Improving data management and decision support systems in agriculture

It is widely recognised that agriculture is a significant contributor to global warming and climate change. Agriculture needs to reduce its environmental impact and adapt to current climate change whilst still feeding a growing population, i.e. become more 'climate-smart'. Burleigh Dodds Science Publishing is playing its part in achieving this by bringing together key research on making the production of the world's most important crops and livestock products more sustainable.

Based on extensive research, our publications specifically target the challenge of climate-smart agriculture. In this way we are using 'smart publishing' to help achieve climate-smart agriculture.

Burleigh Dodds Science Publishing is an independent and innovative publisher delivering high quality customer-focused agricultural science content in both print and online formats for the academic and research communities. Our aim is to build a foundation of knowledge on which researchers can build to meet the challenge of climate-smart agriculture.

For more information about Burleigh Dodds Science Publishing simply call us on +44 (0) 1223 839365, email info@bdspublishing.com or alternatively please visit our website at www.bdspublishing.com.

Related titles:

Advances in crop modelling for a sustainable agriculture
Print (ISBN 978-1-78676-240-5); Online (ISBN 978-1-78676-242-9, 978-1-78676-243-6)

Robotics and automation for improving agriculture
Print (ISBN 978-1-78676-272-6); Online (ISBN 978-1-78676-274-0, 978-1-78676-275-7)

Precision agriculture for sustainability
Print (ISBN 978-1-78676-204-7); Online (ISBN 978-1-78676-206-1, 978-1-78676-207-8)

Chapters are available individually from our online bookshop:
https://shop.bdspublishing.com

climate•SMART•publishing
IN AGRICULTURAL SCIENCE

BURLEIGH DODDS SERIES IN AGRICULTURAL SCIENCE
NUMBER 85

Improving data management and decision support systems in agriculture

Edited by Dr Leisa Armstrong, Edith Cowen University, USA

burleigh dodds
SCIENCE PUBLISHING

Published by Burleigh Dodds Science Publishing Limited
82 High Street, Sawston, Cambridge CB22 3HJ, UK
www.bdspublishing.com

Burleigh Dodds Science Publishing, 1518 Walnut Street, Suite 900, Philadelphia, PA 19102-3406, USA

First published 2020 by Burleigh Dodds Science Publishing Limited

Library of Congress Control Number: 2020934454

British Library Cataloguing in Publication Data
A catalogue record for this book is available from the British Library

ISBN 978-1-78676-340-2 (Print)
ISBN 978-1-78676-343-3 (PDF)
ISBN 978-1-78676-342-6 (ePub)
ISSN 2059-6936 (print)
ISSN 2059-6944 (online)

DOI 10.19103/AS.2020.0069

Typeset by Deanta Global Publishing Services, Dublin, Ireland

Contents

Series list

Title	Series number
Achieving sustainable cultivation of maize - Vol 1 From improved varieties to local applications Edited by: Dr Dave Watson, CGIAR Maize Research Program Manager, CIMMYT, Mexico	001
Achieving sustainable cultivation of maize - Vol 2 Cultivation techniques, pest and disease control Edited by: Dr Dave Watson, CGIAR Maize Research Program Manager, CIMMYT, Mexico	002
Achieving sustainable cultivation of rice - Vol 1 Breeding for higher yield and quality Edited by: Prof. Takuji Sasaki, Tokyo University of Agriculture, Japan	003
Achieving sustainable cultivation of rice - Vol 2 Cultivation, pest and disease management Edited by: Prof. Takuji Sasaki, Tokyo University of Agriculture, Japan	004
Achieving sustainable cultivation of wheat - Vol 1 Breeding, quality traits, pests and diseases Edited by: Prof. Peter Langridge, The University of Adelaide, Australia	005
Achieving sustainable cultivation of wheat - Vol 2 Cultivation techniques Edited by: Prof. Peter Langridge, The University of Adelaide, Australia	006
Achieving sustainable cultivation of tomatoes Edited by: Dr Autar Mattoo, USDA-ARS, USA & Prof. Avtar Handa, Purdue University, USA	007
Achieving sustainable production of milk - Vol 1 Milk composition, genetics and breeding Edited by: Dr Nico van Belzen, International Dairy Federation (IDF), Belgium	008
Achieving sustainable production of milk - Vol 2 Safety, quality and sustainability Edited by: Dr Nico van Belzen, International Dairy Federation (IDF), Belgium	009
Achieving sustainable production of milk - Vol 3 Dairy herd management and welfare Edited by: Prof. John Webster, University of Bristol, UK	010
Ensuring safety and quality in the production of beef - Vol 1 Safety Edited by: Prof. Gary Acuff, Texas A&M University, USA & Prof. James Dickson, Iowa State University, USA	011
Ensuring safety and quality in the production of beef - Vol 2 Quality Edited by: Prof. Michael Dikeman, Kansas State University, USA	012
Achieving sustainable production of poultry meat - Vol 1 Safety, quality and sustainability Edited by: Prof. Steven C. Ricke, University of Arkansas, USA	013
Achieving sustainable production of poultry meat - Vol 2 Breeding and nutrition Edited by: Prof. Todd Applegate, University of Georgia, USA	014

Introduction

The prediction of rapid increase in the world's population and the need to improve food security across the world has led the agricultural sector on a path to improve farm productivity. Farmers are subjected to many risks from changing climate, economic and market volatility. In order to continue to remain viable and increase productivity, they need to be able to make both strategic in-season and long-term decisions. Agricultural decision support systems provide one tool for farmers to improve this decision making.

Agricultural decision support systems (DSS) have made rapid advances in the last 5 years from basic spreadsheet-based tools to sophisticated, interactive systems. Advances in computing techniques, the use of artificial intelligence (AI), geospatial and precision agriculture technologies have provided tools that can improve the granularity of decision making both spatially (to district, farm and field/paddock level) and temporally (from daily to seasonal contexts). The new emerging DSS are data-centric tools which can interface with farm machinery and sensors networks to provide mobile and cloud services with real-time capabilities. The use of these DSS has become widespread across many agricultural enterprises, including broad acre cropping, animal production, horticulture and food supply chains and has shown benefit to both small-scale/subsistence and larger agricultural enterprises.

A number of key factors need to be considered if there is to be continual improvements in these agricultural decision support systems:

1. The increasing quantities of data that have been collected from sources such as satellites, drones, in-field sensors etc. This growth has led to concerns as to how best to collate these enormous quantities of seasonal and historical data to create useful information for making decisions.
2. The need to provide universal data standards for the storage and distribution of these data sets to allow for their effective use.
3. The question as to who owns the data that has been collected and whether open access to data is possible given IP and other concerns such as the cost of collecting data.
4. The increased connectivity of farms through wireless sensor networks, cloud and cloud computing, which has also increased the vulnerability of the data that is being stored and transferred across these networks.
5. The use artificial intelligence (AI) technologies which can now be used in conjunction with more traditional approaches of crop modelling and statistics to develop effective decision support systems.

These issues are discussed in various chapters in the book.

This book reviews and summarises the wealth of research on key challenges in developing better data management and decision support systems (DSS) for farmers and illustrates how those systems are being deployed to optimise efficiency in crop and livestock production. Part 1 reviews general issues underpinning effective decision support systems (DSS) such as data access, standards and security. It also reviews the advances in the use of artificial intelligence (AI), Image processing, GIS and other technologies to improve the effectiveness of these decision support systems. Part 2 contains case studies of the practical application of data management and DSS in areas such as crop planting, nutrition and use of rotations, livestock feed and pasture management as well as optimising food supply chains.

Part 1 General issues

Chapter 1 discusses a key issue in developing decision support systems which is access to good data. The chapter describes various Initiatives which support the sharing of open access data sets. The chapter also discusses challenges such as the need to digitize data as well as issues relating to ownership, accessibility, quality, interoperability and portability which limit the usefulness of data. The chapter shows the various ways these challenges are being addressed in ensuring e.g. that data infrastructures are underpinned by good quality standards. The chapter includes a number of case studies, including the Africa Regional Data Cube and GEOGLAM.

The development of digital agriculture or smart farming highlights the importance of data and data exchange. Chapter 2 discusses how to achieve data standardization as a critical success factor in agricultural decision-making. It emphasizes the importance of understanding business processes and decision-making steps in identifying key issues in data standardisation. The chapter reviews a reference architecture that has been developed in the IoF2020 project to support the interoperability, replicability and re-use of standards and components for integral decision-making in agriculture.

Chapter 3 focusses on data integration, data analytics and decision support methods that can help agriculture to rise to challenges such as climate change adaptation and food security. In this context, smart data acquisition systems, interoperable Information systems, and frameworks for data structuring are required. The chapter describes methods for data identification in phenotyping hybrid information systems (PHIS) and provides recommendations for non-ambiguous universal resource identifiers (URI). The chapter also discusses the enrichment of data with semantics and ways to tag data with relevant ontology. A case study shows these techniques in practice through technologies and methods for plant phenotyping.

Chapter 4 explores how connectivity and information flow are the two key enabling factors for farming enterprise. These elements also represent the most vulnerable aspects open to cyber-attack, with the aim of disrupting food production. The use of a principles-based framework to assess cyber-attack vulnerabilities in precision agriculture technology as well as in the environment to which it is applied can greatly mitigate the risk of cyber-attack. Farmers can then construct a system that is appropriately protected from cyber-attacks and matched to the complexity level of the technology without unnecessary cost. Future areas of research include analyses of cyber-attack consequences through modelling cyber-physical system vulnerabilities, detection design and security architecture.

Chapter 5 reviews developments in the use of artificial intelligence (AI) techniques to improve the effectiveness of decision support systems (DSS) in agriculture. It discusses the use of different AI techniques such as data mining, artificial neural networks (ANN), Bayesian networks (BN), support vector machines (SVM). It includes several case studies of practical application of these techniques to support decision making by farmers, including WAAgData, GeoSense and rice based DSS.

Chapter 6 discusses developments in tools and technologies used in precision agriculture for effective decision making, including remote sensing and geographic information system (GIS) technologies, sensors and sensor networks. There is a particular focus on statistical and crop simulation models for identifying and accounting for variability in crop production systems.

Part 2 Case studies

Chapter 7 offers a comprehensive review of some of the approaches used by decision support systems (DSS) to make fertiliser application decisions. The chapter reviews direct methods and indirect techniques: simulation models, yield forecasts using data-driven approaches and yield forecasts based on water supply. The chapter includes two case studies to estimate season-specific nitrogen requirements of wheat crops at a within-field scale in Australia. These models forecast yield in two key periods of the season in which farmers make decisions for fertiliser applications - pre-sowing, and mid-season.

Crop rotations have formed a fundamental component of agricultural systems for millennia and advanced decision support systems for crop rotations hold great potential for improving soils and agricultural sustainability. Chapter 8 explores new and current opportunities to gather and collect farm data at unprecedented temporal frequency and spatial resolution. These provide adaptive recommendations for multiple sustainability indicators, based on time-varying constraints and real-time data on factors such as climate, markets and pests. At the heart of this new development is the requirement for a better human-centred design with improved efforts to understand how farmers think, aligning their interests between technologists and farming communities, leading to improved access and to facilitate agricultural transformation in the longer term.

Chapter 9 reviews the role of decision support systems (DSS) for pest monitoring and management through information technology like remote sensing, GIS, spectral indices, image-based diagnostics, expert systems and phenology-based degree day models. Applications range from rule-based models to phenology models to help in pest forecasting as well as monitoring, together with intelligent systems able to suggest appropriate integrated pest management (IPM) strategies. The chapter includes case studies of expert systems, crop pest DSS using phenology-based degree day models, and mobile-based artificial intelligence (AI) modules for identification of pests.

Food production takes place in a complex network of enterprises reaching from agriculture to processing, trade and retail. This complex network creates difficulties in the management of data across the food value chain or in coordinating decisions that maintain product supply and quality and serve the needs of all enterprises concerned. Chapter 10 discusses the issues and provides a framework for supporting data management and decision support in food value chains. Starting from a discussion of the decision situation in food value chains, the chapter outlines a selection of tools for decision support and reviews decision problems in data management and food chain organization.

As well as continuing issues about food safety and quality, more concern is being raised about animal welfare and environmental issues caused by livestock. Balancing these conflicting objectives remains an ongoing challenge requiring multidisciplinary research. Chapter 11 reviews the role played by linear and mathematical programming models and other tools in calculating diets for livestock production. Case studies are provided to offer better understanding of the strengths and weaknesses of linear and goal programming. Ultimately, decision support systems (DSS) are essential in balancing economic, environmental and social objectives in order to provide sustainable diets for livestock. Their role is also key in animal product traceability.

Chapter 12 addresses the limited uptake of decision support systems (DSS) in pasture and rangeland farming systems, despite several decades of development. Historically, there has been confusion between use of DSS tools for either farmers, or for learning and research purposes, which has impacted their applicability for farmers. It is highly important to focus on, and involve, the end user early in the development process. Existing DSS development has focussed on relatively static models, that require manual or semi-automated inputs, often needing more time investment from farmers. Opportunities for improvement are focussed on the automation of real time data streams into farm system DSS, linkages to smart middleware and using advanced analytics.

Part 1

General issues

Chapter 1

Improving data access for more effective decision making in agriculture

Ben Schaap, Wageningen University and Research, The Netherlands and Global Open Data for Agriculture and Nutrition (GODAN), UK; and Suchith Anand and André Laperrière, Global Open Data for Agriculture and Nutrition (GODAN), UK

1 Introduction

The Food and Agriculture Organization of the United Nations (FAO) estimates that, by 2017, the number of undernourished people globally will have reached 821 million – around one person out of every nine – with the majority being women and children. Undernourishment and severe food insecurity appear to be increasing in almost all subregions of Africa, as well as in South America, whereas the undernourishment situation is stable in most regions of Asia (FAO, 2019). With the increase of data becoming available, and the adoption of digital technologies to advance precision farming, there is also a growing digital divide emerging between more and less technological advanced farming systems (Van Es and Woodard, 2017).

Today, our society is globally connected, and so is our food system. However, not all the data in our food system is equally accessible in different parts of the world. Farmers across the world will benefit from bridging the digital divide(s) (Jellema et al., 2015; Berdou and Miguel Ayala, 2018). The Global Open Data for Agriculture and Nutrition (GODAN) initiative supports proactive sharing of open data to make knowledge on agriculture and nutrition available, accessible and usable, in an effort to deal with the urgent challenge of ensuring world food security (Schaap et al., 2019; Musker et al., 2018).

http://dx.doi.org/10.19103/AS.2020.0069.02

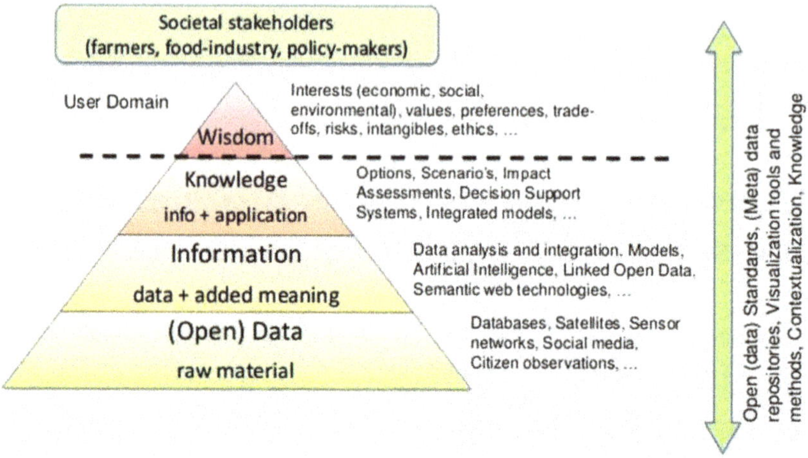

Figure 1 Pyramid of wisdom with open data as raw material. Source: adapted from Janssen et al. (2017).

The modern farmer has access to a growing number and variety of tools to enable better-informed decision making on the farm. Modern farm management information systems (FMIS) allow for better decision making and, thus, more efficient and sustainable agricultural production using an increasing amount of data. With the increased amount of data that is collected on farms, through farm equipment and remote sensing comes the potential for the development of new applications such as precision agriculture (Wolfert et al., 2017).

However, open data by itself will not directly lead to better decision making by farmers (Fig. 1). Generally, open data informs the decision-making tools that farmers use, since applications need to be of value to farm management operations. Additionally, from a user perspective, devices and data need to be trusted in order to make them a reliable part of any decision support system. For any FMIS, good quality data is key to adequate decision support.

2 Key issues in current availability of data

Agricultural reports have been published for a very long time. Agricultural censuses are one of the earliest examples available of data sets in agriculture. They may, in fact, be the first open data sets on record. From the early 2000s, more and more governments have begun to share basic data sets (on land use, for example) that are more or less universally available for public reuse. In many countries in the developing world, records of data such as agricultural statistics have not been fully digitized and thus it is very hard to access the data. Even if the data is made available, there is also sometimes a lack of capacity to be able to utilize the data effectively. One of the key aspects of successful capacity development for utilization of data is localization (e.g. language) to meet local needs and priorities.

Some of the costliest data on farms is soil-related. Due to the technology involved in soil sampling, many of these records are now digitized. Satellite data changed the game in terms of data size and coverage, with a significant amount of this data becoming publicly available. Drones and IoT devices now also produce large amounts of data on a local scale (Liu et al., 2012). These data are often collected by the private sector (farmers and farm machinery manufacturers) and, as a result, the data are not shared widely because of data access rights issues. Besides rights issues, there may also be issues that relate to the quality, interoperability and portability (large size of files) of the data. Figure 2 shows a variety of sensors that can be used on a modern farm. Unfortunately, such variety can create data that is hard to reuse, either due to accessibility problems or due to practical issues such as interoperability.

According to the Open Data Index, most countries provide basic information data sets with creative commons licenses. Much of this data is used by farmers and some is also used by FMIS. For example, national weather data sets and national soil data sets are offered through government open data portals such as www.data.gov. Access to national-level weather data is varied

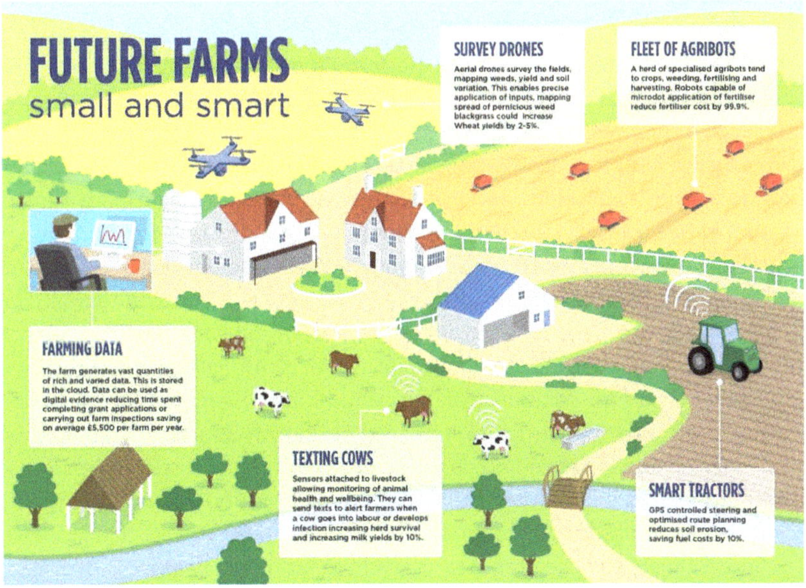

Figure 2 Future farming technologies. Copyright: NESTA. http://nesta.org/precisionfarming.

for different countries. For example, it is difficult to get up-to-date weather data information in some countries in sub-Saharan Africa.

There is a huge global variability in terms of the use of and access to personal computers as well as connectivity to the internet. Many now acknowledge the inability to access and use data as a digital divide (Aaronson and Leblond, 2018). Today's increasing use of available data in the world has been growing at different rates in different regions. It is clear that not everyone is benefitting equally from data, either because they have no access to it, or because there is a lack of capacity to make use of it. In agriculture, we find that this phenomenon is also present (as described by Jellema et al., 2015). In many countries in Africa and Asia, data such as soil data and agricultural statistics have not been digitized at all and thus it is very hard to access. But even if the data is globally offered such as the satellite imagery data from space programs, farmers or extension officers are unable to make use of this data in applications directly due to bandwidth problems. Besides bandwidth, there may also be a lack of storage capacity and general infrastructure to process data. And if the data and technology is available, there is sometimes a lack of capacity to be able to utilize the data effectively. A further digital divide is that women in many situations do not have the same access to technology, predominantly within developing countries. In these environments, a significant part of the farming workload falls to women, whose contribution is too often unreported or ignored. Without access to technology and data, farming work by women is made more difficult, less productive and less competitive.

Most of the data collected on the farm is not stored by the farmer. Soil sample data is stored in a database at the soil testing lab, for example, or sensor data from equipment such as tractors and harvesting machines would be stored by the farm machinery manufacturer. Often this information is available through a cloud application, and sometimes also via an API. Data infrastructures are supported by web standards such as OGC WMS mapping standards for example. However, any data infrastructure needs to be underpinned by good quality standards that are semantically coherent. The Africa Regional Data Cube (ARDC) use case is an example of a data infrastructure that delivers large volumes of earth observation data to decision makers in the region to understand and find solutions for agricultural improvement. An example of a web-based data infrastructure use case is that of Geo-Wiki. Geo-Wiki is a platform for citizen science and crowd-sourced information sources. Use cases presented later in this chapter, like Geo-Wiki, help us understand the potential of initiatives like this for agriculture.

Some initiatives have emerged that aim to facilitate access to farm data on behalf of service providers and app developers. In the United States, OADA provides an open protocol, allowing farmers to share and receive data through cloud technology. In the Netherlands, JoinData is setting up an authentication infrastructure and service for farmers, farm cooperatives and industry partners, providing seamless permission to use each other's data. In France, API-AGRO is another initiative aiming to solve the current lack of trusted authentication infrastructure by issuing permission to use the data (Siné et al., 2015; Rehben et al., 2017).

3 Use of data for decision making: case studies

3.1 Use case 1: *Akkerweb*

Akkerweb compiles a variety of data (closed and open) and information in one central geo-platform, which is re-used across a suite of applications native to the platform (Van Evert et al., 2017). The Akkerweb 'Crop Rotation Application', set up using geo-data, forms the foundation for all the functionalities which provide added value for a range of farm operations such as fertilization and crop protection (Fig. 3). These types of applications allow farmers to make immediate and optimal decisions on the management of crops, without the technical complexity generally associated with big data analysis or dataset management.

3.2 Use case 2: *Africa Regional Data Cube*

Global earth observation will help to promote food security, providing high-quality information derived from satellite imagery that can be used to

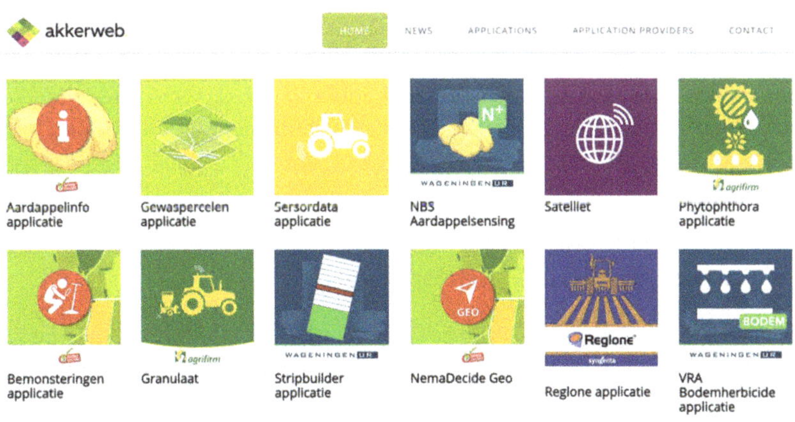

Figure 3 The Akkerweb platform, showing a variety of applications for precision farming, including decision support applications such as a leaf haulm killing application for more precise dosing of the relevant herbicide. Image credits: https://akkerweb.eu/en-gb/.

improve agricultural production solutions. The ARDC is based on satellite and complementary technologies to assist Kenya, Senegal, Sierra Leone, Ghana and Tanzania in addressing food security concerns, as well as issues relating to agriculture such as land degradation, deforestation, crop monitoring and water access. The ARDC was developed by the Committee on Earth Observation Satellites (CEOS), in partnership with the Group on Earth Observations, Amazon Web Services, Strathmore University in Kenya, the Office of the Deputy President of Kenya and the Global Partnership for Sustainable Development Data (ARDC, 2018).

The ARDC is designed to achieve both the historical and multi-sensor data integration impact sought by an increasing number of national data cubes being developed by more than 30 nations across the world. However, it is the first time such a tool has been developed on a regional basis, pointing at new avenues for agricultural improvement; such as the potential to address not only national but also cross-boundary issues related to climate change and pest/disease infestation in particular.

3.3 Use case 3: GEOGLAM

GEOGLAM is the Group on Earth Observations Global Agricultural Monitoring Initiative, set up by the G20 leaders to strengthen global agricultural monitoring, by improving the use of remote sensing tools for crop production projections and weather forecasting. The main objective of GEOGLAM is 'to reinforce the international community's capacity to produce and disseminate relevant, timely and accurate forecasts of agricultural production at national, regional and global scales by using Earth Observation data'. Several regional agricultural productivity monitoring systems are used in GEOGLAM for a global overview of agricultural productivity. The agricultural productivity monitoring systems provide policy makers and industry with information from various data sources (open and closed) on possible unfavourable weather conditions and yield forecasts (Fig. 4). The output information contributes to the transparency of the commodity trade market, and is also used to identify potential food insecurity risks.

3.4 Use case 4: Geo-Wiki platform

The Geo-Wiki platform provides citizens with the means to engage in environmental monitoring of the earth by providing feedback on existing information overlaid on satellite imagery or by contributing entirely new, complementary data. Data can be inputted via a traditional desktop platform or mobile devices, with campaigns and games used to incentivize input. These innovative techniques have been used to successfully integrate citizen-derived

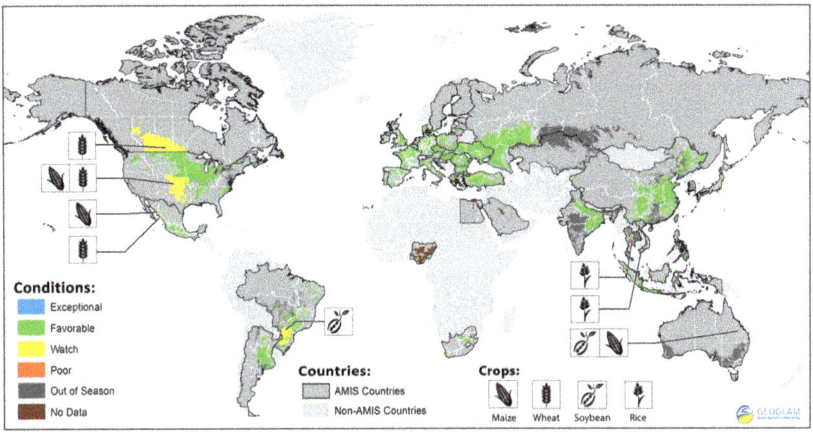

Figure 4 GEOGLAM Synthesis map: Synthesis maps provide a simplified overview of current global crop conditions and trends. The cropped area displayed in the synthesis map is the total area of all four crops (Maize, Wheat, Soybean and Rice) depicted according to the observed crop condition categories (Exceptional, Favorable, Watch and Poor). Crops classified as having other-than-favourable conditions are highlighted with crop symbols, and areas that are out of season are in dark grey. Non-AMIS countries are identified in light grey and are not covered by the Crop Monitor. Dashboard for Maize, Wheat, Soybean and Rice. Image credits: http://www.geoglam.org/index.php/en/global-regional-systems-en/crop-monitor-for-amis.

data sources with expert and authoritative data. Since 2009, Geo-Wiki has grown rapidly, with currently over 15 000 registered users and applications in many successful citizen science campaigns, most recently crowdsourcing global agricultural field-size data. There are many ideas that the global agriculture community can learn and make use of from Geo-Wiki. The work done by Olteanu-Raimond et al. (2018) on integrating citizen and community science into land cover, land use and land change detection processes is also of interest to the agricultural community.

3.5 Use case 5: Sustainable Technology Adaptation for Mali's Pastoralists (STAMP)

Sustainable Technology Adaptation for Mali's Pastoralists (STAMP) is an information service tailor-made to pastoralists' knowledge and decision-making needs, giving them more predictability for their movements. STAMP aims to improve resilience among climate-affected pastoralists, through access to and use of geo-satellite derived data. The service will provide instant access along the points of different transhumance routes to reliable information on biomass availability and quality, surface water availability, herd concentration and market prices for livestock and staple grains (STAMP, 2019).

STAMP is a partnership between various organizations from Mali, the Netherlands, the United States and Spain: SNV Netherlands Development Organisation, ORANGE Mali, Hoefsloot Spatial Solutions (HSS), Action Against Hunger (ACF), Institut d'Economie Rurale (IER) and Project Concern International (PCI). It is an excellent practical example of open data and technology assisting pastoralists in the region: increasing their income from livestock sales, reducing losses, as well as improving overall productivity.

3.6 Use case 6: Farm-Oriented Open Data in Europe (FOODIE)

There are 195 countries in the world today, each with their own environmental and agricultural concerns, each adding to a web of sometimes conflicting rules and regulations. Farm-Oriented Open Data in Europe (FOODIE) has an ambitious aim to bring all this data under one roof. FOODIE is an open data initiative based in Europe with the aim to increase efficiency and open new opportunities for all actors in planning, growing and delivering food to the marketplace. FOODIE aims to change the culture of many large public institutions and government departments, turning them from owners and collectors of data, to organizations that freely share data. This new approach should help to make wiser, more informed decisions, protecting the environment, improving agriculture techniques and driving efficiencies in the food chain. It can be applied across Europe and supports global food production. (FOODIE, 2019; GODAN Success Stories, 2018).

4 Current trends

The vast majority of all the data that is captured in relation to farming is not controlled by the farmer but by third parties. Large companies involved in agriculture are increasingly controlling the data they generate and use. This is typically separated into three categories:

- The first category is public data, open to anyone;
- The second category is proprietary data, either linked to commercial activities, research or product development;
- The third category is farmer data, with companies seeking to safeguard data in order to protect the privacy and intellectual property rights of the farmers who have entrusted them with their data.

This has led to intensive discussions on the subject of data rights and, more fundamentally, data ownership. The challenge is to find the right balance between sharing data that would then benefit the community as a whole without putting at risk or exposing any of its individual members.

For this purpose, privacy regulation has been developed and implemented in a number of countries in Europe and elsewhere, while various codes of ethics have also been developed by governments, consumer groups and associations of various kinds, including equipment manufacturers. For example, farmers organizations such as Copa Cogeco have developed a code of conduct for better and more ethical data sharing (Sanderson and Wiseman, 2018). The idea is that these codes of conduct will help farmers to take into account how manufacturers of farm equipment will deal with permissions to access and reuse the information that is generated by farm sensors.

Unfortunately, most of the data collected is very hard to reuse, either due to being hard to find, and/or poor accessibility, interoperability and licencing problems. There are many examples where the data from farm sensors, such as yield from grain harvesters or milking robots, is not reusable. Reusability is also still a common issue in agricultural research and beyond.

In the Digital Single Markets Strategy (Anon., 2017), the European Commission acknowledged the issue and resolved to set up the European Open Science Cloud (EOSC). The EOSC will function as an open network of heterogeneous data that will be openly available to both researchers and private sector bodies, for the reuse of data for innovation. The Findable, Accessible, Interoperable and Reusable (FAIR) data principles (Wilkinson et al., 2016) are at the centre of the EOSC and have been endorsed in Europe, including by the G20 (Anon., 2016). In order to visualize how the FAIR data principles might work in a farm context, the Farm Data Train animation has been created to show

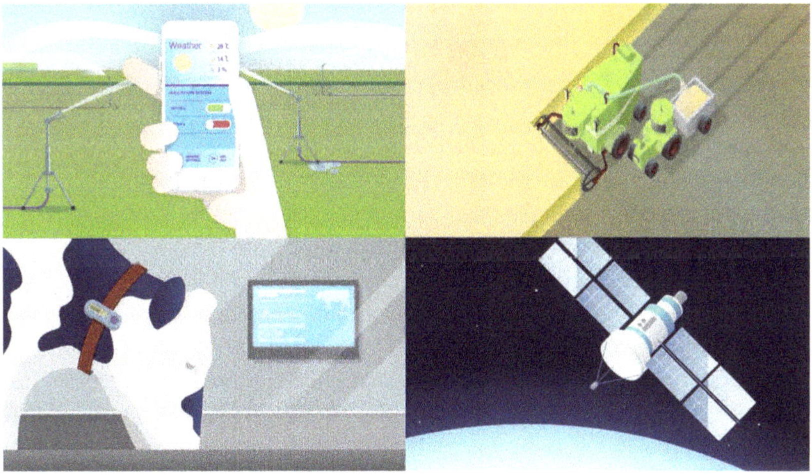

Figure 5 The Farm Data Train visualization of how the FAIR data principles may be implemented for reusability of sensor data by machines in a farm context (Finkers et al., 2017). Image credits: https://www.dtls.nl/fair-data/farm-data-train/.

how reusability of data by artificial intelligence (AI) machines can help farmers, researchers and other stakeholders (Fig. 5; Finkers et al., 2017).

It remains the case that a large amount of data released today is not fully exploited. Arguably, the challenge now resides in finding and accessing the right data for the right purpose, using it for practical applications from IoT monitoring devices to handheld applications. The FAIR data principles provide guidance for both policy makers and data producers in addressing these issues (Wilkinson et al., 2016). However, to make this a reality for our food system, more efforts are needed to achieve interoperability; from common ontologies and vocabularies to technical standards and tools that will help users create machine readable data (Schaap et al., 2019).

5 Conclusions

With the increased amount of data that is collected on agriculture, we live in a time where data seems to be abundant. There is more and more evidence of this data being put to use. Initial findings have also pointed to the usefulness of open data for better decision making in agriculture (Carolan et al., 2015). In order to achieve good availability and reusability of data, we need good policy frameworks that include recommendations on how to open and publish open data (Smith et al., 2017). The GEOGLAM use case, presented earlier in this chapter, is a good example of this. This is an agricultural productivity monitoring system, providing policy makers and industry with information on possible unfavourable weather conditions and yield forecasts, to improve transparency within the commodity trade market and identify food insecurity risks.

In advanced agricultural systems, technology heavily relies on data to make smart decisions to achieve sustainable and healthier food production (Wolfert et al., 2017; Busse et al., 2015). The FOODIE use case in this chapter shows how open data in the food production system is helping various stakeholders along the value chain. On the farm systems and field-level platforms, such as Akkerweb, provide useful precision farming applications for farmers to reduce agricultural inputs and maximize production. There are some promising examples in various parts of the food ecosystem, like the user case on STAMP, which underlines the potential of access to data and technology in increasing income and overall productivity across a range of food production actors.

The uses cases presented show that, with improved access to data, stakeholders in agriculture on various levels will be able to make more effective decisions. This also has to be recognized by policy makers who are starting to invest in policies to support better data sharing and better reuse of existing data. The EOSC is a good example from Europe and the adoption of the FAIR data principles on a global level suggest a significant change in how we currently share data. The current issues with farm data sharing show that there is a need for more clarity on data ethics in order to share data effectively for effective decision making in agriculture. Initiatives on authentication for data sharing show that with the appropriate rights to utilize data, it will be possible to facilitate data sharing.

6 Where to look for further information

- GODAN Gateway on F1000Research: https://f1000research.com/gateways/godan/about-this-gateway.
- Agricultural Information Management Standards: http://aims.fao.org/.
- Global Open Data for Agriculture and Nutrition (GODAN): www.godan.info.

7 References

Aaronson, S. A. and Leblond, P. 2018. Another digital divide: The rise of data realms and its implications for the WTO. *Journal of International Economic Law* 21(2), 245–72. doi:10.1093/jiel/jgy019.

African Regional Data Cube. 2018. https://www.earthobservations.org/documents/meetings/201805_rapp_sdg_odc/201805_odc_announcement.pdf.

Allemang, D. and Teegarden, B. 2017. A global data ecosystem for agriculture and food. *F1000Research*. Available at: https://doi.org/10.7490/f1000research.1114971.1.

Anon. 2016. G20 leaders' communique: Hangzhou summit, 4–5 September 2016. Available at: https://www.consilium.europa.eu/media/23621/leaders_communiquehangzhousummit-final.pdf.

Anon. 2017. EIP-Agri workshop data sharing: Ensuring fair sharing of digitisation benefits in agriculture. Available at: https://ec.europa.eu/eip/agriculture/sites/agri-eip/files/eip-agri_workshop_data_sharing_final_report_2017_en.pdf.

Anon. 2018a. Geo-Wiki: Earth observation & citizen science. Available at: https://www.geo-wiki.org.

Anon. 2018b. The state of food security and nutrition in the world 2018: Building climate resilience for food security and nutrition. Available at: https://www.wfp.org/content/2018-state-food-security-and-nutrition-world-sofi-report.

Anon. 2019. STAMP SNV. Available at: http://www.snv.org/project/stamp-building-success.

Berdou, E. and Miguel Ayala, L. 2018. GODAN Action – a review of relevant methods and frameworks for impact evaluation of open data [version 1; not peer reviewed]. *F1000Research* 7, 809. doi:10.7490/f1000research.1115589.1.

Busse, M., Schwerdtner, W., Siebert, R., Doernberg, A., Kuntosch, A., König, B. and Bokelmann, W. 2015. Analysis of animal monitoring technologies in Germany from an innovation system perspective. *Agricultural Systems* 138, 55–65.

Carolan, L., Smith, F., Protonotarios, V., Schaap, B., Broad, E., Hardinges, J. and Gerry, W. 2015. *How Can We Improve Agriculture, Food and Nutrition with Open Data?* London: Open Data Institute. Available at: http://www.godan.info/sites/default/files/old/2015/04/ODI-GODAN-paper-27-05-20152.pdf.

FAO. 2019. *The State of Food Security and Nutrition in the World 2019*. FAO, Rome, Italy. Available at: http://www.fao.org/state-of-food-security-nutrition/en/.

Ferris, L. and Rahman, Z. 2017. Responsible data in agriculture [version 1; not peer reviewed]. *F1000Research* 6, 1306 (document). doi:10.7490/f1000research.1114555.1.

Finkers, R., Schaap, B. F., Mons, A. and Kok, R. 2017. Farm data train. Available at: https://www.dtls.nl/fair-data/farm-data-train/.

FOODIE. 2019. http://www.foodie-project.eu

García, J. S. 2018. Big Data: A blessing or a curse for agriculture? Available at: https://www.iof2020.eu/blog/2018/03/protocol-of-conduct-on-data-exchange.

GODAN. 2018. Success Stories: Issue 2. https://www.godan.info/files/documents/success-stories-issue-2.

Janssen, S. J. C., Porter, C. H., Moore, A. D., Athanasiadis, I. N., Foster, I., Jones, J. W. and Antle, J. M. 2017. Towards a new generation of agricultural system data, models and knowledge products: Information and communication technology. *Agricultural Systems* 155, 200–12. doi:10.1016/j.agsy.2016.09.017.

Jellema, A., Meijninger, W. and Addison, C. 2015. Open data and smallholder food and nutritional security. CTA Working Paper 15/01. CTA, Wageningen, the Netherlands. Available at: https://hdl.handle.net/10568/75490.

Liu, D., Zhou, J. and Mo, L. 2012. Applications of internet of things in food and agri-food areas. *Nongye Jixie Xuebao/Transactions of the Chinese Society of Agricultural Machinery* 43(1), 146–52.

Maru, A., Berne, D., De Beer, J., Ballantyne, P., Pesce, V., Kalyesubula, S., Fourie, N., Addison, C., Collett, A. and Chaves, J. 2018. Digital and data-driven agriculture: Harnessing the power of data for smallholders [version 1; not peer reviewed]. *F1000Research* 7, 525 (document). doi:10.7490/f1000research.1115402.1.

Musker, R., Tumeo, J., Schaap, B. F. and Parr, M. 2018. GODAN's impact 2014–2018 - Improving agriculture, food and nutrition with open data. *F1000Research*. doi:10.7490/f1000research.1115970.1.

Olteanu-Raimond, A.-M., Jolivet, L., Van Damme, M.-D., Royer, T., Fraval, L., See, L., Sturn, T., Karner, M., Moorthy, I. and Fritz, S. 2018. An experimental framework for integrating citizen and community science into land cover, land use, and land change detection processes in a national mapping agency. *Land* 7, 103.

Rehben, E., Balvay, B. and Haezebrouck, T. P. 2017. Sharing data through an API platform - API AGRO. *Proceedings of the 41st ICAR Conference held in Edinburgh, UK*.

Sanderson, J. and Wiseman, L. 2018. What's behind the Ag-data logo? An examination of voluntary agricultural-data codes of practice. *International Journal of Rural Law and Policy*. Available at: https://doi.org/10.5130/ijrlp.1.2018.6043.

Schaap, B. F., Musker, R., Parr, M. and Laperriere, A. 2018. Open data and agriculture. In: Davies, T., et al. (Eds), *The State of Open Data: Histories and Horizons*. African Minds and International Development Research Centre, Cape Town, South Africa/Ottawa, Canada. Available at: https://docs.google.com/document/d/10kslypUWMfqHguTRRVP1Eod3194yIlBKsAMpfSpart4/view.

Schaap, B., et al. 2019. GO FAIR food systems implementation networks manifesto. Available at: https://www.go-fair.org/implementation-networks/overview/food-systems/.

Siné, M., Haezebrouck, T. P. and Emonet, E. 2015. API – AGRO: An Open Data and Open API platform to promote interoperability standards for Farm Services and Ag Web applications. Available at: http://real.mtak.hu/30159/1/209_1076_1_PB_u.pdf.

Smith, F., Fawcett, J. and Musker, R. 2017. Donor open data policy and practice: an analysis of five agriculture programmes [version 1; not peer reviewed]. *F1000Research* 6, 1900 (document). doi:10.7490/f1000research.1115013.1.

Sonka, S. 2015. Big Data: from hype to agricultural tool. *Farm Policy Journal* 12, 1–9.

Tsiropoulos, Z., Carli, G., Pignatti, E. and Fountas, S. 2017. Future perspectives of farm management information systems. In: Pedersen, S. and Lind, K. (Eds), *Precision Agriculture: Technology and Economic Perspectives. Progress in Precision Agriculture*. Springer, Cham.

Van Es, H. and Woodard, J. 2017. Chapter 4: Innovation in agriculture and food systems in the digital age. In: *The Global Innovation Index 2017: Innovation Feeding the World*. Cornell University. Available at: https://www.wipo.int/edocs/pubdocs/en/wipo_pub_gii_2017.pdf.

Van Evert, F. K., Been, T., Booij, A. J., Kempenaar, C., Kessel, J. G. and Molendijk, P. L. 2017. Akkerweb: a platform for precision farming data, science, and practice. *Proceedings of the 14th International Conference on Precision Agriculture*, 24 June – 27 June 2018, Montreal, Quebec, Canada. Available at: http://edepot.wur.nl/459167.

Wilkinson, M. D., Dumontier, M., Aalbersberg, I. J., Appleton, G., Axton, M., Baak, A., Blomberg, N., Boiten, J. W., da Silva Santos, L. B., Bourne, P. E., Bouwman, J., Brookes, A. J., Clark, T., Crosas, M., Dillo, I., Dumon, O., Edmunds, S., Evelo, C. T., Finkers, R., Gonzalez-Beltran, A., Gray, A. J., Groth, P., Goble, C., Grethe, J. S., Heringa, J., 't Hoen, P. A., Hooft, R., Kuhn, T., Kok, R., Kok, J., Lusher, S. J., Martone, M. E., Mons, A., Packer, A. L., Persson, B., Rocca-Serra, P., Roos, M., van Schaik, R., Sansone, S. A., Schultes, E., Sengstag, T., Slater, T., Strawn, G., Swertz, M. A., Thompson, M., van der Lei, J., van Mulligen, E., Velterop, J., Waagmeester, A., Wittenburg, P., Wolstencroft, K., Zhao, J. and Mons, B. 2016. The FAIR Guiding Principles for scientific data management and stewardship. *Scientific Data* 3, 160018. doi:10.1038/sdata.2016.18.

Wolfert, S., Ge, L., Verdouw, C. and Bogaardt, M.-J. 2017. Big data in smart farming – a review. *Agricultural Systems* 153, 69–80. doi:10.1016/j.agsy.2017.01.023.

Zervas, P., Manouselis, N., Karampiperis, P., Hologne, O., Janssen, S. and Keizer, J. 2018. Roadmap for a pan-European e-Infrastructure for open science in agricultural and food sciences. Available at: http://www.erosa.aginfra.eu/sites/default/files/Roadmap%20Paper.pdf.

Chapter 2

Improving data standards and integration for more effective decision-making in agriculture

Sjaak Wolfert, Wageningen University and Research, The Netherlands

1 Introduction

As sensors and smart machines become more common on farms, and farm data grow in quantity and scope, farming processes will become increasingly data-driven and data-enabled. Rapid developments in the Internet of Things (IoT) and cloud computing enable the development of 'smart farming' (Sundmaeker et al., 2016). Precision agriculture technology has focussed on identifying in-field variability as a basis for more targeted management (e.g. more selective applications of fertilizer, pesticides or herbicides). 'Smart farming' goes beyond that by basing management tasks not only on location but also on other data, enhanced by context awareness and situation awareness, triggered by real-time events (Wolfert et al., 2014). By linking sensors, data acquisition, decision support systems and robotics, smart farming offers the possibility of agile, autonomous systems capable of predicting what is needed, making decisions and implementing them as well as responding intelligently to unforeseen events.

http://dx.doi.org/10.19103/AS.2020.0069.03

Real-time reconfiguration features are required to ensure agility in the system, especially in cases of sudden changes in operational conditions (e.g. changes in weather or a disease outbreak). These features typically include intelligent assistance in the implementation, maintenance and use of the technology. Figure 1 summarizes the concept of 'smart farming' along the management cycle as a cyber-physical system. 'Smart farming' means that smart devices - connected to the Internet - control the farm system. Smart devices enhance the use of conventional tools (e.g. rain gauges, tractors, notebooks). They add autonomous context-awareness using all kinds of sensors. They also add built-in intelligence, capable of executing autonomous actions. Robots can play an increasingly important role. Where humans are still involved, their role in analysis, planning, implementation and monitoring is assisted by machines so that the cyber-physical cycle becomes almost autonomous. Humans will always be involved in the whole process, but increasingly at the level of design and monitoring, leaving most operational activities to machines.

However, the increasing number of devices and related systems needed for 'smart farming' create new challenges for the exchange of data required. There is a clear need for data standards to address this challenge. The objective of this chapter is to explain how standards for data exchange should be developed for effective decision-making in agriculture. I will argue that

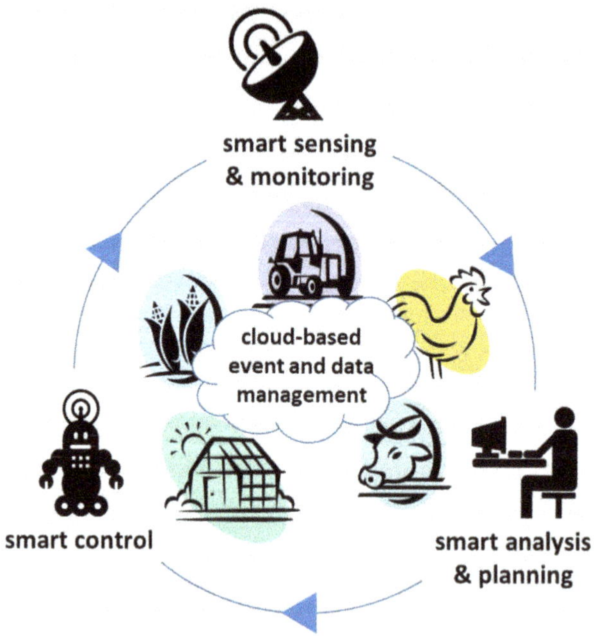

Figure 1 Three-tiered service-oriented architecture with some illustrative examples from the farming sector. Source: adapted from Wolfert et al. (2010).

integration and standardization should not start with the databases themselves, but more from the level of business processes in which data exchange plays a role. This will be illustrated by a precision agriculture example for variable-rate fertilization. Since software increasingly is being developed as smartphone apps and incorporated into a platform, we will describe how standardization plays a role in that context. The chapter will also discuss an overarching reference architecture for interoperability, replicability and re-use that was developed in the European IoF2020 project. This can function as a framework for future standardization in smart-farming applications.

2 Business process modelling to identify data requirements

Information and communication technology (ICT) is often considered as a specialist technical subject to be undertaken by experts such as hardware engineers or software programmers. However, these experts usually do not have a full understanding of the business for which they may be supplying technical products, including the key decision-making processes and systems within the business. This is particularly true for whole supply chain networks that involve different companies. It is very important to start from an understanding of business processes and then see what data and information these processes require, and which need to be exchanged between different systems. This is shown in Fig. 2. Software and other technicians often start at the application service layer at the bottom and see how they can connect different databases so that the data can be used by different users. In our view, it is more important to start the planning and analysis from the top layer, the business process management layer, and then see what data this suggests is needed and how this could be retrieved from the various data sources. In this way actual processes in the business will be supported by applications in a more appropriate way.

An example is a standard crop production process from sowing to harvesting. Production involves several stages at which key decisions need to be made. In the sowing process a farmer decides which crop and which variety has to be grown, on which field and so on. To make these decisions he/she has to get market information, for example, on what markets he/she is growing the crop for, what restrictions there may be in choosing which variety and so on. This information may need to be retrieved from different market information systems and data sources. It would be helpful if the farmer could get this information in a standardized and user-friendly way through one service, for example, showing up as 'get restrictions information' on what varieties he/she is permitted to plant. These data-driven, decision-making tools at the business process management layer of the system may be even more complicated when dealing with an integrated supply chain involving many stakeholders.

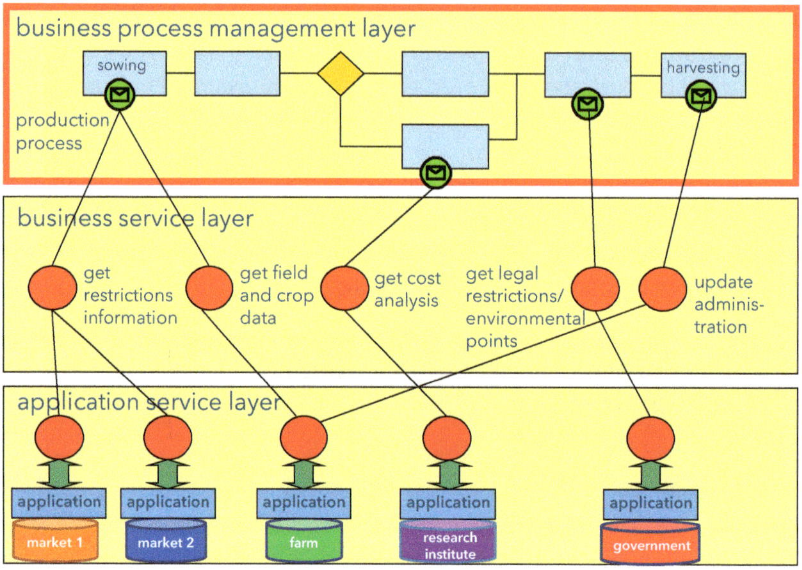

Figure 2 Three-tiered service-oriented architecture with some illustrative examples from the farming sector. Source: adapted from Wolfert et al. (2010).

Figure 2 illustrates other information flows that are needed to support the business processes in the upper layer. The envelope icons (shown in green) indicate when an information flow (from or to the process) is involved. In some cases, external information is not needed to carry out a certain process (in which case there is no envelope icon). In the next section we will illustrate information flows for a specific process: variable rate fertilization.

3 Data flows for a particular process: the example of variable rate fertilization

Variable rate fertilization is a precision agriculture technique in which fertilizer is applied in differing amounts in a field depending on site-specific conditions (e.g. according to the specific nutrient requirements of individual plants). To be able to decide what part of the field needs more or less fertilizers, you need site-specific information about the crop and/or the field. The process is shown in Fig. 3.

The business process starts with a recommendation request from a farm management system based on specific field and crop information. The 'N advice process' (N refers to nitrogen - a key crop nutrient) can be seen as the central process engine that drives the other processes. It will request data about the crop status, for example, from a leaf area index (LAI) map (which uses leaf colour to measure the nutrient status of a plant). This can be combined with

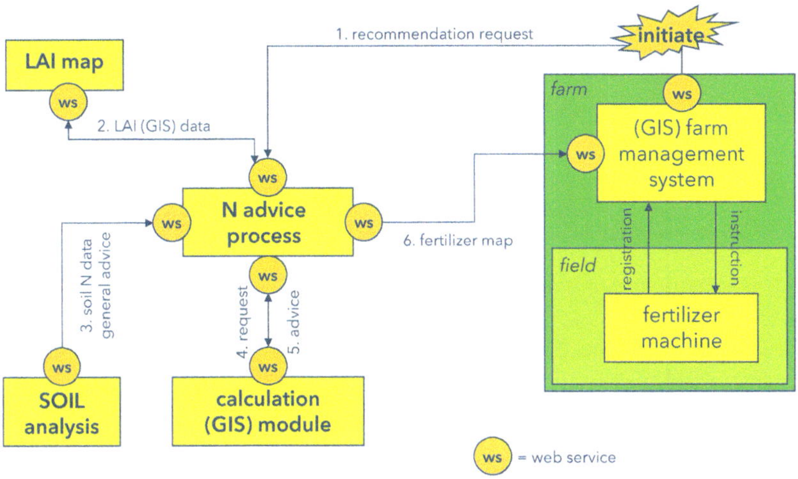

Figure 3 Conceptual graphical representation of the various data flows between different entities for variable rate fertilization. Source: adapted from Wolfert et al. (2010).

an assessment of general nitrogen (N) status for that field based on a standard soil analysis. This information can be combined with a GIS-based calculation model that identifies what fertilizer is needed at any particular point in the field. This is converted into a machine-readable fertilizer map that can be sent to the fertilizer equipment to apply the fertilizer (either directly to the machine or through the farm management system). In the past all this information was provided on paper. A web service (ws) makes it possible for data to be read and understood by different parts of the system. In practice, effective data exchange to enable the system to work is still limited by a lack of standardization in the underlying data.

Figure 3 helps to show what data have to be exchanged and which applications have to be developed. The next step is to develop a business process model as shown in Fig. 4 for the example of variable rate fertilization. The model clearly indicates the interaction between different actors or sub-departments within an organization in the so-called swim lanes. This makes it very clear on who the relevant actors are and where the information interactions take place. There are special software tools available to draw these diagrams, provide more information for each process and define the data flows. These tools make it possible to automatically generate computer code that can then be deployed as a set of web services.

4 Linking platforms and software

When the development of a software application is the responsibility of one software provider, standardization is not difficult because the provider can set

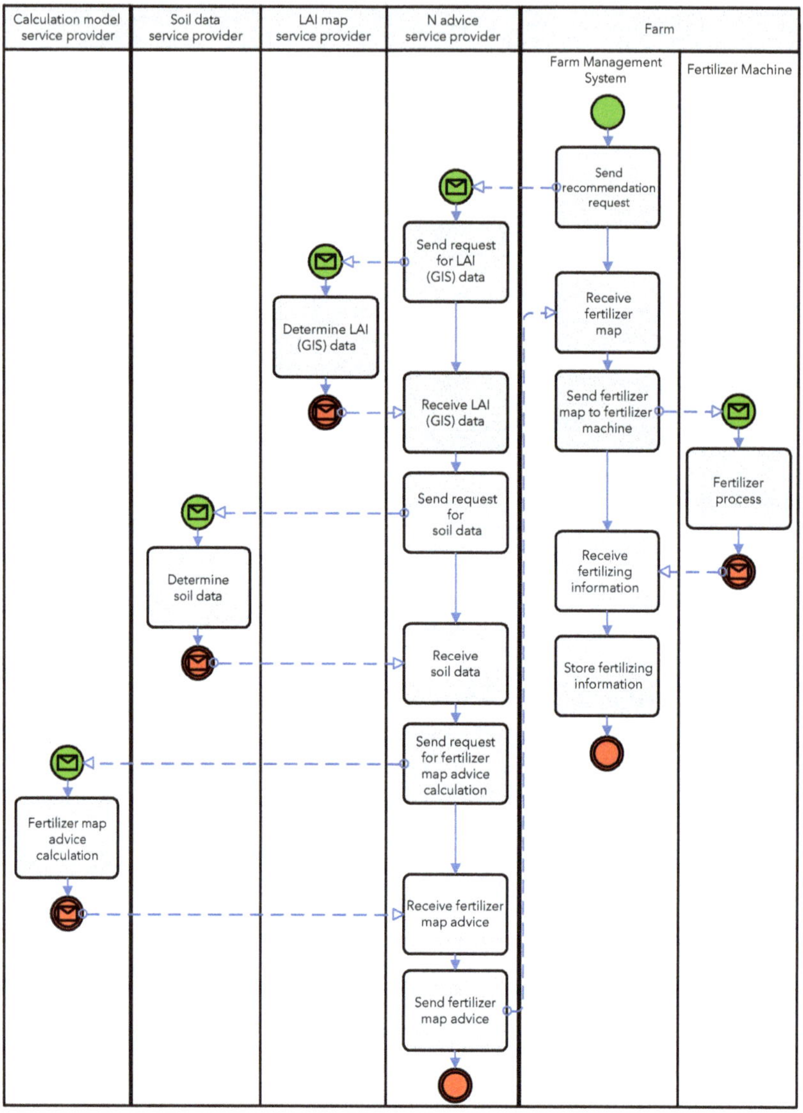

Figure 4 Business process model of the variable rate fertilizing case with various actors and subsystems in 'swimlanes'. Source: adapted from Wolfert et al. (2010).

his/her own standards. This means that every application, module and so on can easily communicate with the others. However, this is often not the case and applications are developed by different providers who may have different development methods and standards. This is especially true in agriculture and other sectors where farmers, for example, will frequently use mobile phone 'apps' which are compact, flexible and easy to use. The key challenge is how to make all these apps, developed from different vendors, work together seamlessly - that is, make them interoperable - in a multi-faceted and multi-stakeholder business process as described for variable rate fertilization.

European researchers have done significant work to develop an open infrastructure for data exchange within the Future Internet programme. This was undertaken through FIWARE, a generic open standard platform for developing open source software (www.fiware.org). A key feature of the programme was the development of a series of Generic Enablers (GEs). GEs are general purpose platform functions available through application programming interfaces (APIs). They are basic building blocks used in the development of smart 'apps' designed to have common shared functions serving multiple-use cases. GEs were developed for applications such as cloud hosting, data and context management services, IoT services, security and big data analysis. The SmartAgriFood project proposed a conceptual architecture for applications for the agri-food domain based on these FIWARE GEs (Kaloxylos et al., 2012). The FIspace project implemented this architecture into a real platform for business collaboration which is shown in Fig. 5 (Verdouw et al., 2016; Wolfert et al., 2014).

FIspace uses FIWARE GEs but has two particular extensions for business collaboration: the app store and the real-time B2B collaboration core. These key components are connected with several other modules to:

- enable system integration (e.g. with IoT applications).
- ensure security, privacy and trust in business collaboration.

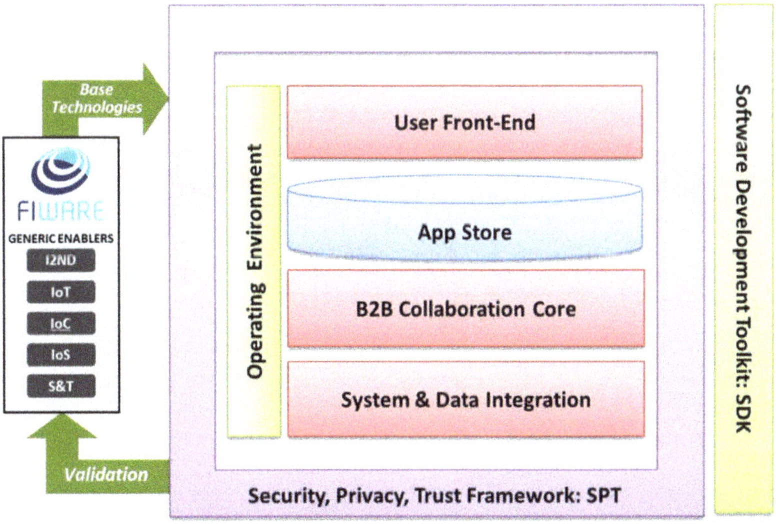

Figure 5 A high-level picture of the FIspace architecture based on FIWARE GEs.

- provide an operating environment and software development kit to support an 'ecosystem' in which apps for the FIspace store can be developed.

The FIspace platform is accessible through various types of front-ends (e.g. web or smartphone) but also allows direct machine-to-machine (M2M) communication.

Figure 6 shows a practical example illustrating the business collaboration concept behind FIspace. It shows a use-case scenario in which a farmer gets expert advice on when to spray his crops to prevent disease, for example, spraying tomatoes in the greenhouse. The process, which involves three different stakeholders, is indicated by the white ellipses. Three apps have to work together:

- a weather information app.
- a spraying certification app.
- a spraying expert advice app.

These are provided by three independent software providers. However, by placing them within the FIspace app store, they can start to work together - interoperate - as if they were one application. This is handled by the FIspace B2B collaboration engine. The FIspace platform takes care of various communication issues with external back-end systems linked by interfaces.

A working prototype of the FIspace platform and several applications, such as spraying decision support system, have been developed in the project.

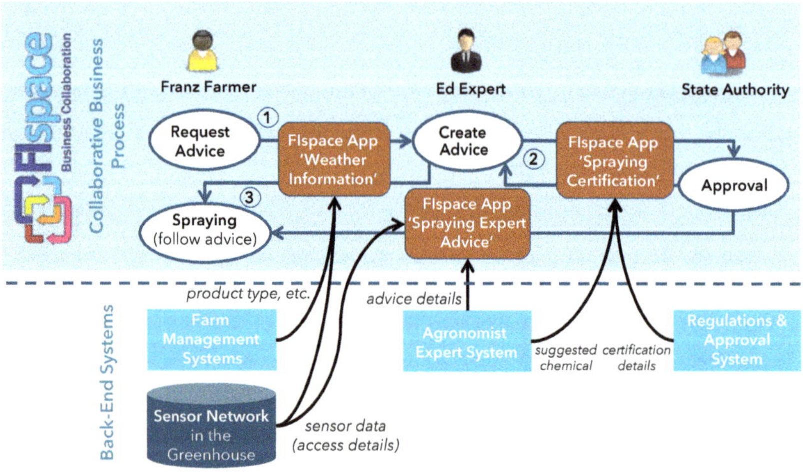

Figure 6 A use-case scenario illustrating the FIspace business collaboration concept. The blue rectangle represents the FIspace platform with all its functionalities that are presented in Fig. 5. Further description in the text.

Although FIspace has not yet developed into a commercial platform, a number of elements are further being developed in commercial applications, especially within the FIWARE business ecosystem and in projects such as Internet of Food and Farm (IoF2020) (Verdouw et al., 2017) (www.iof2020.eu) and SmartAgriHubs, another EU-funded project for the development of digital innovation hubs for digital transformation in agriculture (www.smartagrihubs.eu).

5 Creating a reference architecture for interoperability, replicability and reuse

IoF2020 focusses on applications in arable crops, fruits and vegetables, dairy and meat production. Each use case is an autonomous implementation of an IoT system providing a dedicated solution for a specific domain challenge. However, the project recognizes that it is important to maximize synergies across multiple use-case systems. A core concept of IoF2020 is that the use-case systems function as nodes in a software ecosystem. This means ensuring the interoperability of multiple use-case systems and the reuse of IoT components across them. Figure 7 shows the architectural approach to achieve these objectives during design, development, implementation and deployment.

Developing use-case architectures will be based on a common technical reference architecture to create a shared understanding and to maximize synergies across multiple use-case systems. Each use case will use elements of the reference architecture to address its specific user requirements. The project will provide a catalogue of reusable system components which can be integrated into other IoT systems to facilitate large-scale uptake. This catalogue of software units includes practical guidelines and implementation tools to facilitate uptake. The IoF2020 lab will support the implementation of reusable IoT components in a testbed environment. Finally, IoF2020 will provide

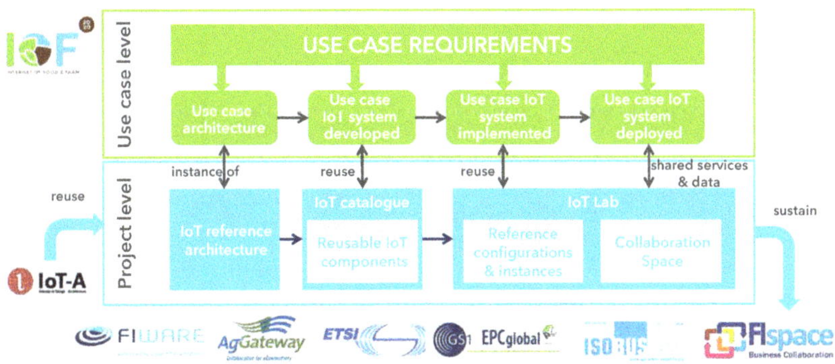

Figure 7 IoF2020 architectural process ensuring re-use and interoperability of IoT systems.

a collaboration space in which services and data can be shared to facilitate interaction between individual use-case IoT systems during deployment. A key feature will be reusing of components and knowledge from previous projects.

IoF2020 has now developed the common reference architecture for the project, though it expects it will be constantly updated and modified in response to changing conditions (Cantera et al., 2018). The technologies and associated standards within this reference architecture address different layers that have to be tackled when developing and deploying a smart food and farming solution, as shown in Fig. 8. The main layers are:

- Physical device layer. This layer consists of IoT devices and agricultural machinery deployed in the field.
- Connectivity layer. This layer enables the bidirectional transmission of data produced by devices and machinery.
- IoT service layer. Using different application-level transport protocols, this layer identifies the raw data generated from IoT devices and, in some cases, actuation commands.

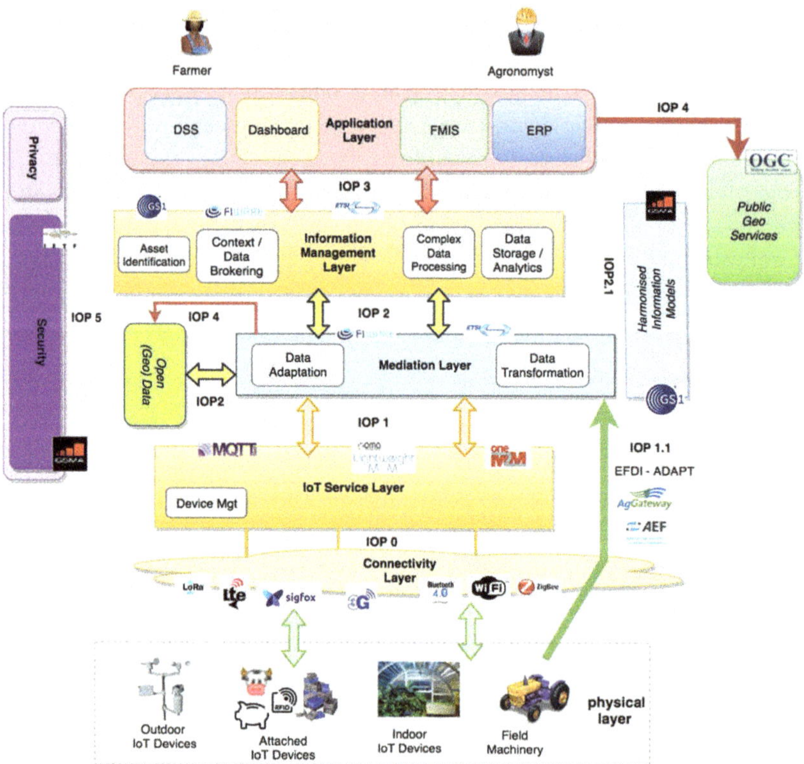

Figure 8 Overview of the IoF2020 reference architecture.

- Mediation layer. This layer transforms the raw data coming from devices or other external services into curated and harmonized data. It also aggregates some data.
- Information management layer. This layer serves mainly as a data hub. It enables the processing of all the information relevant to a food and farming solution.
- Application layer. This layer contains all the different applications such as those related to decision support (DSS), farm management (FMIS) or enterprise resource planning (ERPs).
- Security and privacy layer. This cross-cutting aims to guarantee secure access to information and devices, while protecting the data privacy of farmers and other actors in the system.

Other elements of the system include:

- Open data providers and public geo-services. Examples include public databases offering data in the agricultural domain, geo-services publishing weather or spatial data or even satellite data/image platforms.
- Harmonized information models. These define the structure and representation of the information to be managed, with a view to enable interoperability and portability of solutions in a wider ecosystem.

These interoperable elements are discussed in more detail in the following sections.

6 Key elements in data management

6.1 Interoperability point 0: IoT connectivity layer

The IoT connectivity layer enables communication between IoT devices or agricultural machines (physical device layer) and data gathering platforms. It enables transmission of data from devices (uplink) and reception of actuation commands or task plans by device (downlink). There are three different enabling technologies:

- Short-range communications are based on well-established standards for wireless indoor communications based on local area networks (esp. WiFi and Bluetooth). There are also solutions for outdoor facilities as the ZigBee communication protocols (IEEE 802.15.4). There are challenges related to device battery life, network coverage and operating costs. ZigBee is being used in a solution based on wireless sensor networks for precision agriculture.
- Cellular networks allow data transmission at high speed, long range, with high reliability and autonomy. These include cellular and telco-operated

networks exploiting 3G, 4G and 5G technologies. The main challenge is universal coverage, battery lifetime and device costs. Rural areas still have low connectivity in many areas.
- Low power wide area (LPWA) wireless technology complements existing cellular mobile network and short-range technologies, with lower costs and better consumption characteristics. LPWA networks are well suited for smart farming as they provide long-range communication and low costs for devices together with long battery life. SigFox employs a proprietary technology that enables communication using the industrial, scientific and medical (ISM) radio band. LoRa is intended for wireless battery-operated devices on a regional, national or global network. LoRaWAN targets key IoT requirements such as secure bidirectional communication, mobility and localization services. NB-IoT enables a wide range of devices and services to be connected using cellular telecommunication bands and focusses specifically on indoor coverage, low cost, reliability, long battery life and the ability to connect many devices. LTE-M combines several radio access network and core network features to optimize LTE networks for IoT needs and support of new category of LTE devices.

6.2 Interoperability point 1: IoT service layer

The IoT service layer identifies the raw data generated from IoT devices through different application-level transport protocols based on different paradigms. It also offers interfaces that allow communication with devices for management or actuation purposes. The most relevant technologies enabling this interoperability are:

- MQTT is a lightweight event and message-oriented protocol allowing devices to asynchronously communicate efficiently across contained networks to remote systems. It is a publish/subscribe messaging protocol capable of delivering messages from one publisher to multiple subscribers of a topic. There are multiple MQTT open source applications and client libraries and the protocol has been successfully implemented in the smart farming domain.
- MQTT-SN can be considered a version of MQTT further adapted to wireless communication and optimized for implementation with low-cost, battery-operated devices with limited processing and storage resources. There are different open source applications of MQTT-SN.
- LWM2M is an application layer communication protocol between LWM2M sensors and LWM2M clients. It makes use of a light and compact protocol and an efficient resource data model. It is frequently used with CoAP, which

is a specialized web-transfer protocol for use with constrained nodes and constrained networks in the IoT, particularly designed for machine-to-machine applications. The OMA LWM2M protocol, which enables remote management of M2M devices, has a modern architectural REST-based architecture and is highly adaptable.

- ONEM2M is a horizontal IoT/M2M middleware platform providing common functions, that is, abstract and common APIs, for different vertical service domains. There are commercial and open source versions of this technology.

The main adoption barrier in this area is the existence of diverse standards developed by different bodies. There is a promising initiative called Web of Things, aiming to link IoT devices to the web, regardless of IoT protocols. Although there is no final specification yet, this technology should be developed in the future.

6.3 Interoperability point 1.1: agricultural machinery communication layer

This layer sits between the agricultural machinery (tractors, farm equipment etc.), and the mediation layer, and allows communication of relevant data generated from machinery to the cloud. The most relevant standards are:

- ISOBUS (ISO11783) governs electronics and data exchange between different farm machines (e.g. tractor and farm implement, for example, sprayer) and has been the de-facto standard for decades for tractor manufacturers. A local communication bus system based on CAN bus connects the tractor and other components. Data interchange and process flow are also defined in parts of this standard using ISO-XML and EFDI.
- The ADAPT framework is comprised of an agricultural application data model, a common API, and a combination of open source and proprietary data conversion plugins. Participating FMIS (farm management information system) companies are responsible for completing their own mapping of the agricultural application data model to their FMIS data model.

6.4 Interoperability points 2, 3 and 4: mediation and information management layer

The aim of the mediation and information management layers is to offer the right information to the right application at the right time. These layers are responsible for transforming raw data into information relevant and ready to be consumed by applications, so that smart behaviours are exhibited, enabling the optimization of agri-food processes.

The mediation layer, situated between the IoT service layer and the information management layer (interoperability point 2), is responsible for gathering the raw data coming from devices or other external services. It then curates, harmonizes and, in some cases, aggregates the data so that it can be published as context information, or supplied to upstream data processing algorithms or analytics. In addition, this layer is also capable of sending actuation commands to the IoT service layer. Finally, the mediation layer may also be used for gathering data from other data sources such as agricultural machinery or public geo-services (interoperability point 4).

The information management layer, situated between the mediation layer and the application layer, serves mainly as a data hub to enable consumption of all information relevant to an application (interoperability point 3). Elements include:

- FIWARE NGSI: It is a version of the OMA NGSI-9 and NGSI-10 abstract interfaces for context information management. FIWARE NGSIv2 is based on HTTP/REST and JavaScript Object Notation (JSON), following the usual de-facto industry standards. NGSI supports a powerful yet simple and well-known approach to represent context information, with a meta-model based on entities, attributes and metadata. The most popular version of FIWARE NGSI is the Orion Context Broker, which uses MongoDB as the underlying data source.
- NGSI-LD: It is an evolution of the OMA NGSI information model, designed to better support linked data, property graphs and semantics. It is being developed under the ETSI ISG CIM initiative.
- WFS: It is an interface specified by the Open GIS Consortium to allow exchange of geographic data across the web. It defines the rules for requesting and retrieving geographic information using HTTP. The interface describes the data manipulation operations for geographic features. XML-based geographic markup language is used to exchange information.
- WMS: It is a specification outlining communication mechanisms to allow disjoint software products to request and provide preassembled map imagery to a requesting client. Using WMS, a request results in a ready-made map which can be displayed.

The main adoption barrier in the mediation and information management layers is the proliferation of proprietary APIs, vocabulary and incompatible data models.

6.5 Interoperability point 2.1: harmonized information models

Interoperability in the agri-food sector is not only a matter of harmonized APIs (such as NGSI-LD or WFS). It also requires harmonized domain-specific data models capable of modelling different concepts relevant to different applications. Key elements include:

- CLP.26-IoT big data harmonized data model, published by the GSMA IoT Programme. This is a normative document describing some harmonized data entities used in different IoT domains such as agriculture. The definition of different entity types is based on JSON and the FIWARE NGSIv2 information model, while reusing some parts of schema.org.
- ADAPT framework, developed by AgGateway, includes three core elements: the ADAPT data model (ADM), a plugin manager and a set of plugins. The ADAPT framework includes a proprietary plugin which maps a system's model into the ADM model and serializes the information into a container to share it. The receiver deserializes the container using its own plugin to extract relevant information for its system.
- GS1 core business vocabulary is the data standard used within the electronic product code information service (EPCIS). It is a GS1 standard interface for capturing and sharing event data independent of any data carrier. The core business vocabulary defines elements and their values, for example, for business step identifiers, disposition identifiers, business transactions and respective types, and source/destination identifiers and types.

GS1 enables unique asset identification through different standards:

- GEPIR (Global Electronic Party Information Registry) is a unique, internet-based service that gives access to basic contact information for companies that are members of GS1. These member companies use GS1's globally unique numbering system to identify their products, physical locations or shipments.
- Global Location Number (GLN) can be used by companies to identify their locations, giving them complete flexibility to identify any type or level of location required.
- Global Trade Item Number (GTIN) can be used by a company to uniquely identify all of its trade items. GS1 defines trade items as products or services that are priced, ordered or invoiced at any point in the supply chain.
- GS1 has two GS1 keys for asset identification. The Global Returnable Asset Identifier (GRAI) is especially suitable for the management of reusable transport items, transport equipment and tools and can identify these returnable assets by type and individually for tracking and sorting purposes.

The Global Individual Asset Identifier (GIAI) can be applied on any asset to uniquely identify and manage that asset. This could be, for example, a computer, desk, vehicle, piece of transport equipment or spare part.

- The Electronic Product Code™ (EPC) is a syntax for unique identifiers assigned to physical objects, unit loads, locations or other identifiable entities playing a role in business operations. EPCs have multiple representations, including binary forms suitable for use on radio frequency identification (RFID) tags, and text forms suitable for data sharing among the enterprise information systems (EISs). GS1's EPC tag data standard (TDS) specifies the data format of the EPC, and provides encodings for numbering schemes - including the GS1 keys - within an EPC.

6.6 Interoperability point 5: security and privacy

The digitalization of the agri-food industry to enhance farming processes implies that data is being generated and exchanged throughout the production process. The increasing exchange of data is a major challenge for the sector, and poses questions about privacy, data protection, intellectual property, data attribution (ownership), relationships of trust/power, storage, conservation, usability and security.

An IoF2020 solution should be able to properly react to data privacy and security violations with defined procedures and should incorporate capabilities in order to secure the platform which is going to support the farming services. It needs to provide support for confidentiality, integrity, authentication, authorization, trust and non-repudiation, when needed. IoF2020 has defined a set of IoT Security Guidelines to be followed when implementing use cases and trials. Some relevant initiatives include:

- The EU General Data Protection Regulation (GDPR) was designed to harmonize data privacy laws across Europe, to protect and empower the data privacy of all EU citizens and to reshape the way organizations across the region approach data privacy.
- A coalition of associations from the EU agri-food chain has launched a joint EU Code of Conduct on agricultural data sharing. The Code promotes the benefits of sharing data and enables agri-business models, including agri-cooperatives and other agri-businesses, to swiftly move into digitally enhanced farming. The Code reviews contractual relations and provides guidance on the use of agricultural data, particularly the rights to access and use of data.
- GSMA IoT security guidelines provide an approach to end-to-end security, including 85 detailed recommendations for secure design, development and deployment of IoT services. These guidelines promote best practices

for the secure design, development and deployment of IoT services, and provide a mechanism to evaluate security measures.
- oneM2M develops technical specifications and standards which address the need for a common M2M service layer in IoT. oneM2M has issued two relevant technical specifications, namely TS-0003-V2.12.1 Security Solutions and TR-0008-V2.0.1 Security.

Following these guidelines and specifications will optimize system security and protect the data rights of the different actors in the system.

7 Conclusions

Translating data into useful and actionable information requires appropriate software applications to be developed. This chapter has emphasized the importance of starting from a business process management perspective. Working in a supply chain network context means dealing with multiple stakeholders and their own business processes and objectives. Interoperability, the ability of computers and machines to seamlessly communicate with each other is a key challenge to address in this context. A reference architecture of various layers and interoperability points can help to create a system foundation which allows data to flow easily and effectively through the system. Moreover, it can stimulate re-use and replicability of components and use-case scenarios, leading to an increased adoption of data-driven applications in agri-food business.

8 Where to look for further information

The work in this chapter is largely based on a European innovation ecosystem that was developed through a series of EU-funded projects. A conceptual foundation was laid in a paper that was published in 2010 (Wolfert et al., 2010). The concepts and ideas of this paper were further elaborated and materialized through the following projects that published reports, which are detailed sources of information:

- SmartAgriFood (www.smartagrifood.eu) - Deliverable D600.2 'Plan for standardisation for large scale experimentation' contains a detailed overview of various data standards that are used in agri-food (https://cordis.europa.eu/docs/projects/cnect/6/285326/080/deliverables/001-D6002PlanForStandardiationFINAL2.pdf);
- FIspace (http://www.fispace.eu/) - A series of three deliverables (D500.4.x; x = 1,2,3) deal with various standardization issues can be found at http://www.fispace.eu/publicdeliverables.html;

- FIWARE (www.fiware.org) - this was the overarching project for SmartAgriFood and FIspacein the Future Internet programme that also covered various other sectors (e.g. smart cities, energy, health). FIWARE was turned into a foundation that is supported by various ICT businesses and now has a wide, active network of developers and sources, including a subcommunity on smart agrifood (https://www.fiware.org/community/smart-agrifood/);
- IoF2020 (www.iof2020.eu) - a large-scale pilot on the application of the Internet of Things (IoT) in farming and food production. The reference architecture that is summarized in this chapter is described in more detail in several deliverables of work package three (https://www.iof2020.eu/about/deliverables).

In these projects, there is an active involvement of relevant standardization organizations such as:

- AgGateway Europe (https://www.aggateway.org/AboutUs/Europe.aspx) - part of the AgGateway Global Network. In this context, the ADAPT framework is of particular interest (https://www.aggateway.org/GetConnected/ADAPT(inter-operability).aspx);
- AEF (https://www.aef-online.org/) - the Agricultural Engineering Foundation that amongst others develops and maintains the ISOBUS standard for communication between agricultural machinery;
- GS1 (www.gs1.org) - that global standards organization that provides all kinds of standards for various sectors amongst others agri-food.

9 References

Cantera, J. M., Issa, J. S., van der Vlugt, P., Klaeser, S., Bartram, T., Kassahun, A., Neira, I. and Milin, T. 2018. D3.3 opportunities and barriers in the present regulatory situation for system development. In: IoF2020 (Ed.), *IoF2020 Project Deliverables*.

Kaloxylos, A., Eigenmann, R., Teye, F., Politopoulou, Z., Wolfert, S., Shrank, C., Dillinger, M., Lampropoulou, I., Antoniou, E., Pesonen, L., Nicole, H., Thomas, F., Alonistioti, N. and Kormentzas, G. 2012. Farm management systems and the Future Internet era. *Computers and Electronics in Agriculture* 89, 130–44. doi:10.1016/j.compag.2012.09.002.

Sundmaeker, H., Verdouw, C., Wolfert, S. and Pérez Freire, L. 2016. Internet of food and farm 2020. In: Vermesan, O. and Friess, P. (Eds), *Digitising the Industry - Internet of Things Connecting Physical, Digital and Virtual Worlds*. River Publishers, Gistrup, Denmark and Delft, the Netherlands, pp. 129–51.

Verdouw, C. N., Wolfert, J., Beulens, A. J. M. and Rialland, A. 2016. Virtualization of food supply chains with the internet of things. *Journal of Food Engineering* 176, 128–36. doi:10.1016/j.jfoodeng.2015.11.009.

Verdouw, C. N., Wolfert, S., Beers, G., Sundmaeker, H. and Chatzikostas, G. 2017. IOF2020: fostering business and software ecosystems for large-scale uptake of IoT in food and farming. In: Nelson, W. (Ed.), *The International Tri-Conference for Precision Agriculture in 2017*, Hamilton, p. 7.

Wolfert, J., Verdouw, C. N., Verloop, C. M. and Beulens, A. J. M. 2010. Organizing information integration in agri-food - a method based on a service-oriented architecture and living lab approach. *Computers and Electronics in Agriculture* 70(2), 389–405. doi:10.1016/j.compag.2009.07.015.

Wolfert, J., Sørensen, C. G. and Goense, D. 2014. A future internet collaboration platform for safe and healthy food from farm to fork. *2014 Annual SRII Global Conference (SRII)*. IEEE, San Jose, CA, pp. 266–73.

Chapter 3

Improving data identification and tagging for more effective decision making in agriculture

Pascal Neveu and Romain David, MISTEA, INRAE, Montpellier SupAgro, University of Montpellier, France; and Clement Jonquet, LIRMM, CNRS and University of Montpellier, France

1 Introduction

Global demand for food products is increasing sharply, and the current growth rates in agriculture are clearly inadequate and ill-adapted. Today, an urgent and profound redesign of agriculture is crucial to increase production and reduce the environmental impact. In this context, a major challenge is the shift to digitization - entering the Big Data era - to enable a better understanding of the complex mechanisms underlying the sustainable improvement in crop yields and adaptation. This requires studying not only the genotype, phenotype and environment relationships but also the social or health aspects. This highly interdisciplinary challenge (agronomy, genetics, biology, sociology, etc.) requires intensive data integration. In this context, we shall develop and promote new methods and tools for decision makers, researchers and other agricultural actors, especially in relation to:

- the development of the use of sensors with a smart data acquisition system suitable for the areas such as precision farming,
- the advances in the design of interoperable information systems for agricultural (Big) data, and
- providing data structuring frameworks for visualization, data analytics, knowledge discovery and decision support.

http://dx.doi.org/10.19103/AS.2020.0069.04

However, agriculture researchers face multiple data challenges such as (i) the change of scale in the production and exploitation of data (Big Data), (ii) the need to share and reuse these data with others in an open approach (Open Data) and, finally, (iii) interoperability and the transformation of data into knowledge (Linked Data). Structuring the data is a prerequisite for more effective data exploitation, analysis and decision making. By 'structured', we mean data that are organized into different parts following a specific data model, for example, data contained in databases or spreadsheets or stored in a specific standard format. In Section 2, we discuss two specific aspects related to structuring the data:

- Identification: objects, concepts and data must be clearly identified.
- Semantics and tagging: the meaning of objects and concepts and the relation between them must be clearly formalized.

Additionally, data describing contexts, often from outside producers, are key to interpreting antropic and natural phenomena and effects on agrosystems. Parameters concerning the weather, pedology, hydrology and social environment are produced and banked by different organizations. For instance, social and biodiversity data, which can be essential for developing agro-ecological approaches, are produced and hosted by a multitude of actors (institutes, associations, environmental agencies, etc.). For efficient decision support, these heterogeneous contextual data must enrich and complement agricultural data. In the meantime, both agricultural data and context data are now provided by thousands of various data sources (data repositories, registries and knowledge bases), requiring scientists and stakeholders to develop international recommendations and standards to improve interoperability while ensuring data traceability and ownership. Better semantically described data have proved a source for better decision support systems, including in agriculture (Lousteau-Cazalet et al., 2016; Guillard et al., 2015). Stakeholders are now embracing the FAIR data principles (Findable, Accessible, Interoperable, and Reusable) to allow the implementation of integrative analyses and multisource decision support systems, based on better data structuring, analysis and curation (Wilkinson et al., 2016).

High-throughput phenotyping (phenomics), the plant selection process that aims to identify the most adapted genotypes, is a good illustration of the data challenges faced by the agricultural research community. For example, in plant sciences, phenomics platforms produce huge complex datasets (images, spectrum, human readings, soil analysis) from different scales (molecular to plant population) in various contexts of strongly instrumented installations (field, greenhouse). Phenomics datasets must be accessible to the scientific communities (geneticist, bioinformatician, ecophysiologist, agronomist, statistician, sociologist, etc.) who have intensive data integration needs in order to help them in their selection. This case study will be further detailed in Section 3.

2 Structuring the data

In agriculture, observation and management systems, developed and used in many settings, produce a large volume of heterogeneous data, which are difficult to aggregate since they focus on specific issues. There are various data sources in agriculture that require miscellaneous knowledge and skills to be used together accordingly. For instance, agricultural data sources can be related to agricultural production, farm practices, transformation, distribution and so on. Since a few years, another important sources of data are not only connected objects in agriculture (Tzounis et al., 2017) - weather stations, insect traps, soil moisture sensors and water meters connected to irrigation - but also various sensors installed on animals to evaluate their conditions (health measures, temperature, movement), milking robots (quantity and quality of milk) or feeding automata. Agro-equipment is increasingly enriched with sensors, for precision farming (e.g. provide the plant exactly what it needs) and predictive maintenance. Satellite images are another example: the Sentinel constellation delivers free images at a very high temporal frequency (every 5 days), which opens up new research and business opportunities. Agricultural production traceability requirements are now supported, in part, by automated reading systems, with radio frequency identification (RFID) and NFC chips, or by the manual input of agricultural interventions from smartphones with direct transmission to the applications software. The challenge is to automate data acquisition so that it has virtually no cost and is not an additional charge for farmers or scientists (Wolfert et al., 2017). Finally, high-throughput phenotyping methods, essential for shortening the production cycle of new seeds, are also sources of massive data (e.g. phenotype-monitoring platforms produce thousands of images per day) to link with genotypic data (Halewood et al., 2018).

Organized and structured access to primary agricultural data is a sine qua non condition for building efficient decision support systems to achieve the conservation of biodiversity and sustainable development. Organizing, managing and storing of various data require new approaches. Proper data structuring enables to organize data to suit a specific purpose so that they can be accessed and worked with in appropriate ways. The better the data structure, the better we will be able to group them with other data and learn from them.

2.1 Identification

An identifier is a sort of name that identifies a specific object (digital or not) in a set of objects. In an ideal world, identifier should be unique for each object (bijection); in practice this is rarely the case. In most cases a resource (object) can have several (not all unambiguous) identifiers depending on the context. An identifier is unambiguous if it makes it possible to identify an individual in a specific context in a safe way (McMurry et al., 2017). An unambiguous identifier, which cannot refer two different objects, is called a globally unique identifier (GUID) or a universally unique identifier (UUID); irrespective of whatever the database or source, all disciplines taken together, no other object will be designated identically, for example, ISBN for books. For software objects, GUIDs are typically randomly generated 128-bit codes. There are several specifications for identifiers, for example, UUID, LSID, ARK, DOI, URI, RFID, XRI. The relevance of these different mechanisms depends on the context and, of course, of the characteristics of objects to identify. Data identification also depends on the range of the use of the resource. If the resource shall be referenced only within a limited range or system, it could be assigned a local identifier. But if it shall move to another system (e.g. for the purposes of expert measures such as chemistry of soil and water quality) or if it shall be reused and aggregated with data of different provenances or contexts, a 'reliable' global and long-term identification mechanism is necessary.

Long-term structuring of data requires to reliably identify all the concepts, objects and their properties described in the information systems. A persistent identifier is an identifier that is permanently assigned to an object (ideally usable in several decades). For example, once an ISBN is assigned to a particular book, that number is always associated with that book and no other book will ever receive the same number. Likewise, identifiers must be persistent and shall not change. The problem is that during periods of decades, many changes can occur not only within databases but also in institutions or organizations in charge of the data. It is thus necessary to preserve and recover dependencies between these elements, in time and in localization. Persistent identifiers play a key role in adopting Open Science (Dappert et al., 2017). The reliability of this identification depends on some essential qualities described, for instance, in W3C Recommendations (https://www.w3.org/TR/cooluris/) and must assure persistent security, traceability and reusability of data. The key to rich integration is a commitment to deploy and reuse globally unique, shared identifiers and to implement services that link those identifiers (Page, 2008). The major persistent identification system appears in chronological order: Handle (1994), Persistent URL (1995), Uniform Resource Name (URN; 1997), Archival Resource Keys (ARKs; 2001) and eXtensible Resource Identifier (XRI; 2005).

For instance, persistent GUIDs are usually generated as groups of dash-separated hexadecimal characters, for example, 120a-e29f-a861-12f5-5a52. Their three main qualities are: 1) to be generated in a non-centralized way, 2) to make extremely improbable the random generation of two identical identifiers and 3) to be completely opaque and not sensitive to the changes of authorities or names of authorities. These automatically generated GUIDs can be used as a basis for the construction of other identifiers, for example, by adding a prefix (URI, URL, domain name, authority name). A GUID, when it is integrated in an URI, can be dereferenceable as explained hereafter.

Uniform resource identifier (URI) is defined by the RFC 3986 standard[1] provided by the W3C, which specifies: 'An URI is a compact sequence of characters that identifies an abstract or physical resource. This specification defines the generic URI syntax and a process for resolving URI references that might be in relative form, along with guidelines and security considerations for the use of URIs on the Internet. The URI syntax defines a grammar that is a superset of all valid URIs, allowing an implementation to parse the common components of a URI reference without knowing the scheme-specific requirements of every possible identifier. This specification does not define a generative grammar for URIs; that task is performed by the individual specifications of each URI scheme'. All Web hyperlinks (URLs) are expressed as URIs. Dereferencing is also an important aspect: an URI is said to be dereferenceable if it is possible to obtain all the digital contents describing the referenced resource (e.g. URL). The act of retrieving an information of a resource identified by a URI is known as dereferencing that URI. To summarize on URI/URL one can say that:[2]

- URL identifies what exists on the Web;
- URI identifies, on the Web, what exists; and
- IRI identifies, on the Web, in any language, what exists.

Life science identifiers (LSIDs) are represented as an URN with the following format: urn:lsid:<Authority>:<Namespace>:<ObjectID>[:<Version>]. But LSIDs are not strictly URIs, and so are not always dereferenceable. Bioinformatics and biodiversity communities use them as a way of identifying species in global catalogs. LSIDs have been criticized as violating the Web architecture's good practice of reusing existing URI schemes.

Digital Object Identifier (DOI), initially used in bibliographic databases, allows the identification of digital resources, such as a report, scientific articles or any other type of digital object objects. The purpose of the DOI is to associate metadata describing the object, for example, in bibliography, to produce more reliable, unambiguous and longer-lasting citations. DOIs are issued by DOI agencies, part of the DataCite consortium. A DOI is a special case of Handle ID[3] with the following format: doi:10.<Naming Authority>:<Registry_Number>, and it contains a link to the metadata (restrictions of use or copyright and

1 https://tools.ietf.org/html/rfc3986
2 Credit to Fabien Gandon's (INRIA).
3 The Handle System is a technical specification for assigning, managing and resolving persistent identifiers assigned to digital objects and other internet resources.

naming authority among others), described by a data model common to all DOIs, the indecs Data Dictionary, an address or physical location for the digital object (usually a URL) that the DOI translator will use to redirect. For instance, prefixing a DOI with https://doi.org/ allows to dereference the identifier into a landing page storing or describing the object identified. A DOI provides a good frame for a persistent identification of agricultural datasets.

ARK is a perennial identifier system based on the URI standard. ARK is designed to ensure long term identification of a resource, scalability and independence. An ARK contains a portion impervious to changes and a flexible portion, which designates a shape of the object or a mode of access thereto. An ARK URL is subdivided into two URLs: the first, optional, gives the addressing authority Name Mapping Authority (NMA), while the second is the ARK URL, fixed and proper, which includes a Name Assigning Authority Number (NAAN) and the name given to the object.

An XRI is a schema and resolution protocol for abstract identifiers compatible with URIs. The goal of XRI is to provide a universal format for abstracts, structured identifiers that are independent of domains, locations and transport applications, so that they can be shared across a large number of domains, directory and protocols.

Identifying samples and real objects with a persistent identifier is possible with several standardized methods that can be linked with previous persistent identifiers (e.g. bar code that is a visual, machine-readable representation that describes something about the object that carries the barcode). It can have one or two dimensions and represents a numerical identifier. For instance, Universal Product Code from industrial sector is a worldwide retail, GS1-approved international standard (ISO/IEC 15420).

The identification of real objects has been increasing since the appearance of the internet of things (IoT). Between RFID chips, naming solutions and middlewares, the IoT is composed of many complementary elements, each having its own specificities.

For real objects, RFIDs are based on radio tags that can be pasted or embedded in objects or products and even implanted into living organisms (animals, human body). This identification method can be used to identify objects, such as those with a barcode (electronic label), people (being integrated in passports, transport card, payment card or domestic carnivores by implantation under the skin), cats, dogs and so on. The RFID identification of pets is mandatory in many countries. For traceability purposes, this is often the case for farm animals. The International Geo Sample Number (IGSN) retains the identity of a sample even if it is transferred from one laboratory to another, and the data appear in different publications, thus eliminating any ambiguity from similar names of other terrestrial samples. It allows researchers to reconstruct the analytical history of a sample. The IGSN, developed as part of System for

Earth Sample Registration (SESAR), is a nine-character identifier. It is designed to ensure backward compatibility with previously collected data as new techniques are developed. The IGSN network allows to link data generated by researchers and published in different scientific articles. Research Resource Identifier (RRID) aims to authenticate the key resources: antibodies, model organisms and tools (software, databases, services). But it is dedicated to the medical domain, and it could be relevant to extend it to agriculture.

The persistence of an identifier relies on the durability of the system to provide the identifier and its capacity to dereference. It is clear that this sustainability is not always a strong priority of institutions. Durability also depends on the durability of the organizations themselves. The missions and perimeters of national and international identifying organizations are regularly re-evaluated and modified. Several global database organizations were created, to catalog and monitor research organizations worldwide, such as the Global Research Identifier Database (GRID), Ringgold IDs, International Standard Name Identifiers (ISNIs) or the Research Organization Registry. For instance, most URIs, regardless of their type, include the name of the institute that generates or host them. These changes needed to be tracked so that identifiers stay valid; this is the role of so-called data authorities and a prerequisite to the consensual adoption of a truly perennial and shared global identification system. Today, many self-established or defecto reference identifier generators, some of them proprietary, co-exist (e.g. Pensoft, Zenodo, PubMed, ResearchGate, ResearcherID, HAL-ID). However, the existence of systems established and promoted by economic actors is questionable, either for ethical or economical reasons, for example, what if Google-Bing-Yahoo-launched Schema.org reference system was stopped because it is unprofitable? Therefore, identifier governance and management should be based on a system equivalent to the management of domain names and supported by a standard Web organization like W3C.[4]

We clearly established identifiers that are used in schemas or standard vocabularies and ontologies (cf. Section 2.2) to provide information (properties, relations) about the object (e.g. responsible organization, type of object, definition, labels). As ontologies are changing with digital objects, the persistent identification method must support different versions of an object. Versioning becomes then an important aspect when building identifiers, for example, predefined period and important update releases (curation of data, campaign of collection). On the other hand, some versioning processes must trace all the transformations made on the data for history management. Services such as B2HANDLE can allow to support this.

4 The World Wide Web Consortium is led by three organizations: the MIT Computer Science and Artificial Intelligence Laboratory (the United States), Keio University (Japan), the National Institute of Research in Computer Science and Automation (France). Its role is only advisory.

Today, the URI system is a standard used in a large variety of domains: genetic, chemistry, IoT, life sciences and so on. As an identifier, the URI must have some properties: non-ambiguousness, unicity, persistence, stability and resolvability.

Here is an example of an URI: http://www.phenome-fppn.fr/m3p/arch/2017/c17000915 (which is not dereferenceable). It uses the following pattern: http://subdomain.yourdomain.topdomain/path/identifier.

Properties of URIs are:

- Non-ambiguousness: the URI must be associated with only one resource.
- Unicity: only one URI for one resource.
- Persistence: once a resource is given an URI, one should not replace or delete the URI.
- Stability: URI has to remain the longer possible (at least 20 years) and should not be reassigned to another resource. The definition is close to the persistence; stability is persistence over a long time.
- Resolvability: URI should be used through internet browser to find information about the resource or the resource itself (also called dereferenceable).

When these principles are not respected, one may encounter several issues. Usually non-ambiguousness and unicity are usually not a problem as everyone understands their importances. However, stability and persistence are much more difficult to get: typical case is when part of this URI is changing. For example, the domain name www.phenome-fppn.fr later becomes phenotyping.fr. Thus, the unicity of phenotyping.fr/m3p/arch/2017/c17000915 is not guaranteed; there could be two different resources identified with the same URI, the phenotyping.fr ID and another with the phenome.fppn.fr ID.

In summary, few rules must be followed to create good URIs in agriculture: i) use minimal information and do not use everything that may change, ii) use persistent URL, iii) provide multiple output format - content negotiation - and link them together, iv) request on the external identifier (identifiers.org, n2t.net, w3id.org, ePIC[5]), v) integrate and reuse already-existing identifiers. Things to avoid include: i) avoid file extension in the URI, ii) avoid query-specific characters (e.g. '?' or '&'), iii) use auto-incrementation carefully (not for versioning purpose). To conclude, identification is a first step to go further in order to improve decision support. Once the objects have been identified, it is necessary to specify how they are used and the role of each in relation to the others, in other words, to allow the implementation of methods related to semantics and tagging.

5 https://www.pidconsortium.eu/

2.2 Semantics and tagging

2.2.1 Interoperability with ontologies and the Semantic Web

Data structuring and understanding need metadata for their description and use. Several categories of metadata must be provided such as descriptive, administrative, technical and provenance. Too often, metadata are poor and incomplete, hampering effective data reuse. In some virtuous situations, these data and associated contexts can be informed, very precisely, but not in a machine-readable format. Metadata are often simple (wording, date of creation, contact point, cartographic projection, size, etc.) or very detailed (data quality measurements for each data element, provenance, versioning or historical maintenance service of the measurement instruments, constraints of use, etc.). Structuring the data and providing metadata are essential for the understanding and good use of data in decision processes. It is therefore important to pay special attention to the semantics of the data. Earlier we get metadata better they are.

Semantic interoperability enables data integration and fosters new scientific discoveries by exploiting various data acquired from different perspectives (e.g. agricultural and context data). For instance, a scientist experimentally measures the sensitivity of a plant to a disease (agronomy vision), whereas a farmer concretely observes the leaves of the plant turning brown (agriculture vision). Both are phenotypes, or traits, information, but they come from two different worlds that must yet be more connected. This shall be possible only through lifting the data into meaningful knowledge for humans, yet exploitable by machines.

A researcher studying a certain plant trait (e.g. resistance to a disease) is interested in the gene that controls this phenotype, the expression of this trait in different crop varieties observed in different environments and, of course, its effect on the crop yield or for associated needs such as the use of pesticides. The information we need to answer such questions is available in multiple datasets expressed using various ontologies (Crop Ontology, Plant Ontology, Trait Ontology, etc.) and at various levels (e.g. population, individual, organ); the issue is finding that information and combining it in a meaningful way for researchers, breeders and ultimately farmers, consumers or any stakeholders of the value chain.

Ontology engineering is a sub-domain of knowledge engineering that deals with knowledge representation and reasoning. An ontology is described as a 'formal specification of conceptualization' (Gruber, 1993); it 'defines the terms to describe and represent an area of knowledge'. Ontologies are composed of concepts, relations and instances. For example, if you want to define a car, you should say: 'a car is a transportation object, with four wheels, and one needs a licence to drive it. My blue Ford Mustang is a car'. 'Car' is a concept, 'is a' is a relation and 'My blue Ford Mustang' is an instance.

The Semantic Web is the area in which ontologies are used to structure data into formal knowledge. The Semantic Web provides the necessary techniques and technologies to build a Web of data (or Linked Open Data (LOD)) as well as reasoning on ontology concepts and mapping between ontologies. The Semantic Web relies on a set of core technologies such as RDF, RDF-S, OWL, SPARQL and SKOS; all of them are built on top of the notion of URIs, which are employed to formally identify objects and remove ambiguity. The Resource Description Framework (RDF) is the W3C language to describe data. It is the backbone of the Semantic Web. SPARQL is the corresponding query language. Complementary, RDF Schema (RDF-S), the Web Ontology Language (OWL) and the Simple Knowledge Organization System (SKOS) are languages to build schemas, ontologies and vocabularies/thesaurus. Figure 1 (credit to http://lod-cloud.net) illustrates, as of the beginning of 2017 and previously, part of the

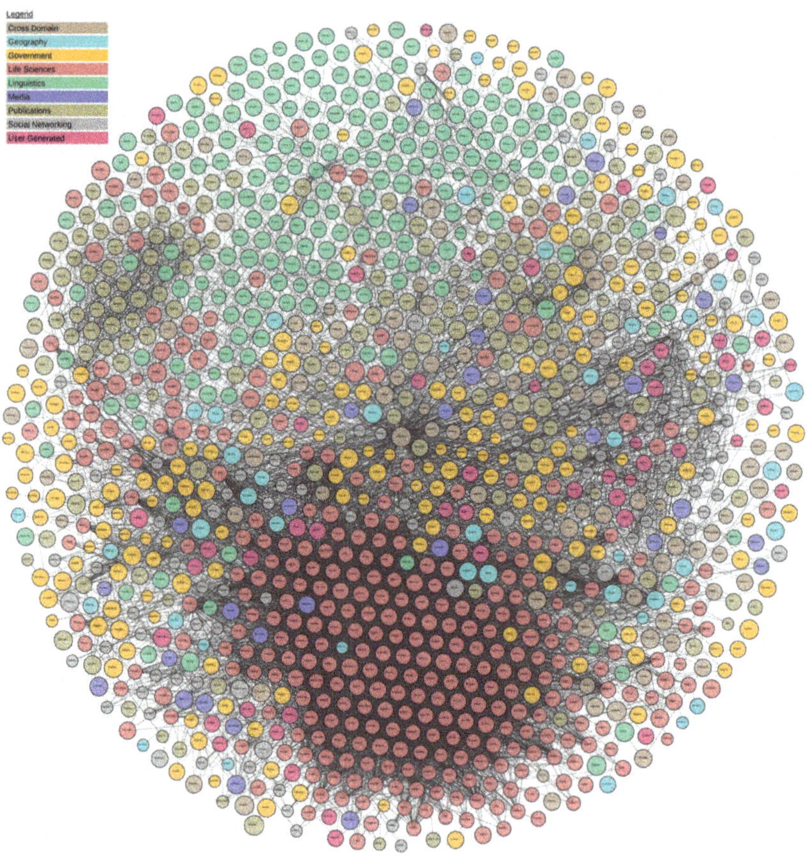

Figure 1 Linked Open Data cloud diagram (version 2019-03-29), CC-BY-SA by A. Abele, J. P. McCrae, P. Buitelaar, A. Jentzsch and R. Cyganiak (cf. http://lod-cloud.net).

amount available as LOD and the importance of ontologies/vocabularies (most of the life sciences section in pink are ontologies listed in the NCBO BioPortal; Noy et al., 2009).

The Semantic Web offers the methods and technologies to extract/transform Big Data into actionable knowledge (Antoniou and Van Harmelen, 2004).[6] It relies on standard vocabularies and ontologies to formally capture the knowledge of a domain into semantic resources that computers use to index, search or reason on the data.

Tim Berners-Lee, the inventor of the Web and initiator of the Linked Data project, suggested a five-star deployment scheme for Linked Data. The five-star Linked Data system is cumulative. Each additional star presumes the data meet the criteria of the previous step(s).[7]

☆ *Available on the Web, in whatever format*
☆☆ *Available as machine-readable structured data (i.e. not a scanned image)*
☆☆☆ *Available in a non-proprietary format (i.e. CSV, not Microsoft Excel)*
☆☆☆☆ *Published using open standards from the W3C (RDF and SPARQL)*
☆☆☆☆☆ *All of the above and links to other LOD*

The purpose of the Web of data is not to create another Web, since it is based on its current architecture (the URI system and the HTTP protocol), but to create an extension. RDF is to structured data what HTML is to documents, an interoperability framework that ensures consistency in the handling and processing of these data by machines.

2.2.2 Ontologies and semantic tagging in agriculture

In recent years, we have seen an explosion in the number of semantic resources (thesauri, terminologies, vocabularies and ontologies) being developed in agronomy and agriculture, for instance, the Plant Ontology, Environment Ontology, Crop Ontology or Agronomy Ontology, which opened the space from various types of semantic applications to data integration or decision support. Ontologies in agriculture are spread out around the Web (or even unshared), in many different formats and artifact types, and with different structures. Agronomy (and its related domains such as food, plant sciences and biodiversity) needs a one-stop shop, allowing users to identify and select ontologies for specific tasks, as well as offering generic services to exploit them in search, annotation or other scientific data management processes. The need is also for a community-oriented platform that will enable ontology developers and users to meet and discuss their respective opinions and wishes. And, with such a number of ontologies, new problems have raised such as describing, selecting,

6 See the MOOC Web of Data for a quick introduction: https://www.coursera.org/learn/web-data
7 https://www.w3.org/2011/gld/wiki/5_Star_Linked_Data

evaluating, trusting and interconnecting ontologies. For this reason, ontology repositories, such as AgroPortal (Jonquet et al., 2018), are offering a reference point of entry for interconnected vocabularies and ontologies in agronomy and agriculture. AgroPortal offers a robust and reliable service to the community that provides ontology hosting, search, versioning, visualization, comment and recommendation; enables semantic annotation; stores and exploits ontology alignments; and enables interoperation with the Semantic Web.

One important use of ontologies is for annotating and indexing text data. Indeed, ontologies allow representing data with clear semantics that can be leveraged by computing algorithms to search, query or reason on the data. One way of using ontologies is by means of creating semantic annotations or semantic tags. An annotation is a link from an ontology term to a data element, indicating that the data element (e.g. article, experiment, observation, medical record) refers to the term. When doing ontology-based indexing, we use these annotations to 'bring together' the data elements from these resources. However, explicitly annotating data is still not a common practice for several reasons (Jonquet et al., 2009):

- Annotation often needs to be done either manually by expert curators or directly by the authors of the data.
- The number and format of ontologies available for use are large, and ontologies change often and frequently overlap.
- Users do not always know the structure of an ontology's content or how to use the ontology to do the annotation themselves.
- Annotation is often a boring additional task without immediate reward for the author.

Semantic annotation is an important research topic in the Semantic Web community. Tools vary along with the types of documents that they annotate (e.g. image annotation). For an overview and comparison of semantic annotation tools, the reader may refer to the study by Uren et al. (2006).

Previous work has encouraged and exalted the use of ontologies for annotation at various levels (Rhee et al., 2008). For a while, the prevalent paradigm in the use of ontologies was that of manual annotation and curation. However, several researchers have shown that such manual annotation, though highly desirable, will not scale to the large amounts of data being generated (e.g. in the life sciences; Baumgartner et al., 2007). If one examines the reasons for the low adoption of ontology-based annotation methods among database providers, the high cost of manual data curation remains the main obstacle. In light of this situation, researchers have called for the need of automated annotation methods (Dowell et al., 2009) and for leveraging natural language processing tools in the curation process (Altman et al., 2008).

An example of such service for agronomic/agricultural data is the AgroPortal Annotator (http://agroportal.lirmm.fr/annotator), a Web service that provides a mechanism to employ ontology-based annotation in curation, data integration and indexing workflows, using any of the ontologies in the AgroPortal repository.[8] The Annotator tags raw text descriptions with relevant ontology concepts and returns the annotations to end users. Those annotations are mainly made of an URI identifying the annotating concept. Services like the AgroPortal Annotator can be used to structure data into unambiguous and semantically identified parts, hence contributing to the process of transformation/extraction of data into knowledge.

3 Case study: plant phenotyping

In this section, we provide an example of the use of some of the technologies and methods presented in Section 2 in a case study of plant phenotyping and agriculture (Neveu et al., 2019).

Plant-derived products are at the center of challenges posed by increasing requirements for food, feed and raw materials. Integrating approaches across all scales from molecular to field applications is necessary to develop sustainable plant production with a higher yield and using limited resources. While significant progress has been made in molecular and genetic approaches in recent years, the quantitative analysis of plant phenotypes - the structure and function of the plant - has become the major bottleneck. Plant phenomics is an interdisciplinary science that links genomics with plant ecophysiology and agronomy. The functional plant body (phenotype) is formed during plant growth and development from the dynamic interaction between the genetic background (genotype) and the physical world in which plants develop (environment). These interactions determine plant performance and productivity, measured as accumulated biomass and commercial yield and resource use efficiency.

Phenomics platforms produce huge complex datasets (images, spectrum, human readings) from different scales (genetic to plant population). Phenomics datasets need to be accessible to the large scientific community (genetician, bioinformatician, ecophysiologist, agronomist, etc.). Their reanalysis requires tracing relevant information on thousands of plants, sensors and events. The open-source Phenotyping Hybrid Information System (PHIS - http://www.phis.inra.fr) is proposed for plant phenotyping experiments in various categories of installations (field, glasshouse).

8 This service is AgroPortal's version of the NCBO Annotator described in (Jonquet et al., 2009).

3.1 Identification in PHIS

Tracking all objects involved in a phenotyping experiment (e.g. plants, pots, sensors) and representing relationships between them are essential in a high-throughput context where thousands of plots, plants or sensors are involved. This requires a proper strategy that allows to individually identify each specific object as well as semantic properties for creating relationships between such objects.

For instance, the replacement of a sensor at a given position (e.g. meteorological sensor or soil tensiometer) is not obvious in the outputs of an environmental database. In greenhouse experiments, a plant can be replaced by another plant at the same position and vector (e.g. pot, cart) during an experiment, potentially generating confusion. All objects therefore need to be identified in order to keep the necessary information associated to them (e.g. positions over time, successive calibration for sensors, origin for plants).

In the following text, we illustrate PHIS's identification system. PHIS object identification is based on URIs. This ensures traceability in space and time, while a typical identification by numbers (e.g. 'plant 736') refers to different plants in different experiments and installations. URIs are generated automatically for each object via the user interface and implemented by QR codes, creating a set of connected objects that can be accessed, along with all their properties, from any terminal (e.g. mobile device, barcode reader).

What are the things to identify? Ideally, we want to identify everything, but we have very different resources - do we identify them the same way? Are URIs the best option to identify every resource? Those are questions one should ask before designing an URI scheme. For example, measures collected by a sensor can be gathered in a dataset and require only one URI for the dataset, or even be aggregated in a database. Then, the measures per day are identified with a primary key or an incremental ID.

3.1.1 How to make non-ambiguous URI

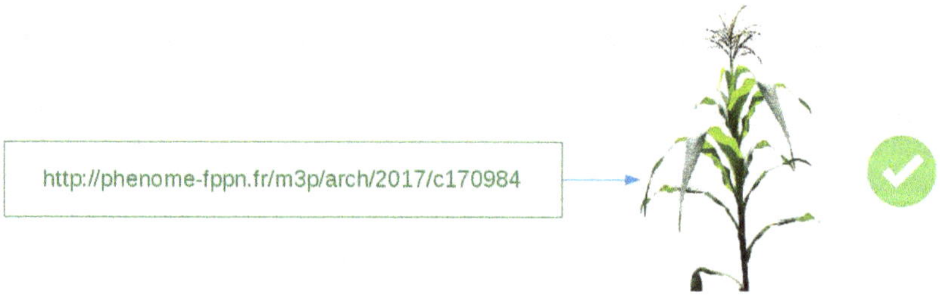

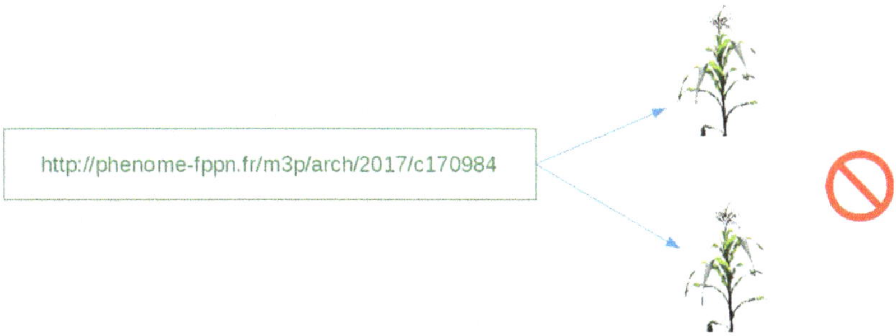

PHIS's non-ambiguous identifiers use an incremental number (the number of plants), prefixed with a letter that helps human manipulate the URI and real objects.

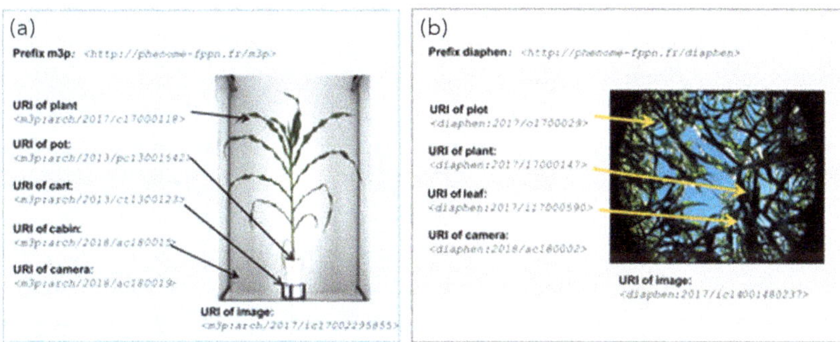

In PHIS, the semantic implementation is realized by a set of standardized ontologies written in OWL2. Based on these ontologies, the first step is to organize objects and concepts with a specialization hierarchy (sort of). For instance, corncob is a sort of a plant organ, that is, corncob is subClassOf plantOrgan. The description of this object (metadata) is formalized as properties. These properties can be values (dataProperty) or objects (objectProperty). Semantic links between objects, between events and between traits used in PHIS are realized through the annotation ontology and some specific application ontologies (such as Ontology of Experimental Events (OEEV)).[9]

In order to integrate data, the relations between objects need to be represented adequately in a high-throughput context. For instance, if thousands of sample tissues have been collected on different leaves of different plants, the information 'sample 884 belongs to the leaf 7 of plant 736' may be lost if kept in a spreadsheet. The links between objects are based on

9 http://agroportal.lirmm.fr/ontologies/OEEV

two OWL application ontologies. The Ontology for Experimental Phenotypic Objects (OEPO)[10] describes objects involved in phenotyping experiments (e.g. infrastructure, devices, germplasm, scientific objects) and defines specialization hierarchy between them according to the specificities of the installations and experiments. OEEV characterizes events that occur during an experiment, for example, moving of plants, dates of sowing, application of a given treatment, harvesting, measurements or sampling for -omics measurements or any category of technical problems. For instance, the Trouble concept distinguishes Breakdown (sensor or conveyor), Dysfunction (sensor fault, irrigation trouble) and Incident (a pot falls down, a leaf is blocked in an imaging cabin, lodging of a plot, human error, etc.). As described in the associated semantic graph, an event can be associated with objects (e.g. plant, plot, sensor) and with the user who has annotated the event, and the occurrence date can be tracked along with every relevant detail.

The use of ontologies allows to deal with the complexity of phenotyping data in order to link a large number of different data sources. Data integration process can be done automatically:

- Concept mapping is one of the approaches for data integration from different sources. Ontologies will help for concept mapping. For instance, the 'field' is equivalent to 'cultivated land'.
- Data-linking approach is based on the use of common standardized RDF properties in several data sources. It allows to identify common individuals in different sources. For instance, GPS coordinate values and the plant species name allow to know common plots of different datasets.

Ontology-driven approach for data management allows to deal with the same system data from greenhouse or fields, thanks to a precise formalization of agricultural objects. This approach makes easier the data integration process. In other words, by connecting greenhouse and field experiments, the decision-making process is strongly improved.

Other uses are made. Indeed, this approach based on ontology-driven information systems can facilitate decision making for many agricultural applications such as agroecological system design, precision agriculture and breeding. For instance, in agroecology we formalized bioagressors, lifecycles and impacts. All these applications require interdisciplinary work and intensive data integration. The formalization of concepts, the links between concepts and tagging are fundamental and constitute a crucial step. This information system generation encourages the production of FAIR data that can be used across disciplines.

10 http://agroportal.lirmm.fr/ontologies/OEPO

4 Conclusion and future trends

As we have seen earlier, the structuring of data in order to make them reusable is based on their identification in the long term (beyond the decade) and on the reuse of ontologies and interdisciplinary standards. In practice, organization evolutions and staff turnover have important effects for the long-term data management. In too many cases, data are often produced and designed for 'immediate consumption'. Reusing ontologies is the way that we must choose, but efficient tools for improving reuse are needed. Data come from various devices; simulation, observation or crowdsourcing and too often data repeatability/reproducibility is not well known or impossible. Structuring data will be a significant advance. For projects, institutions and companies, Data Management Plans (DMPs) are a sine qua non condition for the evaluation of produced data in agriculture. DMPs will allow the development and improvements of methods for the identification of agricultural objects and the associated data semantics. An interesting example is the world of software where many developers do not hesitate and are very active to share their production. Data papers improve the process of data sharing and data indexing.[11] A citation mechanism is designed to reward the efforts of people and institutes that collect and manage data. But recognizing data sharing is still in its infancy, and the generalization of persistent identifiers, data papers and the Web of data could help change things. Part of the answer is also in the availability of integrative data tools for visualization, analysis, prediction and decision support. Access to a new generation of tools can motivate communities of agriculture. It will support agriculture to raise challenges.

The Web of data - built out of Linked Open Data (LOD) - is the concrete and most salient outcome of 20 years of Semantic Web research. Ontologies and vocabularies are its backbone as they are used to semantically annotate and interlink datasets. Methods and techniques have recently been developed, allowing the massive publication of structured data on the Web. Yet, that vision has not been fully applied to agronomy and agriculture for which data have some specificities that require new models (e.g. spatio/temporal dimension, complex and multi-scale data (from gene to environment), data streaming from IoT devices in precision agriculture). And agronomy and agriculture mix data from different disciplinary fields and scientific perspectives, making the integration even more challenging. Despite recent initiatives like the Agronomic Linked Data RDF knowledge base (AgroLD - www.agrold.org) (Venkatesan et al., 2018), we have not seen integrated semantic resources that have had a major impact in agronomy and agriculture such as the ones that have been developed in biomedical and health sciences (e.g. Bio2RDF.

11 https://freshwaterblog.net/2012/06/29/what-does-a-data-paper-look-like/

org, EBI RDF). Indeed, despite the large adoption of some semantic resources, like AGROVOC – the most widely adopted vocabulary to index, retrieve and organize agricultural system data – we cannot yet measure their impact in terms of LOD produced and made available to the rest of the world. Some may ask: where are agronomy and agriculture in the famous LOD cloud diagram previously illustrated? This question is at the center of a the D2KAB project (hereafter referenced).

Data curation needs to be developed and to go further than 'cleaning' its imperfections. The curation of data, from Latin *curare*, which means 'to take care', is essential before any process of analysis or decision. It consists of improving the capacity of the data to describe a system in an unambiguous and explicit way. It is essential to prepare a dataset for a large set of analysis methods, given the opportunity to aggregate different datasets of different provenances, structures and semantics.

To meet the agricultural challenges, well-structured and described data are essential, but how to use them better? Ideally, we shall have powerful tools to automatically select and integrate huge datasets from various sources (agriculture, environment, social, health, etc.). A first stage is more reasonably to use semi-automatic tools, in order to produce the most complete knowledge that constitutes the decision support material.

The structuring of data must be accompanied and allow the construction of different kinds of decision support tools. The main goal is to promote the adoption of increasingly decision-making and 'smart' decision support tools in the agricultural domain. These systems will use not only more data but also better data, updated, cheaper to produce, more standardized and more efficient for decision making.

Produced data should meet FAIR principles; if widely adopted, the connections they enable will result in improved access to information, opportunities for collaboration, reduced administrative overhead and, ultimately, increased trust in studies and research (Meadows et al., 2019).

However, ensuring long-term persistence of identification is a challenge: it is theoretically easy to install and practically very difficult to maintain. Reuse, obsolescence and updating standards (semantic resources, formats, access protocols, etc.) are another main challenge to the sustainable interoperability of information systems, especially in areas handling heterogeneous data such as agriculture. For these reasons, interoperability level is strongly linked to the quantity of work insured at a long-term scale and to the capacity of different data authorities to build together community-approved ontologies and data schemes. Interoperability at a human level as well as a machine level is a key to integratively analyzing environmental data and building decision support systems that are relevant to address the future challenges of agriculture.

5 Where to look for further information

Further reading and references on identification:

- ARK: https://tools.ietf.org/html/draft-kunze-ark-18.
- B2HANDLE: https://eudat.eu/catalogue/B2HANDLE.
- DOI: https://www.doi.org/.
- Datacite: https://datacite.org/.
- ePIC: https://www.pidconsortium.eu/.
- GRID: https://www.grid.ac/.
- Handle System Namespace and Service Definition: http://www.ietf.org/rfc/rfc3651.txt.
- Handle System Protocol: http://www.ietf.org/rfc/rfc3652.txt.
- IGSN | SESARSystem for Earth Sample Registration: www.geosamples.org/igsnabout.
- IRI: https://www.ietf.org/rfc/rfc3987.txt.
- ISNI: http://www.isni.org/.
- LSID: http://www.lsid.info/.
- ORCID: https://orcid.org/.
- RINGGOLD: https://www.ringgold.com/ringgold-identifier/.
- RRID: https://scicrunch.org/resources.
- UUID: https://www.w3.org/wiki/UriSchemes/uuid.
- XRI (OASIS): https://www.oasis-open.org/committees/xri/x.

Research data organizations such as the Research Data Alliance[12] (RDA) or Force11 are developing. They coordinate actions, research or communication focusing on the structuring of the data for the next tens. Force11[13] is a community of scholars, librarians, archivists, publishers and research funders that has arisen organically to help facilitate the change toward improved knowledge creation and sharing. The RDA was started in 2013 by the European Commission and several governments with the goal of building the social and technical infrastructure to enable open sharing and reuse of research data. RDA supports interest groups and working groups all along the range of research data issues from DMPs to data repository, identification, standardization and more. Agriculture is especially well represented at the RDA with the Interest Group on Agricultural Data (IGAD). This group gathers several working-groups such as on wheat data interoperability (https://ist.blogs.inra.fr/wdi/), rice data interoperability and Agrisemantics (https://agrisemantics.org/) which is focus on data management and interoperability with adopting semantic resources and tools.

12 https://rd-alliance.org/
13 https://www.force11.org/

Additionally, there are a number of current research projects designed to support data management issues in agronomy or agriculture such as:

- French ANR project (*Data to Knowledge in Agronomy and Biodiversity* - www.d2kab.org), which goals is to create a framework to turn agronomy and biodiversity data into knowledge - semantically described, interoperable, actionable, open- and investigate scientific methods and tools to exploit this knowledge for applications in science & agriculture.
- Big Data Grapes H2020 project (*Big Data to Enable Global Disruption of the Grapevine-powered industries* - http://www.bigdatagrapes.eu) which aims to help European companies in the wine and natural cosmetics industries become more competitive in the international markets. This project helps companies across the grapevine-powered value chain ride the big data wave, supporting business decisions with real time and cross-stream analysis of very large, diverse and multimodal data sources.
- EMPHASIS ESFRI and the EPPN H2020 (*European Plant Phenotyping Network* - https://emphasis.plant-phenotyping.eu/) research infrastructure projects aim to address the technological and organizational limits of European plant phenotyping to make the most of genetic and genomic resources available and essential for crop improvement in times of a changing climate.

6 Acknowledgements

This work was partly achieved within the Data to Knowledge in Agronomy and Biodiversity (D2KAB - http://www.d2kab.org/) project that received funding from the French National Research Agency (grant ANR-18-CE23-0017), Phenome 'Infrastructure Biologie Santé' PHENOME-EMPHASIS project (ANR-11-INBS-0012) and EPPN2020 (UE H2020 grant agreement No 731013).

7 References

Altman, R. B., Bergman, C. M., Blake, J., Blaschke, C., Cohen, A., Gannon, F., Grivell, L., Hahn, U., Hersh, W., Hirschman, L., Jensen, L. J., Krallinger, M., Mons, B., O'Donoghue, S. I., Peitsch, M. C., Rebholz-Schuhmann, D., Shatkay, H. and Valencia, A. 2008. Text mining for biology - the way forward: opinions from leading scientists . *Genome Biology* 9(Suppl. 2), S7. doi:10.1186/gb-2008-9-s2-s7.

Antoniou, G. and Van Harmelen, F. 2004. *A Semantic Web Primer*. MIT press.

Baumgartner, W. A., Cohen, K. B., Fox, L. M., Acquaah-Mensah, G. and Hunter, L. A. 2007. Manual curation is not sufficient for annotation of genomic databases. *Bioinformatics* 23(13), i41-8. doi:10.1093/bioinformatics/btm229.

Dappert, A., Farquhar, A., Kotarski, R. and Hewlett, K. 2017. Connecting the persistent identifier ecosystem: building the technical and human infrastructure for open research. *Data Science Journal* 16, 28. doi:10.5334/dsj-2017-028.

Dowell, K. G., McAndrews-Hill, M. S., Hill, D. P., Drabkin, H. J. and Blake, J. A. 2009. Integrating text mining into the MGI biocuration workflow. *Database: the Journal of Biological Databases and Curation* 2009, bap019. doi:10.1093/database/bap019.

Gruber, T. R. 1993. A translation approach to portable ontology specifications. *Knowledge Acquisition* 5(2), 199–220. doi:10.1006/knac.1993.1008.

Guillard, V., Buche, P., Destercke, S., Tamani, N., Croitoru, M., Menut, L., Guillaume, C. and Gontard, N. 2015. A decision support system to design modified atmosphere packaging for fresh produce based on a bipolar flexible querying approach. *Computers and Electronics in Agriculture* 111, 131-9. doi:10.1016/j.compag.2014.12.010.

Halewood, M., Chiurugwi, T., Sackville Hamilton, R., Kurtz, B., Marden, E., Welch, E., Michiels, F., Mozafari, J., Sabran, M., Patron, N., Kersey, P., Bastow, R., Dorius, S., Dias, S., McCouch, S. and Powell, W. 2018. Plant genetic resources for food and agriculture: opportunities and challenges emerging from the science and information technology revolution. *The New Phytologist* 217, 1407–19. doi:10.1111/nph.14993.

Jonquet, C., Shah, N. H. and Musen, M. A. 2009. The open biomedical annotator. In: *American Medical Informatics Association Symposium on Translational BioInformatics, AMIA-TBI'09*, San Francisco, CA, March 2009, pp. 56–60.

Jonquet, C., Toulet, A., Arnaud, E., Aubin, S., Dzalé Yeumo, E., Emonet, V., Graybeal, J., Laporte, M., Musen, M. A., Pesce, V. and Larmande, P. 2018. AgroPortal: a vocabulary and ontology repository for agronomy. *Computers and Electronics in Agriculture* 144, 126–43. doi:10.1016/j.compag.2017.10.012.

Lousteau-Cazalet, C., Barakat, A., Belaud, J., Buche, P., Busset, G., Charnomordic, B., Dervaux, S., Destercke, S., Dibie, J., Sablayrolles, C. and Vialle, C. 2016. A decision support system for eco-efficient biorefinery process comparison using a semantic approach. *Computers and Electronics in Agriculture* 127, 351-67. doi:10.1016/j.compag.2016.06.020.

McMurry, J. A., Juty, N., Blomberg, N., Burdett, T., Conlin, T., Conte, N., Courtot, M., Deck, J., Dumontier, M., Fellows, D. K., Gonzalez-Beltran, A., Gormanns, P., Grethe, J., Hastings, J., Hériché, J. K., Hermjakob, H., Ison, J. C., Jimenez, R. C., Jupp, S., Kunze, J., Laibe, C., Le Novère, N., Malone, J., Martin, M. J., McEntyre, J. R., Morris, C., Muilu, J., Müller, W., Rocca-Serra, P., Sansone, S. A., Sariyar, M., Snoep, J. L., Soiland-Reyes, S., Stanford, N. J., Swainston, N., Washington, N., Williams, A. R., Wimalaratne, S. M., Winfree, L. M., Wolstencroft, K., Goble, C., Mungall, C. J., Haendel, M. A. and Parkinson, H. 2017. Identifiers for the 21st century: how to design, provision, and reuse persistent identifiers to maximize utility and impact of life science data. *PLoS Biology* 15(6), e2001414. doi:10.1371/journal.pbio.2001414.

Meadows, A., Haak, L. L. and Brown, J. 2019. Persistent identifiers: the building blocks of the research information infrastructure. *Insights the UKSG Journal* 32(9), 1–6. doi:10.1629/uksg.457.

Neveu, P., Tireau, A., Hilgert, N., Nègre, V., Mineau-Cesari, J., Brichet, N., Chapuis, R., Sanchez, I., Pommier, C., Charnomordic, B., Tardieu, F. and Cabrera-Bosquet, L. 2019. Dealing with multi-source and multi-scale information in plant phenomics: the ontology-driven phenotyping hybrid information system. *The New Phytologist* 221, 588–601. doi:10.1111/nph.15385.

Noy, N. F., Shah, N. H., Whetzel, P. L., Dai, B., Dorf, M., Griffith, N. B., Jonquet, C., Rubin, D. L., Storey, M.-A., Chute, C. G. and Musen, M. A. 2009. BioPortal: ontologies and integrated data resources at the click of a mouse. *Nucleic Acids Research* 37, W170-3. doi:10.1093/nar/gkp440.

Page, R. D. M. 2008. Biodiversity informatics: the challenge of linking data and the role of shared identifiers. *Briefings in Bioinformatics* 9, 345–54. doi:10.1093/bib/bbn022.

Rhee, S. Y., Wood, V., Dolinski, K. and Draghici, S. 2008. Use and misuse of the gene ontology annotations. *Nature Reviews. Genetics* 9(7), 509–15. doi:10.1038/nrg2363.

Tzounis, A., Katsoulas, N., Bartzanas, T. and Kittas, C. 2017. Internet of things in agriculture, recent advances and future challenges. *Biosystems Engineering* 164, 31–48. doi:10.1016/j.biosystemseng.2017.09.007.

Uren, V., Cimiano, P., Iria, J., Handschuh, S., Vargas-Vera, M., Motta, E. and Ciravegna, F. 2006. Semantic annotation for knowledge management: requirements and a survey of the state of the art. *Journal of Web Semantics* 4, 14–28. doi:10.1016/j.websem.2005.10.002.

Venkatesan, A., Tagny Ngompe, G., Hassouni, N. E., Chentli, I., Guignon, V., Jonquet, C., Ruiz, M. and Larmande, P. 2018. Agronomic Linked Data (AgroLD): a knowledge-based system to enable integrative biology in agronomy. *PLoS ONE* 13(11), e0198270. doi:10.1371/journal.pone.0198270.

Wilkinson, M. D., Dumontier, M., Aalbersberg, I. J., Appleton, G., Axton, M., Baak, A., Blomberg, N., Boiten, J. W., da Silva Santos, L. B., Bourne, P. E., Bouwman, J., Brookes, A. J., Clark, T., Crosas, M., Dillo, I., Dumon, O., Edmunds, S., Evelo, C. T., Finkers, R., Gonzalez-Beltran, A., Gray, A. J., Groth, P., Goble, C., Grethe, J. S., Heringa, J., 't Hoen, P. A., Hooft, R., Kuhn, T., Kok, R., Kok, J., Lusher, S. J., Martone, M. E., Mons, A., Packer, A. L., Persson, B., Rocca-Serra, P., Roos, M., van Schaik, R., Sansone, S. A., Schultes, E., Sengstag, T., Slater, T., Strawn, G., Swertz, M. A., Thompson, M., van der Lei, J., van Mulligen, E., Velterop, J., Waagmeester, A., Wittenburg, P., Wolstencroft, K., Zhao, J. and Mons, B. 2016. The FAIR Guiding Principles for scientific data management and stewardship. *Scientific Data* 3, 160018. doi:10.1038/sdata.2016.18.

Wolfert, S., Ge, L., Verdouw, C. and Bogaardt, M. 2017. Big data in smart farming - a review. *Agricultural Systems* 153, 69–80. doi:10.1016/j.agsy.2017.01.023.

Chapter 4

Advances in data security for more effective decision-making in agriculture

Jason West, University of New England, Australia

1 Introduction

Precision agriculture (PA) is playing an increasingly vital role in meeting the future global need for food production (Godfray et al., 2010). The interplay between technology applications including hardware in the form of sensors and transmitters and software in the form of data systems and 'artificial intelligence' is driving capacity toward data-driven decision-making. Coupled with enhanced machinery, irrigation, and farming networks, these efforts seek to optimize production and reduce input-output variability in agricultural systems. A particular strength of PA applications is the moderation of important resources used in farming systems including water, fertilizers, and pest control to prevent over-application or misuse, based on understanding highly granular sensitivities to current and expected agronomic conditions. The promise of PA in optimizing investment returns under a range of economic and environmental

http://dx.doi.org/10.19103/AS.2020.0069.05

conditions is a critical enabler across both the primary production of food and in all areas of the value-chain.

A key feature of PA systems also serves as its primary weakness. The operation of PA relies heavily on the capacity for data to transfer between sensory systems and processing hubs, as well as potentially wide arrays of sensors connected to wireless networks. This necessarily involves the need for remote transmission and system actuation, particularly in large farming operations, and must be able to operate in real time to take full advantage of updated information and data analytics. This means that PA architecture and network structures will be highly vulnerable to all forms of cyberattack, either through the network itself or in the architecture used as its support. This poses a significant risk to the future use of PA applications. PA is undergoing a radical transfer of mechanical activities combined with labor-intensive ones towards a network of mostly mechanical activities controlled by a data-driven network, which are necessarily comprised of online connections. This dramatically increases the attack space available to threat actors, both functionally and geographically. What are normally trivial threats are beginning to have unique and far-reaching consequences on the agricultural industry and may act as a barrier to further investment in productivity-enhancing technologies, if the risk of cyberattack remains high without risk mitigation through protection, surveillance, rectification, recovery and, in the worst case, insurance.

Cyberattacks specifically aim to exploit computing systems and networks driven by any number of personal or professional motivations. Cyberattacks come in many forms, and we are witnessing their growth in both scale and complexity from sources ranging from individuals with activist motives to infiltration, theft and extortion by nation states, and organized crime. This trend is entrenched in other industrial sectors such as finance and manufacturing, and it is only a matter of time before agriculture becomes a more attractive target. Some attacks use malicious codes to alter algorithms, logic, or data housed in integrated systems. Other attacks steal data for commercial reasons. A small but growing number of attacks seek to debilitate the wider network of data collection devices to cripple an entire management system or supply chain. Agricultural operators using PA to capture positive but often marginal profits, or to improve productivity and safety, are easy targets for cyber-crime and generally have limited capacity to recover from a widespread attack.

Large firms have historically been an obvious mark for cybersecurity attacks and are generally targeted with a highly focused campaign aimed at disruption or data theft. In contrast, small- and medium-sized enterprises are highly vulnerable to broad-scale attacks, where the same algorithmic process is applied regardless of target.

Since 2014, nearly two-thirds of all targeted attacks were aimed at small- to medium-sized enterprises, including primary producers, agricultural suppliers, and agribusiness firms (Symantec, 2019). One in two firms in manufacturing and financial sectors has been the direct target of a cyberattack, and while over three-quarters of spear-phishing attacks that occurred in 2015 targeted small businesses, many of these included farms. Critically, in 2016 alone, almost two-thirds of small businesses, including farmers who became cyber-crime victims, were forced to exit the industry within 6 months of an attack due to the irrecoverable failure of their systems (Federal Bureau of Investigation, Cyber Division, 2016). While an attack itself may be confined to a specific system within a wider operation, the costs of recovery and usually replacement are prohibitively high relative to the size of the business. This risk is largely uninsurable and represents one of the greatest threats to small enterprises.

It is relatively easy to corrupt data through certain cyberattack methods. Centrally controlled systems can become infected to the point of catastrophic failure. Even when fully prepared for cybersecurity, the chances of a breach are substantial. A study by Crowe (2016) conducted on 60 cybersecurity companies who were the victim of a successful ransomware attack showed that 100% of attacks bypassed the companies' antivirus software systems, 95% of attacks bypassed the firewall, 77% of attacks bypassed email filtering, 52% of attacks bypassed anti-malware programs, and one-third of attacks were successful, even if the companies' employees had extensive cybersecurity training.

It is prudent for businesses to address the specific security challenges facing PA. This chapter provides a simple risk-based framework to help understand and counter the challenges of mobile computing and data-driven decision-making in food production, while maintaining adequate consideration of subsequent sacrifices in efficiency and production effectiveness. After all, process simplicity and productivity will suffer from the intervention of any additional system restrictions, but minimizing disruption for a given level of protection yields the greatest benefit.

The framework developed here is based on the threat prediction model that is derived from the Common Vulnerability Scoring System (CVSS) to form the basis for an assessment. Each threat assessment is tested against network traffic and vulnerability characteristics of the system, which allows cyberattack solutions to then be customized. Adopting the framework will allow precision agriculture systems to achieve an appropriate level of protection from each category of cyberattack without unnecessarily compromising internal process efficiencies, upon which a PA solution is usually selected for in the first place.

We also wish to ensure that this framework is fit for purpose. Specifically, we need a prediction model coupled with a decision framework that can fully understand the set of vulnerabilities facing PA systems, particularly for farmers in remote areas who are heavily reliant on such technologies for water management, crop production, and livestock tracking.

2 Security challenges in PA systems

2.1 Network characteristics

Sensor networks to monitor crops using wireless technologies, infrastructure, and integration mechanisms offer some valuable benefits to agricultural enterprises. Bespoke networks can greatly assist farming operators with monitoring and automation which are useful for activities in remote areas. They also permit insights with greater precision of specific zones within the span of their operations. PA sensor technologies have been successfully deployed across a range of high-value agriculture operations including irrigation control at vineyards, pregnancy window forecasting for dairies, and pest control in fruit-tree orchards.

New types of sensors now exist that accurately capture field heterogeneity to an extremely fine degree. Spatial and temporal heterogeneity for both crop and livestock production typically use wireless sensor networks (WSN) (e.g. Chawla and Hussain, 2015) which dominate most PA activities and delivers the greatest economic advantages to a farming enterprise. Not only are direct measures such as temperature, water balance, rainfall, and disease used to make automated decisions, but advances in the analysis of 'unstructured' data are leading to better decisions and actions. Unstructured data includes ancillary parameters collected by sensors, imagery, ad-hoc data reports, and other types of information, especially when collected from third-party sources. Despite the often nonlinear and multivariate interactions that affect inputs and monitoring processes, the processing of unstructured data using neural networks and deep learning algorithms can lead to important predictive capabilities that may not be available using highly structured, one-dimensional data feeds. Development of nonlinear dynamic models whose inputs rely on newer technologies such as unmanned aerial vehicles (UAVs), multi-spectral imaging methods, and other automated mobile sensors are also gaining traction. The integration of these data types represents a powerful source of momentum in the ability for farming operators to conduct business using automated decision systems.

A necessary component of PA systems is the need for substantial computing power that does not need to be located on-site. However, the need for on-site (usually temporary) and off-site (usually permanent) data storage capacity represents a natural point of weakness in PA system architectures. Although the computational resources for semi-automated decision-making already exist, they do not currently reside within the secure confines of the farm and for a variety of reasons, including security and privacy, are unlikely to in the future. Data acquisition, transmission, computation, reporting, and actuation require an interconnected range of storage nodes and communication links. These links constitute the most vulnerable points in Internet of Things (IoT) technologies and generally constitute the initial targets for cyberattack. The remoteness and geographical span of many agricultural operations accentuates this risk even further, particularly when little or no redundancies are made available to sustain the network.

In almost all settings, we define a WSN as a complex dynamic system of interacting sensor nodes that aggregate in some form to collect and collate data. In agricultural settings, such systems can perform relatively advanced analytics from raw data, which are then typically transformed into automated decisions using computational and control functions. An example is the use of a sensor network for receiving real-time data from food production equipment, along the entire supply chain (from field to fridge) in a fruit-packing facility. Plant operators can be highly reactive by avoiding the need for manual checks, visual inspection or data loggers and the use of IoT monitoring tools, industrial wireless sensors, and vibration monitors to moderate the production process.

The sensors themselves are today available in various forms, several of which are capable of actuating controls to perform an activity (e.g. activating a valve, reducing production throughput, or shutdown procedures in an emergency). For instance, a hydrometer may continuously detect humidity at a given location in an orchard or crop. The central processor analyses humidity data and, once the cumulative total of dry conditions reaches a threshold, it activates an irrigation actuator to precisely distribute water to the field in a controlled way. This process can be fully automated and often is.

Continued advances in micro-electromechanical systems (MEMS) and distributed computing permit a wider use of WSNs in multiple applications, including the usual need for monitoring weather conditions and air quality, along with more difficult applications such as soil behavior, tree health, ecological habitats, water balance, livestock tracking, and livestock health monitoring. Sensor nodes can be equipped with modules for sensing, computing (to a limited degree), powering, and communicating to monitor specific phenomena via self-organizing protocols. This is feasible because node positions do not need to be predetermined and can change location without loss of quality.

In terms of system architecture, WSNs comprise a sensor node where the microcontroller or computing module processes the data observed by a sensing module(s). Data transmission to data warehouse is enabled via a wireless link with a communication module. Pre-processed and post-processed data can subsequently shift to other locations to enable further processing by third parties or for data storage purposes. Enterprise data, in either raw or processed form, can be especially valuable. Access to valuable data by external parties may allow them the capability to perform their own research on plant biology, genetics, economic viability, supply forecasts, veracity of farm equipment, and productivity. Data protection is therefore a vital component of the system; there is little point investing in a PA system to improve productivity only to witness the theft of valuable system data which undermines the economic advantage gained by an enterprise.

Downstream participants in the value chain, direct competitors, and substitute producers can greatly benefit from access to unprotected or poorly protected information and may be tempted to pay for such insights if offered by a cyberattacker. For instance, a premium craft beer producer operating on tight margins coupled with a need for continual shipments of high-quality grain as an input to their production process may seek a renegotiation in contractual terms if high-frequency produce quality data were made available to a third party that implied substantial variability through time.

Facilities where data are aggregated, stored, and analyzed, along with the data themselves, provide a very high-value target for corporate, nation state, and/or activist adversaries with the goal of data theft and/or sabotage and misinformation. Figure 1 describes the typical approach for cybersecurity attacks on a network-based agriculture system with third-party access (e.g. ag-tech firm and agronomy analytics provider) using malicious software (malware). In this instance, third-party access to on-farm systems is used as the gateway through which the malware infiltrates the network and re-transmits data to dedicated servers that mimic the access characteristics of the third party.

A drawback from this example is that it does not highlight the interference capability of many cyberattack sources. For instance, many actors are able to remotely control irrigation or nutrient delivery systems through supervisory control and data acquisition (SCADA) system architectures, which are commonplace in many intensive agricultural operations. But the example does highlight that what are commonly referred to as 'man in the middle' attacks are very difficult to detect and cannot be defeated using a 'security by obscurity' response within the network. Recent attacks have demonstrated that infiltrations will grow in sophistication and can easily escape detection over a substantial time horizons.

Applying the Confidentiality, Integrity, and Availability (CIA) model of information security to the enterprise risks associated with PA systems offers a neat perspective on the range and variety of threats. More widely, CIA principles form the basis of information security analysis within this framework. The CIA principles are well-understood and commonly applied by cybersecurity proponents, and so it is reasonable to continue their use when adapting systems for PA.

2.2 Key confidentiality threats

Data privacy tends to be a critical issue for PA applications. Highly productive farmers are naturally protective of their information, particularly data related to yield and crop quality, land prices, and herd health. Misuse of these data can

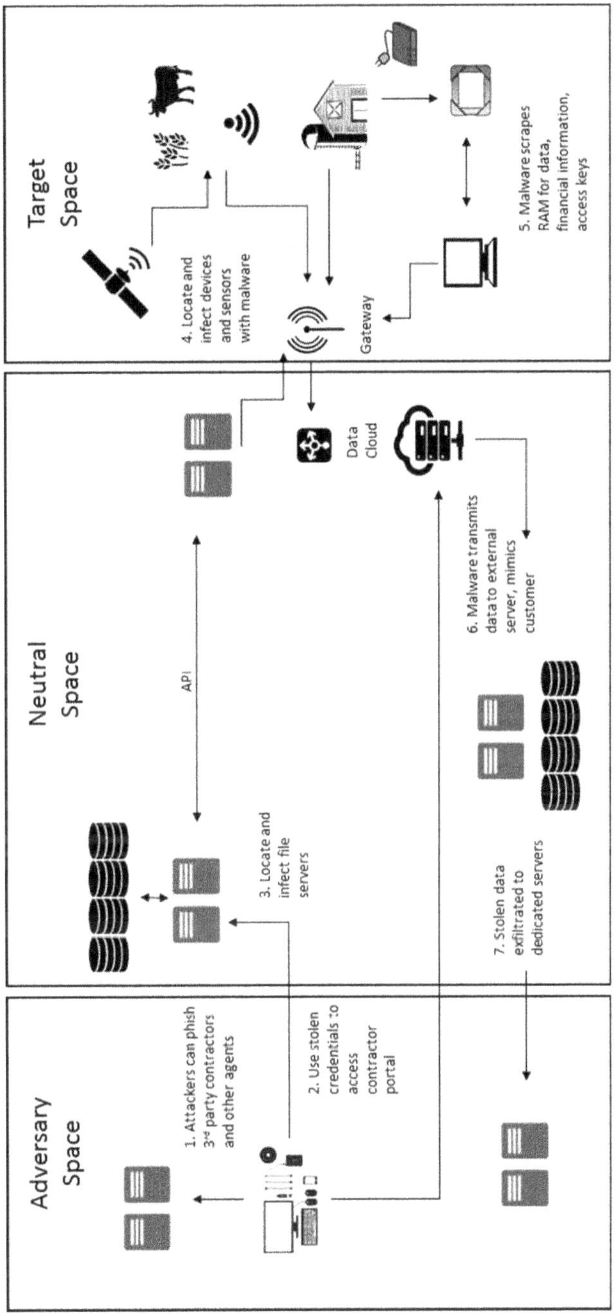

Figure 1 Example malware attack on farm systems using third-party access. Source: West (2018).

have dramatic financial impacts on farmers, due to the highly sensitive nature of reputations in high-value produce. Confidentiality threats are multifaceted.

Firstly, the intentional theft of data collected through decision support systems (DSS) or the unintentional leakage of data to third parties is a major risk. While there has been a rapid increase in the number of DSS and farm information management systems (FIMS) that use mobile apps to deliver information to farmers, many were built by newly minted technology firms or university extension programs with outsourced programming which rarely include code updates or security patches. As a result, privacy controls, user agreements, third-party applications, and system updates do not exist. In fact, there have been circumstances where some DSS applications are malicious, by design, to steal data.

Secondly, the disclosure of confidential information to cause damage to an enterprise is an important risk. No major cyberattacks have occurred in the agricultural sector as yet but the public release of confidential pricing and market data could have catastrophic impacts for suppliers. This possibility is causing lingering mistrust among some consumers, particularly those with extensive trading links to international buyers.

Finally, the sale of confidential data to third parties to act against farmers in commodity markets can result in inefficient price discovery and unfair market outcomes. Production volume and quality data, along with supply chain costs, provide valuable insights, which can be exploited during price negotiations. Lack of data protection acts as a threat not only to individual producers, but to entire sectors of the food economy.

2.3 Key integrity threats

PA's aggressive transition to 'smart farming' using wide-scale sensor nets for crop and livestock monitoring enables real-time decision-making. This has helped reduce enterprise risks in various activities such as the mistiming of crop planting and harvests, pesticide application, and livestock rotation decisions. This necessarily involves automation, robotics, machine learning, and edge computing which means that threats to data integrity will be manifest in ways previously uncontemplated. There are three main issues here.

Firstly, deliberate falsification of data can greatly disrupt crop and livestock markets, particularly any association with virulent disease or uncontrolled genetic modification. This threat has always existed. Malicious actors have been found to falsify information which can take many months (typically years for livestock disease outbreaks) to detect and then restore. Protecting the veracity of vast volumes of agricultural data is therefore a critical mission for PA developers to maintain data integrity.

Secondly, data corruption of existing sensor networks has been found to damage high-value crops and livestock operations. Reliance on 'edge computing' to make decisions at the source, as opposed to making it at a central hub, allows hackers to corrupt data networks. Examples of this include the destruction of crops through the under/over-application of water and nutrients, as well as threats to livestock systems, particularly assets that control HVAC automated feeding systems.

Thirdly, many analytical capabilities that rely on machine learning are still in their infancy. While these will improve over time, the use of automation coupled with machine learning activation can increase risks to an enterprise. For instance, underdeveloped machine learning algorithms for advanced PA applications may overtly bias data outliers in Bayesian predictive models, which can have potentially adverse effects in control systems.

2.4 Key availability threats

Farming enterprises are increasingly using equipment as a system of systems. The result is an operation comprising of complex embedded tools, communication networks, and guidance systems. Clearly, cyberattack threats that limit or stall equipment availability can result in unpredictable outages, which are always costly.

The first and major threat relates to timing. Crops experience very narrow and precise windows for planting and harvesting. Access to functioning equipment during these windows is essential. Enterprises may miss optimal planting windows or delay harvest which can result in yield reduction or quality loss. A delay in corn planting by two weeks can result in a negative yield deviation of 20% (Long et al., 2017). This threat may come not only from malicious actors, but also from poorly tested PA algorithms that disrupt many machines simultaneously. It can also be caused by poorly timed system updates.

Second, disruption to positioning, navigation, and timing (PNT) systems is probably a greater threat to the wider industry rather than to specific enterprises. The increasingly crowded signal spectrum may disrupt PA operations through signal loss or interference. Malicious actors attacking one system may inadvertently affect others. Some guidance systems rely on GLONASS and other foreign systems, as well as GPS, which may be denied during a crisis or conflict. This is not unique to agriculture but is worthy of consideration as an enterprise risk. Disruption to ground-based real-time kinematic (RTK) positioning systems will be a more common feature for PA activities to provide centimeter-scale precision for seed planting, input application, and pest control. RTK systems can be particularly vulnerable to attack where only a limited number of base stations are available in remote areas.

Third, geographical constraints and cost control means that PA systems are usually constructed using highly distributed sensor networks that depend on reliable wireless data transfer capacity. Access to broadband services is a limiting factor for many PA applications, especially in remote areas. The loss of any available signal will directly stall equipment availability for these systems. By necessity, farmers currently rely on a mixture of radio, cellular, RTK, and wireless networks with some degree of physical data transfer. While each of these technologies represents a major point of vulnerability for PA systems operating far from a base station, the combination of them offers redundancies to many enterprises. Shifting toward greater reliance on a single system may dramatically increase vulnerability to attack.

Finally, smart livestock production facilities are complex, network-reliant structures. They are especially vulnerable to malicious cyberattack because corrupted data or system malfunction may not be known until livestock losses have occurred. Production systems therefore need to be coupled with highly reliable monitoring systems. This increased system integration as a protection against cyber threats may be too costly for many enterprises.

3 System architecture and legal recourse

The data collection, processing/computation, and storage infrastructure for PA systems have not traditionally been designed with security as a top priority. Systems have naturally evolved from bespoke single-use applications toward hub-spoke wide-scale systems that integrate data from many single-use systems. While the confidentiality of data needs to be maintained, the system must also be able to fully use internal applications that depend on interrogating aggregated data. This is becoming of great importance for large-scale prediction and region-wide analytics. Crop models with accuracies great enough to support PA decision-making, when driven by stolen data, might well enable within-season, regional crop performance projections of relevance to economic markets (Kshetri, 2006).

The treat of a legal response to cyberattackers is unlikely to ever be a deterrence against hackers. Litigation against data invaders, if they are ever caught, is certainly not a guaranteed path for recovery. In terms of liability, it is not unreasonable to assume that PA operators would naturally wish to maintain direct 'ownership' of their data. From a legal perspective, ownership refers to a bundle of rights whose composition depends on the nature of the object involved. Of the four major types of intellectual property, namely patent, copyright, trademark, and trade secrets (Michalowski and Pfuhl, 1991), it is the latter which has been shown to form the 'best' legal basis in the protection of PA data (Sinrod and Reilly, 2000).

However, a legal defense that relies on trade-secret doctrine requires that data owners must demonstrate they are active in avoiding disclosure of secrets

(Can, 2015). Evidence demonstrating systems of documented data handling and protection goals, objectives, and methods of implementing procedures are needed to cover all parties in the data chain. This could be extensive for a PA system. Without this, a legal challenge against external users of the information, even those who exploit the data though they were not involved in the attack, can be very limited.

3.1 Old dogs, new tricks

SCADA is a control system architecture that uses computers, networked data communications, and graphical user interfaces. It can also integrate peripheral devices such as programmable logic controllers and discrete proportional-integral-derivative (PID) controllers to integrate process plant and machinery. An operator accesses the system through an interface and issues process commands (which can also be automated). Real-time control logic or controller calculations are performed by networked modules which connect to the field sensors and actuators. The advantage of SCADA systems is that all collected information from the farm is centralized and operated remotely.

Over 1 million SCADA systems and other industrial control systems (ICS) are connected to the internet in the United States alone (IBISWorld, 2015). Similarly, large numbers of connections also exist in Europe, Asia, Canada (Cardwell and Shebanow, 2016), and this figure is anticipated to be growing by 2000–8000 additional systems per day (Goodman, 2017). Since most of these devices directly face the internet, they are highly vulnerable.

An example can be useful to demonstrate the potential. An attacker could use Shodan, which is a search engine designed specifically for internet-connected devices, to locate a device. It can then obtain its software version, retrieve a viable exploit code for that device using penetration testing software, establish a proxy connection that is difficult to trace, and then exploit the remote system through roaming, data downloads, corrupting systems, data encryption, or another manner.

The robustness of SCADA/ICS systems under direct cyberattack is poor. Many of these systems were never intended to be connected directly to the internet. Newer PA systems have recognized this and are starting to be redesigned with multiple redundant security controls (e.g. firewalls/data-diodes and identity/access management systems) between the real-time system and the internet, segregation between system architectures and minimal network-facing remote sensor usage (Gronau, 2015).

Many older SCADA/ICS systems, including existing PA systems, are so vulnerable that they are unable to withstand even simple security scans. Some have 'backdoor' administrative accounts and, in some cases, authentication credentials are hardcoded which guarantees hacker access. Basic fuzz testing (automated software that provides invalid, unexpected, or random data as an input to test for memory leaks or program crashes) causes many SCADA systems to crash, and buffer overflows remain a serious issue. Some current systems contain no password timeouts which allows hackers to login using brute-force key searches. Once the base level of a system has been breached, such as the industrial processes that contain programmable logic controls (PLC), hackers can then simply upload new ladder logic which may cause systems to ignore multiple safety interlocks. This grants hackers complete control over the system, and no method of retrieval is possible.

While there have been attempts to counter this, the construction of more robust SCADA/ICS products to append existing systems is not practical. Protection against cyberattacks now requires complete segregation of vulnerable systems from high-risk networks, which means that device IP addresses must be prevented from being directly accessible from the internet. This greatly limits system flexibility.

While cyberattack threats escalate linearly, the number of vulnerable systems will grow exponentially. Each new attack will gain the capacity to infiltrate a broad array of older architectures.

4 Security framework considerations for PA systems

A cybersecurity framework for PA needs at least the following three components to defend against the most basic attacks:

1 abnormal measurement detection;
2 access control; and
3 encryption.

4.1 Detection

Anomaly detection searches for unusual patterns that deviate from expected data behavior. This is needed to eliminate false alarms that trigger unnecessary emergency interventions. Faulty diagnosis results, false alarms, and unreliable monitoring will eventually undermine system confidence. Sensors can be configured to report abnormal values for various infractions such as data corruption, energy depletion, electromagnetic interference, variable signal quality, connectivity interception, severe weather, malfunction, and false inputs. The detection of measurement outliers and distinction between sensor faults and emergency situations are a primary system feature.

4.2 Access control

Sharing sensor data is required for WSN domains, but if the sharing process is overly complex or involves many external parties, then the integrity of system is easily undermined, especially for dynamic equipment activities. Access control is needed to guarantee access to sensitive information by entities with legitimate reason for doing so, along with control mechanisms that support selective sharing.

Controls range from being fully centralized to fully distributed, with well-known benefits and trade-offs in computational power and security. The increased use of 'edge computing' and 'fog computing' ensures that as much processing as possible is performed at the edge (i.e. on-site), while deeper levels of analysis and data backups are offloaded to a computing facility via farm-office gateways.

4.3 Encryption

Encoding data is a powerful form of protection but remains subject to exploitation. Sensor data that is not encrypted at the point of capture but is replicated in a central location is very vulnerable to cyberattack. For instance, yield maps generated by harvesters derived from collected data would be encrypted by third-party vendor hardware using a middle-box either on-site or at an external location. But the raw collected data may remain on the collection device itself and therefore be open to exploitation.

The increased use of external advisory services by producers greatly increases this vulnerability, particularly when autonomous activities occur at the farm based on decision logic using the analysis results. Currently, around half of the users employ mobile devices to access information from cloud systems and, over time, they will access WSN information in this manner as well. Ensuring that such users will employ access controls and encryption in proper ways will continue to be a major challenge to PA technologists.

5 Modern cyberattack methods

There are several elements required for a cyberattack.

1. The target of the attack (also known as the endpoint) is generally motivated by a purpose, which is to either control, corrupt, or disable.
2. The level of vulnerability is defined as the weakness that permits the endpoint to be infiltrated. Common vulnerabilities include software flaws, system design weaknesses, insecure configurations, and human errors.
3. Malicious software, known as malware, often involves more than one single type which are classified by their purpose or intended effect.
4. A delivery vehicle is needed to deliver the malware to the victim's system through a medium including social engineering (phishing and whaling), network infiltration, or direct access.

5 The method of execution defines the means through which attackers get the resources necessary (access, processing time, data, etc.) to execute the attack.

For security protection against a cyberattack, it needs to prevent malware from achieving its purpose without necessarily preventing a breach or eradicating malware code from the system. There are different types of malware, as follows:

1 Depositors are malware whose primary purpose is to 'land and expand.' They conceal the introduction of malware by separating the exploit of the system from the execution and installation of the malicious program. Depositors are dangerous because they allow attackers to sneak malware past traditional security before transforming it into its executable form. For example, the Cryptowall 2.0 code used a sophisticated dropper that embedded multiple potential exploits to infiltrate target systems. It performed anti-VM and anti-emulation checks prior to decrypting and executing the malware, decreasing the likelihood of identification and detection. This is a powerful mechanism to infiltrate PA systems without ever revealing its presence to system operators. Two basic types of depositors are downloaders and droppers.
   a A downloader is a type of Trojan program that, once running, fetches and installs the malware executables.
   b A dropper is a type of Trojan executable that combines both the installation functionality and the actual malware program within the same object. As a result, a dropper does not necessarily require a permissive network connection.
2 Ransomware is a form of malware that threatens to destroy data unless a victim makes payment. Its primary purpose is to extort money from victims. Two basic types of attacks are encrypting ransomware and destructive ransomware. The encryption approach encrypts critical assets after collating an inventory of system files. Victims are shown a screen with instructions on how to pay an untraceable ransom to decrypt the files. The destructive approach demonstrates its presence on the system to the user by disabling administrator access and demands payment to avoid the automatic destruction of data or malicious remote system operation. This form of attack is a relatively easy and lucrative way to monetize the exploitation of a system. Quite inexpensive and extensible ransomware packages are common and are freely available, so cyber protection must be able to defeat these simple attacks. Ransomware can lie dormant for long periods of time, avoiding detection and making it difficult to trace. The most notorious recent examples of ransomware are the Cryptowall and Cryptolocker attacks which have reportedly cost

businesses around $50 million in direct payments between 2016 and 2017 (ISACA, 2017).

3 Backdoor malware provides at-will unauthorized remote access or command and control. These are programs created to provide repeated and unfettered access to systems for reconnaissance or to leverage systems for botnets to launch attacks against others. Backdoors find ways to make their existence persistent and undetected across reboots, providing attackers the ability to tap back into the system and reconnect. In 2014, it was reported that, on average, 18 systems are infected with botnets *every second* (ISACA, 2017). Today, it is expected that this number is exponentially higher, although no one really knows the likely scale.
4 Viruses and worms are self-replicating, self-perpetuating malware that infect host systems and then spread to infect others autonomously. Once on a system, viruses and worms insert copies of themselves into programs, files, and drives. Worms can then execute actions to spread onto other computers via their network. Worms and viruses can also carry additional 'payloads' designed to perform harmful or disruptive activity on their infected hosts. Worms enable attackers to create a network of hijacked machines (a botnet) for use as distributed denial-of-service (DDoS) attack centers. In contrast, viruses are often shared by unaware users who go on to infect others.
5 Vandalizers are malware that is purely destructive in nature. Once a vandalizer has access to a vulnerable site, it can cripple a system completely. Vandalizer attacks are usually very public.

The above classes of malware rarely operate in isolation. Good cyberattacks will involve multiple classes of malware to increase effectiveness, enable rapid propagation, or obfuscate underlying operations. As an example, attackers use a downloader to establish a command and control infrastructure for ransomware, then insert a worm to propagate a vandalizer against monitoring systems, and finally steal credentials to a hosted email account to spread the ransomware further.

The interconnected nature of IoT PA systems and the requirement to access data generated in real time are the key challenges in cybersecurity. One reliable approach in defending against threats is the use of zones that segregate system architectures. This method is usually feasible for PA systems. Highly secure controls limit the spread of attacks and prevent access to real-time processing.

The escalation in system segregation architectures is illustrated in Fig. 2. The efficiency of a system containing widespread but secure ICS/SCADA infrastructure is compromised by the volume of remote sensors linked to secured conduits (WAN and VPNs).

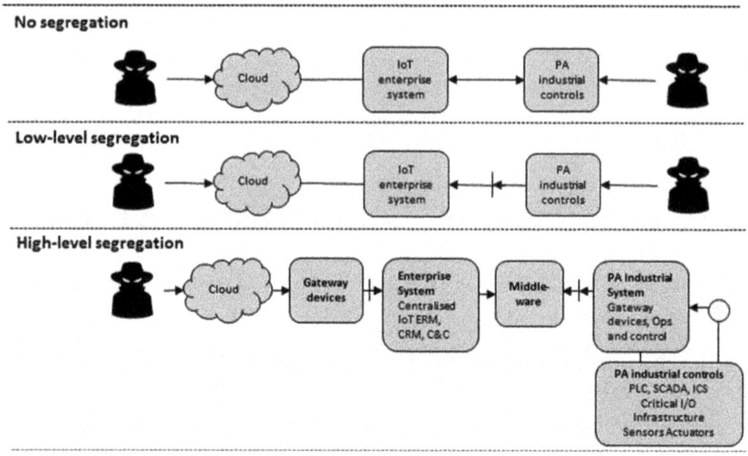

Figure 2 Segregating data and real-time network architecture. Source: adapted from West (2018).

Users can safely assume that new systems booting for the first time are hacked by automatic malware algorithms within around 60 seconds. The use of dedicated modules that disperse the command and control functions for a PA system vastly improve the depth of the defense against cyberattack. The costs associated with segregation are generally modest but a protective suite should be integrated as a complete system to prevent any compromise during system construction.

6 Classifying cyberattack source psychology

While the channels available to cyberattack are well-understood, the motivations behind attacks are more difficult to quantify. Some understanding of these motives can be useful when preparing a defense.

Criminology scholars suggest that criminal intent is a function of the difference between the combined likelihood-benefit attributes of an attack (Fishbein and Ajzen, 1974) and the combined outcome-disadvantage costs anticipated by the hacker (Voiskounsky and Smyslova, 2003). This will vary between hacker types but the anticipated likelihood-benefit/outcome-disadvantage value is attributable to each hacker type with a single order of magnitude. Figure 3 presents a general summary of hacker types categorized by motives, actions, and general features.

Beginners (known as script-kiddies) are generally the most common form of cyberattackers who deploy simple hacking tools, have low persistence levels, and are concerned with building their skills. The attacks create a lot of noise in

Source	Motives	Actions	Persistence	Skill	Greed	Stealth
Script-kiddy	Curiosity	Loud, no specific target, many attempts	2	3	0	0
Malicious insider	Revenge	Stealing information/ damaging internal systems	2	5	7	4
Activist	Revelation	Revealing trade secrets, exposing a cause	9	3	1	4
Spy	Espionage	Accessing state secrets, embarrass governments	10	9	5	9
Terrorist	Destruction	Infiltrate, discredit or destroy data/systems	8	7	0	7
Organized crime	Wealth	Making money without destroying computer infrastructure	8	5	10	3

Figure 3 Threat profile of cyberattack sources. Source: adapted from West (2018).

perimeter and boundary intrusion detection and are easily remedied through SIEM analysis (security information and event management analytics) and boundary defense to disrupt common threats and hacking enumeration and scanning techniques. In contrast, malicious insiders use their access to systems under their responsibility to steal data or inflict damage to control systems. The number of attacks from this source has been increasing steadily and represent a particular risk to stand-alone PA systems.

Pure activists prefer the use of denial-of-service (DoS) attacks or the alteration of public-facing media, as well as accessing information to protest or manipulate data. Remedies are largely reactive, and locating the source of traffic to cause the DoS can take time. The best defense is typically to conduct filtering and baseline reviews of network traffic to predict DoS attacks with sufficient time to filter the traffic based on the attacker's IP.

Spies target all industries, especially high-value sectors like finance and defense manufacturing. Agricultural system attacks from this type have been few but attacks may well increase between nation states in the future. The industrial espionage ecosystem provides channels that are not easily infiltrated and detection may never be achieved. If an exfiltration has been discovered, the usual response is to review the degree of compromise and the structure of the channel of attack and trace system activities that permitted the attack. Defense in depth is compromised at a system's weakest link, so the construction of security layers is the best approach.

Terrorist attacks are based on enumeration espionage to find system weaknesses and are persistent in probing security infrastructure until an entry point is located. For these attacks, all critical security controls should be used which severely limits system flexibility (e.g. remote networks, automated vehicles are an easy target and robust protection is difficult). Remedies are reactive and business continuity followed by disaster recovery are the priority. Attacks of this sort have not directly affected agriculture, but, as automation and remote observation increase, they will likely present an 'easy' target to networks in areas that rely heavily on the sector for income and employment.

Organized crime deploys phishing or malware or directly infiltrates networks through under-protected system features (e.g. reporting and control devices). The source is highly persistent and motivated by extortion. Immediate remedies are to close all network connections to disrupt the kill chain. Often, only external parties identify the compromise rather than it being detected internally. The source of the attack is motivated by financial gain rather than system or data destruction, so, once the infiltration has occurred, a financial payment may be the only remedy available. Several agricultural entities have suffered losses from these types of attacks in recent years, which may escalate with more targeted attacks in the future (Rogers et al., 2006).

7 Cybersecurity frameworks for PA

7.1 System vulnerability

To help enterprises address cyber concerns, the U.S. National Institute of Standards and Technology (NIST) has developed a set of voluntary, best practices for threat assessments (ISACA, 2016). Enterprises using PA systems can be drastically different, which requires the use of frameworks which are flexible enough to meet varying needs. One such framework consists of five fundamental functions (ISACA, 2017):

- identify,
- protect,
- detect,
- respond, and
- recover.

Each function is associated with activities, desired outcomes, and applicable references. A breakdown of the categories and subcategories of the framework is available in the Appendix with application to four types of PA activities to demonstrate its flexibility: (1) IoT distributed network/mesh with remote access, (2) automated vehicle (Agbots) monitoring, data collation, (3) ICS with PLC, and (4) integrated farming.

For PA systems with predominantly online services, web traffic has been shown to serve as a good proxy for usage rates (Fich and Fich, 2004). Using this notion, we can employ the above framework to derive the probability of a cyberattack incident by monitoring web-traffic behavior as part of the detection function. There are other relatively simple web-traffic monitoring processes that can help identify vulnerabilities to memory corruption and malware (Houmb et al., 2010).

A popular method of quantifying vulnerability for a system is the use of the CVSS. The CVSS provides an objective way to capture the principal characteristics of a vulnerability and produce a numerical score reflecting its severity, as well as a textual representation of that score (Houmb et al., 2010). The score is then translated into a qualitative category (e.g. low, medium, high, or critical) to assess and prioritize vulnerability management activities. The CVSS approach is useful because it provides a standardized vulnerability score, and it is developed from an open framework, and it directly helps identify and prioritize risks. The CVSS scoring rubric is relatively straightforward and is available in several formats.

Referring to Table 1, the CVSS approach uses three groups of parameters to score potential risks: basic parameters, temporal parameters, and

Table 1 CVSS scoring overview

Basic parameters	Temporal parameters	Environmental parameters
Access vector	***Exploitability***	***Collateral damage potential***
Local	Unproven	None
Adjacent network	Proof-of-concept	Low
Network	Functional	Low-Medium
	High	Medium-High
	Not Defined	High
		Not Defined
Access complexity	***Remediation***	***Target distribution***
High	Official Fix	None
Medium	Temporary Fix	Low
Low	Workaround	Medium
	Unavailable	High
	Not Defined	Not Defined
Authentication	***Report confidence***	***Security requirements***
Multiple	Unconfirmed	Low
Single	Uncorroborated	Medium
None	Confirmed	High
	Not Defined	Not Defined
Confidentiality impact		
None		
Partial		
Complete		
Integrity impact		
None		
Partial		
Complete		
Availability impact		
None		
Partial		
Complete		

environmental parameters. Each group is represented by score compound parameters ordered as vectors. Basic parameters represent the intrinsic specifications of the vulnerability, while temporal parameters represent the specifications of a vulnerability that might change over time due to technical changes. These two categories will be governed by the type of technologies and products supplied for the PA system. Environmental parameters represent the specifications of vulnerabilities derived from the site-specific environment. These will be best specified by the farmers who have intimate knowledge of their environments and perhaps also of known vulnerability impacts. The main benefit of the CVSS approach is that the score will remain unaffected when omitting environmental metrics, and thus the basic and technical vulnerabilities of the system remain common regardless of location (Hur and Noh, 2011).

8 Case study: PA system assessment

We examine a sample IoT PA system integrated using a wireless network for a diversified farming operation that includes cattle backgrounding paddocks and a feedlot, horticultural fields, an equine unit with co-located equine paddocks, a wildlife enclosure, a piggery, a dairy, and a disassociated infrastructure to test for radiance, dust, noise, and so on. The network forms a mesh that includes sensors for water-levels, chemical content of ring tanks, dams, and piggery effluent, as well as for a waste-water management lake system. A similar sensor network has been discussed in López et al. (2015). A second multisensory mesh network for a 180 ha beef facility is located around 10 km from the first IoT network acting as a remote operation. For this facility, the multi-sensor mesh network has been set up to monitor improved and unimproved pasture where a herd of drought-resistant cattle are fattened. Communication is achieved via low-power communication radios within a mesh-network topology. Data at both sites are recorded at the nodes in a point-to-point, star-mesh topology and pre-programmed timeframe, depending on the application, and then transferred wirelessly (cellular or Wi-Fi) to a data server in a third-party provider's cloud.

For this PA system, we prepare a CVSS score to demonstrate the level of overall system vulnerability aggregated across the three parameter categories. Table 2 provides the scoring solution and justification for each component's vulnerability level.

The CVSS formula provides a mathematical approximation of all possible metric combinations ranked in order of severity using a vulnerability lookup table. Metrics are assigned to real vulnerabilities and a severity rating (low, medium, high, or critical). There are a limited number of numeric outcomes (101 in total), with scores ranging from 0.0 to 10. Multiple scoring combinations can produce the same numeric score. The qualitative severity rating scale for the CVSS score describes a rating of low risk for scores of 0.1–3.9, medium risk for scores of 4.0–6.9, high risk for scores of 7.0–8.9, and critical risk for scores of 9.0–10.

Variations in the level of vulnerability to cyberattack can be easily quantified using CVSS as the primary metric. For instance, if the attack vector can be conducted through the network from a remote location then the base score increases to 7.5, placing the PA system at high risk. Similarly, if the three key aspects of systems (confidentiality, integrity, and availability) are all needed to be high (for an operation critical to the farm's profitability), then the environmental CVSS score spikes to a high level of 8.1. The CVSS score permits PA system integrators to determine how best to consolidate system needs and protections against potential attacks.

Representing four generic types of existing PA systems (IoT distributed network/mesh with remote access, automated vehicle, ICS with PLC, and

Table 2 CVSS scoring breakdown for a sample PA system

PA system	Basic	Temporal	Environmental
IoT distributed network/mesh with remote access over two locations	**Basic Score: 4.6**	**Temporal Score: 4.2**	**Environmental Score: 6.7**
	Attack Vector: Physical. Attacker requires physical access to the device	***Exploit Code Maturity:*** Proof-of-concept. Network structure in design phase and testing underway.	***Confidentiality Requirement:*** Low. IP protection needed for some elements.
	Attack Complexity: Low. Attack steps are simple	***Remediation Level:*** Workaround. Full security suite not yet installed but several workarounds in place.	***Integrity Requirement:*** Medium. System and data integrity is important but can be reduced if system efficiency is affected.
	Privileges Required: None. Worse-case scenario device is not protected with a PIN	***Report Confidence:*** Confirmed. High degree of confidence known vulnerabilities and technical details.	***Availability Requirement:*** Medium. Continuous interoperability needed but can be reduced if system efficiency is affected.
	User Interaction: None. No user interaction is required		***Modified Attack Vector:*** Network. Mesh represents most vulnerable element of IoT system.
	Scope: Unchanged. Vulnerable and impacted components are the same		***Modified Attack Complexity:*** Low. Relatively simple attacks will hack mesh system, compromise of data is important but not critical. No PLA systems.
	Confidentiality Impact: None. Confidentiality impact is secondary		***Modified Privileges Required:*** Low. Access to third parties is needed for functional operation of system.
	Integrity Impact: High. Importance (security) of this feature.		***Modified User Interaction:*** None. Users can track data only. No actuators exist in system.

PA system	Basic	Temporal	Environmental
	Availability Impact: None. Any availability impact is secondary.		***Modified Scope:*** Unchanged. Hardware and software as installed represents. ***Modified Confidentiality:*** Low. Some IP is confidential but most sensor data are not. ***Modified Integrity:*** Low. Tamper-evident data storage needs to be minimally secure. ***Modified Availability:*** High. System interaction needed for continuous monitoring. High vulnerability. Critical to maintain.

integrated farming) in the two dimensions of base-temporal score and environmental score yields important insights into the vulnerabilities and protections required by each. Figure 4 illustrates the relative CVSS for each system.

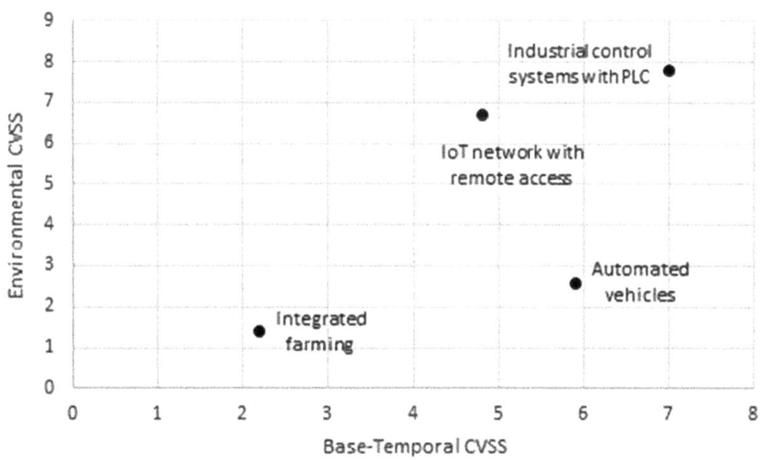

Figure 4 Base-temporal CVSS against environmental CVSS for four types of PA systems. Source: adapted from West (2018).

This representation demonstrates that the trade-offs between technology maturity and local adaptations of the technology can diverge and thus expose systems to compromise (West, 2018). For instance, automated vehicles have been operating in many industries for many years and the technology is relatively mature. But in remote farming operations with intermittent access to wireless data transfer links and potentially harsh conditions, such technology requires added features and system modifications that expose the technology to attack. This is represented via the CVSS score which quantifies the degree to which each of these emerging technologies are vulnerable to future cyberattack.

9 Future trends

9.1 Research pathways

While the agricultural sector faces multiple risks on a regular basis, the traditional threats related to weather and market availability/stability will be increasingly coupled with emerging risks related to cyberattack as agricultural systems develop and become more automated. A more technology-dependent and connected farm is a more vulnerable one. This is especially true for those operating on the fringe of research into PA opportunities that do not possess necessary security protocols. This threat is not trivial. Cyberattacks in the agricultural sector could easily decimate crops. This would have severe consequences for the quality of life of many rural and urban populations. Cyber security will need to be at the forefront of 'agritech' business development planning.

It is difficult to predict what the future holds for IoT technologies in the development of PA systems, but the growth in prevalence of cyber-physical systems (CPS) that integrate cyber and physical components using modern sensor, computing, and network technologies will increase. The development of CPS architecture is shifting toward the integration of a cyber-layer with a physical-layer, to enable the process of (1) monitoring; (2) networking; (3) computational processing; and (4) actuation.

Future research into cyber security of CPS-PA systems appears to be developing along four major themes:

1 Estimation of cyberattack consequences: Attacks are designed to cause significant damage to the cyber and physical layers comprising CPS. Adequate assessment of the impact of attacks on the normal functioning of physical processes needs to be better understood. This analysis can be used to evaluate the destructive impact of cyberattacks and quantify attack consequences so that the availability of cyber activities is known. Research conducted in this area by Genge et al. (2015), Sicari et al.

(2016), and Wu et al. (2015) is uncovering more efficient methods for assessment of the different CPS layers in PA systems.

2 Modeling CPS attacks: Attack and vulnerability models can better identify weaknesses in CPS systems to support their search strategy and understand the nature of future attacks. The use of attack models enables the development of countermeasures to ensure CPS security. This can uncover the range of failure conditions for equipment, control principles, and process behavior of CPS systems. Advanced research in this area includes visual flowcharting in Khalil (2016) and Mitchell and Chen (2015) who have developed a suite of analytical models based on stochastic Petri nets for modeling and analysis of attacks and countermeasures for CPSs.
3 Attack detection: The development of detection algorithms and countermeasures for well-known attacks in advance can greatly reduce the impact of attacks. Areas of research along this path include the classification of external attacks and intrusion detection in sensor networks by Finogeev and Finogeev (2017), distributed host-based collaborative detection (DHCD) methods to identify and mitigate false data injection attacks in smart grids by Li et al. (2017), and design approaches to enable real-time attack detections to supplement quality control systems by Vincent et al. (2015).
4 Security architecture development: Future development in advanced CPS solutions for PA systems will be constrained by security factors. A key component in the design of more complex CPS architectures is to test and validate 'secure design' to ensure the security and reliability of both physical and cyber components. New reliable control and evaluation algorithms under development that consider more realistic attack models from a security perspective include research by Yoo and Shon (2016) and Venkitasubramaniam et al. (2015).

For PA systems designers, the growth of development in successful IoT solutions necessarily requires growth in network security at a greater rate. Strategic partnerships with third parties who have the expertise in securing connectivity of individual devices and network structures is the highest priority.

10 Conclusion

Agricultural decision systems that integrate real-time environmental and farm-status data streams from in-situ sensors to actuate irrigation, pest control, and forecast data generate substantial efficiencies. Systems with high-frequency, real-time decision recommendations based on best management practices are especially appealing to the agriculture sector. However, system integrity will be a limiting factor in the widespread adoption of PA systems.

There are two types of PA systems: those that have been hacked and those that will be. This study represents a basic framework to help understand cyberattack vulnerabilities in the technology itself, as well as in the environment to which it is adapted. The principles-based framework addresses the need to capture and encrypt data, store data only in appropriate locations, limit access to actuation systems and automated vehicles through strictly controlled channels of communication, and only act on data-driven decisions when the system is deemed secure. The level of system vulnerability can be quantified which better allows for a measured approach to improving the integrity of complex PA systems.

11 Where to look for further information

A first place to start gaining greater familiarity with cyber threats is with a non-technical text titled *Cybercrime and Digital Deviance* by Roderick S. Graham and Shawn K. Smith, published by Routledge in September 2019. The following articles provide a good overview of the subject from a number of different perspectives:

- Goode, S., Lin, C., Tsai, J. C. and Jiang, J. J. 2015. Rethinking the role of security in client satisfaction with Software-as-a-Service (SaaS) providers. *Decision Support Systems* 70, 73–85.
- Herath, T. H. and Rao, R. 2009b. Protection motivation and deterrence: a framework for security policy compliance in organizations. *European Journal of Information Systems* 18(2), 106–25.
- Mathieson, K. 1991. Predicting user intentions: comparing the technology acceptance model with the theory of planned behavior. *Information Systems Research* 2(3), 173–91.
- Workman, M., Bommer, W. H. and Straub, D. 2008. Security lapses and the omission of information security measures: a threat control model and empirical test. *Computers in Human Behavior* 24(6), 2799–816.

Key research in the application of cybersecurity to agriculture can be found at the following organizations:

- Agricultural Aerial Remote Sensing Standards Council: https://agrscouncil.org/.
- American Farm Bureau Federation, Privacy and Security Principles for Farm Data: https://www.fb.org/issues/technology/data-privacy/privacy-and-security-principles-for-farm-data.

- CIS - Center for Internet Security: https://www.cisecurity.org/controls/.
- NASA Earth Observatory Precision Agriculture: https://www.earthobservatory.nasa.gov/Features/PrecisionFarming/.
- Open Ag Data Alliance: http://openag.io.
- SANS: https://www.sans.org/critical-security-controls.

12 References

Can, N. 2015. Legal issues concerning the cyber security of GNSS. *7th International Conference on Recent Advances in Space Technologies (RAST) Proceedings*. IEEE Recent Advances in Space Technologies, Istanbul, Turkey, pp. 861-4. Available at: http://ieeexplore.ieee.org/document/7208461/.

Cardwell, L. and Shebanow, A. 2016. The efficacy and challenges of SCADA and smart grid integration. *Journal of Cyber Security and Information Systems* 1(3), 1-7.

Chawla, P. and Hussain, R. 2015. WSN application: insect monitoring through their behavior. *International Journal of Advanced Research in Computer Science* 6(6), 129-32.

Crowe, J. 2016. Survey: ransomware vs. traditional security. *Barkly Stats and Trends*. Available at: https://blog.barkly.com/ransomware-attacks-bypassing-antivirus.

Federal Bureau of Investigation, Cyber Division. 2016. Smart farming may increase cyber targeting against US food and agriculture sector (Private Industry Notification No. 160331-001). Available at: https://info.publicintelligence.net/FBI-SmartFarmHacking.pdf.

Fich, R. and Fich, E. M. 2004. Effects of web traffic announcements on firm value. *International Journal of Electronic Commerce* 8(4), 161-81. doi:10.1080/10864415.2004.11044312.

Finogeev, A. G. and Finogeev, A. A. 2017. Information attacks and security in wireless sensor networks of industrial SCADA systems. *Journal of Industrial Information Integration* 5, 6-16. doi:10.1016/j.jii.2017.02.002.

Fishbein, M. and Ajzen, I. 1974. Attitudes towards objects as predictors of single and multiple behavioral criteria. *Psychological Review* 81(1), 59-74. doi:10.1037/h0035872.

Genge, B., Kiss, I. and Haller, P. 2015. A system dynamics approach for assessing the impact of cyber-attacks on critical infrastructures. *International Journal of Critical Infrastructure Protection* 10, 3 17. doi:10.1016/j.ijcip.2015.04.001.

Godfray, H. C., Beddington, J. R., Crute, I. R., Haddad, L., Lawrence, D., Muir, J. F., Pretty, J., Robinson, S., Thomas, S. M. and Toulmin, C. 2010. Food security: the challenge of feeding 9 billion people. *Science* 327(5967), 812-8. doi:10.1126/science.1185383.

Goodman, K. 2017. Cyber maturity will help ensure cyber security. *Crain's Cleveland Business*. Available at: http://www.crainscleveland.com/article/20170417/BLOGS05/170419840/cyber-maturity-will-help-ensure-cybersecurity.

Gronau, I. 2015. Implementing precision data privacy, security and ownership policies. *Precision Farming Dealer*, June 29. Available at: https://www.precisionfarmingdealer.com/articles/1516-implementing-precision-data-privacy-security-andownership-policies.

Houmb, S. H., Franqueira, V. N. L. and Engum, E. A. 2010. Quantifying security risk level from CVSS estimates of frequency and impact. *The Journal of Systems and Software* 83(9), 1622–34. doi:10.1016/j.jss.2009.08.023.

Hur, J. and Noh, D. K. 2011. Attribute-based access control with efficient revocation in data outsourcing systems. *IEEE Transactions on Parallel and Distributed Systems* 22(7), 1214–21. doi:10.1109/TPDS.2010.203.

IBISWorld. 2015. Precision agriculture systems & services in the US (IBISWorld Industry ReporOD4422). Available at: https://www.ibisworld.com/industry-trends/specialized-market-research-reports/technology/scientific-systems-devices/precision-agriculture-systems-services.html.

ISACA. 2016. Implementing the NIST cybersecurity framework using COBIT 5: a step-by-step guide for your enterprise. Available at: http://www.isaca.org/Knowledge-Center/Research/ResearchDeliverables/Pages/Implementing-the-NIST-Cybersecurity-Framework-Using-COBIT-5.aspx.

ISACA. 2017. State of cyber security 2017: Part 2: current trends in the threat landscape, June. Available at: https://cybersecurity.isaca.org/state-of-cybersecurity#0-part-2-june.

Khalil, Y. F. 2016. A novel probabilistically timed dynamic model for physical security attack scenarios on critical infrastructures. *Process Safety and Environmental Protection* 102, 473–84. doi:10.1016/j.psep.2016.05.001.

Kshetri, N. 2006. The simple economics of cybercrimes. *IEEE Security and Privacy Magazine* 4(1), 33–9. doi:10.1109/MSP.2006.27.

Li, B., Lu, R., Wang, W. and Choo, K. R. 2017. Distributed host-based collaborative detection for false data injection attacks in smart grid cyber-physical system. *Journal of Parallel and Distributed Computing* 103, 32–41. doi:10.1016/j.jpdc.2016.12.012.

Long, N. V., Assefa, Y., Schwalbert, R. and Ciampitti, I. A. 2017. Maize yield and planting date relationship: a synthesis-analysis for us high-yielding contest-winner and field research data. *Frontiers in Plant Science* 8, 2106. doi:10.3389/fpls.2017.02106.

López, J. A., Navarro, H., Soto, F., Pavón, N., Suardíaz, J. and Torres, R. 2015. GAIA2: a multifunctional wireless device for enhancing crop management. *Agricultural Water Management* 151, 75–86. doi:10.1016/j.agwat.2014.10.023.

Michalowski, R. J. and Pfuhl, E. H. 1991. Technology, property, and law – the case of computer crime. *Crime, Law and Social Change* 15, 255–75.

Mitchell, R. and Chen, I. R. 2015. Modeling and analysis of attacks and counter defense mechanisms for cyber physical systems. *IEEE Transactions on Reliability* 65(1), 350–8. doi:10.1109/TR.2015.2406860.

Rogers, M. K., Seigfried, K. and Tidke, K. 2006. Self-reported computer criminal behavior: a psychological analysis. *Digital Investigation* 3(Suppl.), 116–20. doi:10.1016/j.diin.2006.06.002. Available at: http://www.dfrws.org/2006/proceedings/15-Rogers.pdf.

Sicari, S., Rizzardi, A., Miorandi, D., Cappiello, C. and Coen-Porisini, A. 2016. A secure and quality-aware prototypical architecture for the Internet of Things. *Information Systems* 58, 43–55. doi:10.1016/j.is.2016.02.003.

Sinrod, E. J. and Reilly, W. P. 2000. Hacking your way to hard time: application of computer crime laws to specific types of hacking attacks. *Journal of Internet Laws* 4(3), 1–14.

Symantec. 2019. Internet security threat report (vol. 24). Available at: https://www.symantec.com/security-center/threat-report.

Venkitasubramaniam, P., Yao, J. and Pradhan, P. 2015. Information-theoretic security in stochastic control systems. *Proceedings of the IEEE* 103(10), 1914–31. doi:10.1109/JPROC.2015.2466089.

Vincent, H., Wells, L., Tarazaga, P. and Camelio, J. 2015. Trojan detection and side-channel analyses for cyber-security in cyber-physical manufacturing systems. *43rd SME North American Manufacturing Research Conference*, Charlotte, North Carolina.

Voiskounsky, A. E. and Smyslova, O. V. 2003. Flow-based model of computer hackers' motivation. *Cyber Psychology and Behavior: the Impact of the Internet, Multimedia and Virtual Reality on Behavior and Society* 6(2), 171–80. doi:10.1089/109493103321640365.

West, J. 2018. A prediction model framework for cyber-attacks to precision agriculture technologies. *Journal of Agricultural and Food Information* 19(4), 307–30. doi:10.1080/10496505.2017.1417859.

Wu, W., Kang, R. and Li, Z. 2015. Risk assessment method for cyber security of cyber physical systems. *1st International Conference on Reliability Systems Engineering*, Beijing, China.

Yoo, H. and Shon, T. 2016. Challenges and research directions for heterogeneous cyber-physical system based on IEC 61850: vulnerabilities, security requirements, and security architecture. *Future Generation Computer Systems* 61, 128–36. doi:10.1016/j.future.2015.09.026.

13 Appendix

A framework for PA system control is presented as a five-step process

13.1 Identify cyber threats

Function	Subsystem	IoT distributed network/ mesh with remote access	Automated vehicle (Agbots) monitoring, data collation	Industrial control systems, PLC	Integrated farming
IDENTIFY	**Asset Management:** Data, personnel, devices, systems, facilities	Classify remote devices Assign security roles to network access Track third-party network access	Classify in-vehicle devices Assign security roles to assets Track third-party access to data streaming	Classify assets and devices Assign security roles to linkages Limit third-party access to PLA actions	Classify all devices and assets Assign security roles where feasible Track third-party physical and remote access

Function	Subsystem	IoT distributed network/ mesh with remote access	Automated vehicle (Agbots) monitoring, data collation	Industrial control systems, PLC	Integrated farming
	Business Environment: Objectives, stakeholders, activities and risk management	Identify dependencies and critical network functions Resilience testing to support critical services for each state (normal operations, attack, recovery)			
	Governance: Farm regulatory, legal, risk, environmental, and operational requirements	Set security policy Identify roles/ responsibilities Privacy requirements Legal and regulatory obligations Risk management standards			
	Risk Assessment: Conduct cybersecurity risk assessment	Map network vulnerabilities Set risk tolerance at a single level across the network Map threats, impacts, likelihood at each node Compute risk exposure using CVSS processes	Test asset vulnerabilities Set risk tolerance Map threats, impacts, likelihood Compute CVSS risk exposure	Map full set of asset and network vulnerabilities Set multiple levels of risk tolerance for each Use CVSS measure to mitigate total risk level	Map physical and remote asset vulnerabilities Set risk tolerance for physical and remote systems Map threats, impacts, likelihood Use CVSS score to mitigate total risk level
	Supply Chain Risk Management: Constraints, risk tolerances, and assumptions	Identify supply chain links Map threats to each link Audit third-party access and monitor Build recovery plan and test			

13.2 Protect systems, networks, actuators, and automated devices

Function	Subsystem	IoT distributed network/mesh with remote access	Automated vehicle (Agbots) monitoring, data collation	Industrial control systems, PLC	Integrated farming
PROTECT	**Identity Management and Access Control:** Physical and logical access to assets	Penetration testing of remote access Track access permissions Network integrity audit Third-party network access audit Identities logged and mapped to credentials	Penetration testing Test physical access Third-party data streaming access audit Identities logged and mapped to credentials	Penetration testing of remote access Test physical access Track access permissions Network integrity audit Third-party network access audit Identities logged and mapped to credentials	Penetration testing of remote access Test physical access Track access permissions Network integrity audit Third-party network access audit Identities logged and mapped to credentials
	Awareness: Cybersecurity awareness education and training	Users to positively confirm roles and responsibilities Response and recovery plans tested			
	Data Security: Information and data remain confidential but available	Test integrity of data-at-rest and data-in-transit Sensors managed through transfer and disposal Network capacity testing Software and firmware testing Production and development environments segregated	Test integrity of data-in-transit Assets managed through transfer and disposal Software and firmware testing	Test integrity of data-at-rest and data-in-transit Assets managed through transfer and disposal Capacity testing Software and firmware testing Production and development environments segregated	Test integrity of data-at-rest and data-in-transit Software and firmware testing Production and development environments segregated

Function	Subsystem	IoT distributed network/mesh with remote access	Automated vehicle (Agbots) monitoring, data collation	Industrial control systems, PLC	Integrated farming
	Information Protection Process: Security policies and procedures	Baseline configuration of network controls Develop system development lifecycle Change control configuration Test information and controls back-ups Protection measures upgraded with system augmentation Develop vulnerability management plan Develop business continuity plans	Baseline configuration of automation controls Test control back-ups Protection measures upgraded with system augmentation Develop vulnerability management plan Develop business continuity plans	Baseline configuration of ICS, PLA, industrial controls Construct system development lifecycle Change control configuration Test information and controls back-ups Protection measures upgraded with system augmentation Develop vulnerability management plan Develop business continuity plans	Change control configuration Test information and controls back-ups Asset protection measures upgraded with system augmentation Test vulnerability management plan Develop business continuity plans
	Maintenance: Maintenance and repairs of industrial control and ICS	Asset maintenance and repairs logged with approved controls Remote access maintenance and repairs logged, on site testing supported			

Function	Subsystem	IoT distributed network/mesh with remote access	Automated vehicle (Agbots) monitoring, data collation	Industrial control systems, PLC	Integrated farming
	Protective Technology: Technical security solutions to ensure the security and resilience of systems	Apply principle of least functionality to ensure systems provide essential data/control Test communication and control networks Systems capable to operate in pre-defined functional states (normal operations, attack, recovery)	Test communication and control networks Systems capable to operate in pre-defined functional states (normal operations, attack, recovery)	Apply principle of least functionality to ensure systems provide essential data/control Test communication and control networks Systems capable to operate in pre-defined functional states (normal operations, attack, recovery)	Test communication and control networks Systems capable to operate in pre-defined functional states (normal operations, attack, recovery)

13.3 Detect infiltration

Function	Subsystem	IoT distributed network/mesh with remote access	Automated vehicle (Agbots) monitoring, data collation	Industrial control systems, PLC	Integrated farming
DETECT	**Anomalies and Events:** Timely detection and potential impact understood	Map baseline of network operations and expected data flows Establish incident alert thresholds Record and analyse event data including impacts	Record and analyse event data including impacts	Map baseline of network operations and expected data flows Establish incident alert thresholds Record and analyse event data including impacts	Establish incident alert thresholds Record and analyse event data including impacts

Function	Subsystem	IoT distributed network/mesh with remote access	Automated vehicle (Agbots) monitoring, data collation	Industrial control systems, PLC	Integrated farming
	Security Continuous Monitoring: Monitored at discrete intervals to identify cybersecurity events, verify the effectiveness of protection	Monitor remote network functionality continuously Track and log external service provider activity Log unauthorized access to devices and system Develop and run automated vulnerability scans	Monitor physical assets Monitor personnel activity Track and log external service provider activity Log unauthorized access to assets, devices, systems Develop and run automated vulnerability scans	Monitor physical assets and remote network functionality continuously Monitor personnel activity where possible Track and log external service provider activity Log unauthorized access to assets, devices, systems Develop and run automated vulnerability scans	Monitor physical assets Monitor personnel activity where possible Track and log external service provider activity Log unauthorized access to assets, devices, systems Develop and run automated vulnerability scans
	Detection Processes: Detection process maintained and tested	Test detection processes Develop and test event detection escalation procedures			

13.4 Incidence response

Function	Subsystem	IoT distributed network/ mesh with remote access	Automated vehicle (Agbots) monitoring, data collation	Industrial control systems, PLC	Integrated farming
RESPOND	**Response Planning:** Response process and procedures executed and maintained	Test and execute response plans			

Function	Subsystem	IoT distributed network/ mesh with remote access	Automated vehicle (Agbots) monitoring, data collation	Industrial control systems, PLC	Integrated farming
	Communications: Response activities coordinated with internal and external stakeholders	Develop internal and external Information sharing procedures Stakeholder coordination response Single point of authority delegations	Personnel duties developed and tested Stakeholder coordination response Single point of authority delegations	Personnel duties developed and tested Develop internal and external Information sharing procedures Stakeholder coordination response Single point of authority delegations	Personnel duties developed and tested Develop internal and external Information sharing procedures Stakeholder coordination response Single point of authority delegations
	Analysis: Ensure adequate response and support recovery	Record detection system notifications Impact testing and recovery outcomes Audit response plan effectiveness to incident			
	Mitigation: Prevent expansion of event, mitigate effects minimize incident impact	Identify containment, mitigant responses Log new vulnerabilities and map to cyber-security plan upgrades			
	Augment: Response activities incorporate lessons from previous detection and response	Develop lessons learned profile Update response plans			

13.5 Recovery activities

Function	Subsystem	IoT distributed network/ mesh with remote access	Automated vehicle (Agbots) monitoring, data collation	Industrial control systems, PLC	Integrated farming
RECOVER	**Recovery Planning:** Recovery process executed and maintained for system restoration	Test recovery planning integrity and execution			
	Improvements: Recovery planning uses lessons learned	Test lessons learned and strategy upgrades			
	Communications: Restoration coordinated with internal and external parties	Audit of public response and reputation management Communicate recovery activities			

Chapter 5

Advances in artificial intelligence (AI) for more effective decision making in agriculture

L. J. Armstrong, Edith Cowan University, Australia; N. Gandhi, University of Mumbai, India; P. Taechatanasat, Edith Cowan University, Australia; and D. A. Diepeveen, Department of Primary Industries and Regional Development, Australia

1 Introduction

The rapid advance in computing hardware and infrastructure, along with the adoption of machine learning, provides the means for large repositories of historical and real-time seasonal data to be leveraged to improve production and sustainability in the agriculture sector. The potential of artificial intelligence (AI) technologies to improve processes and optimize production has been demonstrated in many industries over the last decade. These techniques can complement or improve traditional statistical and modeling approaches. A number of researchers have highlighted how AI can be used to enhance decision making in agriculture and provide the means for farmers to make better decisions in crop and livestock production.

http://dx.doi.org/10.19103/AS.2020.0069.06

A number of factors have influenced the adoption of AI technologies to improve agricultural decision making. There has been an exponential increase in the amount of data available due to developments such as the increased use of sensors, better access to satellite imagery, cheaper data loggers to collect in-field measurements, the increased use of drones to collect crop image data and open access to government data sets. Despite this increase in potentially useful data, its volume; the need to deal with missing, incomplete or poorly formatted data; and the need to collate and process it effectively mean that it is of little or no use to farmers in its raw form. There are also issues as to how far this wealth of data can be made relevant at national, regional, district or farm level. There is a need to understand how AI can be used to complement and inform other technologies such as robotics, cloud and data analytics and bioinformatics as well as plant breeding, sowing, cultivation, pest management harvesting and other farm management practices.

This chapter outlines the main issues and challenges in use of AI technologies in agriculture, the data used for AI applications and the core AI technologies which are emerging in the area of decision making for agriculture (e.g. data mining and machine learning). The chapter provides a comparison with other approaches and how AI is being integrated with other technologies such as robotics and drones to develop state-of-the-art decision tools for agriculture. The chapter also provides some case study examples as well as discusses future trends in the subject.

2 Agricultural DSS using AI technologies: an overview

There have been a number of studies that have reported on the use of decision support tools (DSSs) to improve agricultural production and farm management (Soltani and Sinclair, 2012a; Mahmoud et al., 2015; Giusti and Marsili-Libelli, 2015). Table 1 provides an overview of some of these studies. Hochman and Carberry (2011) have proposed a minimum set of requirements for agricultural DSSs. These include the need to 'better match farmer's naturalistic decision making processes and ... to educate the farmer's intuition, not to replace it with optimized recommendations; enabling users to experiment with options that satisfy their needs rather than attempting to present optimised solutions; embedding a support network consisting of farmers, consultants and researchers ...; requiring on-going improvement, testing and validation; and providing to the decision maker new and useful information that would not be otherwise accessible.'

Armstrong et al. (2010) have suggested a framework for an information technology-based agriculture DSS. This framework incorporates AI techniques as an integral part of the system to improve the effectiveness of decision making, using data capture, processing and analysis followed by the delivery

Table 1 Overview of the application of DSS using AI technologies

Category	Description	References
Farm planning	Production planning of cropping activities, which will help in building profitable cropping plans, process control and integration; economic and market analysis	Dath and Balakrishnan (2013)
Crop management	Crop management category of expert system provides decision support to farmers for growing any particular crop	Prasad and Babu (2006), Naidu et al. (2013), Negid (2014)
Soil and water conservation	Soil and water conservation over soil erosion and the selection of water pumps for agricultural irrigation	Seflek and Carman (2010)
Pest and disease management	Pest and disease management weed seedling identification, manage diseases and infections	Prasad and Babu (2006), Shafinah et al. (2013), Naidu et al. (2013), Roseline et al. (2012), Ravisankar et al. (2010), Bayu and Aris (2011), Derwin Suhartono et al. (2013), Kaura et al. (2013), Bergsma et al. (2013), Hassan et al. (2011)
Horticulture	Orchard and fruit production	Sridhar et al. (2010), Selvakumar et al. (2011), Naderloo et al. (2012), Naderloo et al. (2012), Guerrero et al. (2013), Hu et al. (2010), Yadav et al. (2014)
Fertilizer management	Manage crop and pasture fertilizer management	Jadhav et al. (2011), Arshad et al. (2012), Hitoshi et al. (2012), Romeo et al. (2013), Montalvo et al. (2013), Ni (2015)

of targeted information to the farmer. Data is collected from various sources, validated and categorized following a defined set of rules and data mining tools to process and integrate the data into a format that contributes to the knowledge base used by the farmer.

Early attempts to develop an effective DSS failed to deliver effective decision making support and benefits such as improved productivity (Parker and Campion, 1997; Long and Parton, 2012). More recent DSSs have improved functionality (Mahmoud et al., 2015; Giusti and Marsili-Libelli, 2015). Some of these have used fuzzy analytical processes to facilitate decision making (Jaiswal et al., 2015). Other DSS have integrated crop models and data from geographical information systems (GISs) to analyze and predict crop yields (Kadiyala et al., 2015). Mirschel et al. (2016), for example, developed a spatial model-based DSS for evaluating agronomic practices to adapt to climate change such as crop rotation, soil tillage, fertilization and irrigation at the farm level. Hardjomidjojo et al. (2014) have developed a prototype spatial model-based DSS for food security policy analysis.

Andrew et al. (2013) have reported on the number of DSS applications for crop production. These systems included 3-PG, APSIM, CABALA, GrassGroTM, GrazFeedTM, MetAccessTM and Yield Prophet. Other examples reported included MyCrop, ROOTMAP, Yield Calculator and SPLAT. Garg et al. (2016), for example, have developed a simple farmer-friendly water impact calculator (WIC) DSS for enhancing water use efficiency in agriculture using a WIUE-based model. Neuhaus and Bowden (2014) have developed a fertilizer decision tool using models based on yearly field trial data, with the more recent inclusion of neural networks to improve the quality of fertilizer recommendations. Chen et al. (2019) have used simulation models to predict yields of crops such as wheat.

There are a few examples of agricultural DSSs which have fully utilized AI techniques. An early study by Zhang et al. (2012a,b) proposed a DSS that comprised different modules including a transaction processing system, database, data mining modules and user interface (see Fig. 1). Another example of an agriculture DSS architecture design was the GeoSense DSS reported by Adinarayana et al. (2012). This was an information, communication and dissemination system and consisted of five modules: crop water requirements, rice yield simulation, energy balance, weather profile studies and crop pest and disease prediction (Tripathy et al., 2012). This DSS-integrated wireless sensor networks (WSNs) and cloud services to provide real-time information to users (see Fig. 2). Tanaka and Kiura (2014) described a similar approach with a DSS, which incorporated a meteorological data acquisition function and a weather data generator, crop models and a crop model execution engine as well as display and comparison functions. Another study by Ramteke and Dhande (2013) suggested ways to improve DSSs based on a traditional relational database model by using data mining, data warehousing, on-line analytical processing (OLAP) and on-line transactional processing (OLTP) technologies.

Armstrong and Nallan (2016) proposed a DSS that could be used by farmers to improve decision making in crop variety selection for different climatic and agronomic scenarios in western Australia. The AgMine DSS comprises six major components, including data input, data mining, statistical analysis, a database, prediction and visualization modules. They show how visualization and data mining modules can be used to improve farmer decision making. Different geospatial Kriging and association rule mining techniques were also used to determine the relationship between crop variety, rainfall and soil at the district level. The authors combined different techniques for decision making by providing visualizations of seasonal patterns of rainfall for individual

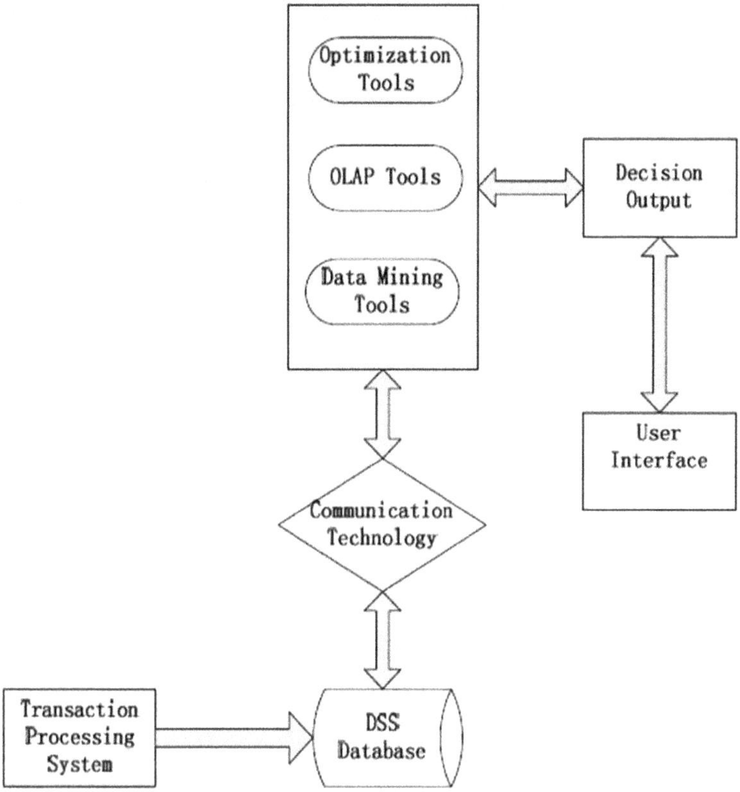

Figure 1 A general framework of DSS. *Source*: Zhang et al. (2012a,b).

districts and showing the effect of various scenarios of dry and wet years on crop production.

Gandhi et al. (2016a,b, 2017) proposed a rice yield DSS which incorporated neural networks and association rule mining to predict rice yield for different agroclimatic zones of India. A historical data set was used to predict rice crop yield for the kharif season of the humid subtropical climatic zone of India. The results showed that the classification techniques such as J48 and LADTree performed well in predicting crop yield. A DSS was developed that was able to predict rice crop yields accurately for different agroclimatic zones and districts across India.

Another study by Czimber and Gálos (2016) outlined a DSS using maximum likelihood and fuzzy logic techniques to analyze data on topography, vegetation, climate, soils and hydrology. It processed time-series data such as meteorological variables and used geospatial, spatial and climate space distribution and yield data to provide yield predictions.

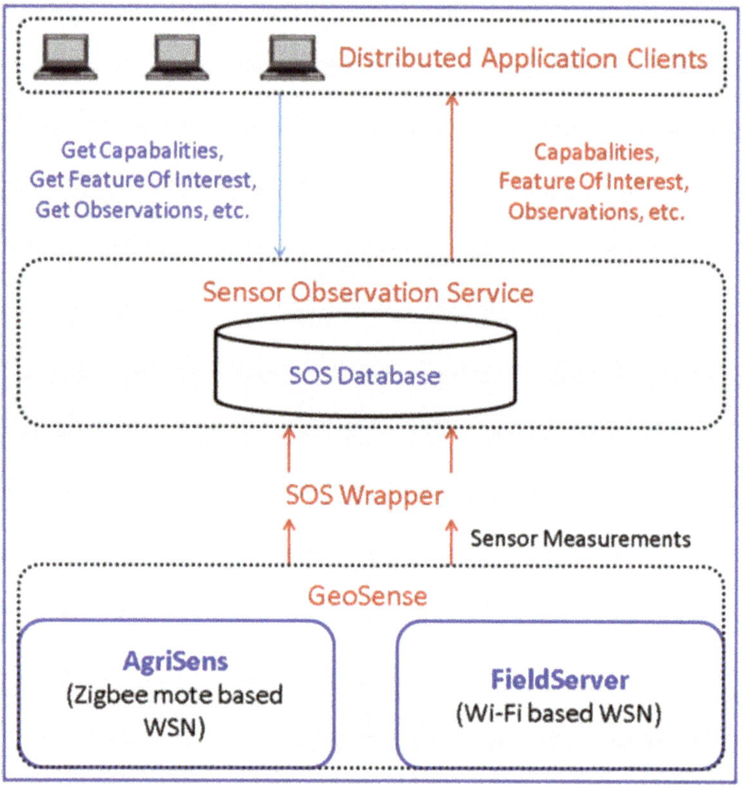

Figure 2 Oriented architecture for GeoSense. *Source*: Adinarayana et al. (2012).

3 Data and image acquisition

The acquisition of data can be considered to be the first stage in developing a DSS. An important consideration in the development of agricultural DSS using AI is the acquisition, pre-processing and quality of data. Types of data used for agricultural DSS include various weather indices, soil and crop/pasture growth parameters and irrigation and other agronomic inputs. Data can be collected through sources such as research trials, plant breeding and variety trials. Advances in sensor hardware, use of satellites, drones, proximal and handheld sensors together with sensors mounted on farm machinery, as well as the increasing use of cloud technologies to store vast amounts of data acquired from these sensors and systems has resulted in more robust and accurate DSS. The development of internet of things (IoT) technologies and cloud computing has provided an opportunity for more data-driven approaches to agricultural management (Wolfert et al., 2017; Da Paz et al., 2019). The need for DSS has been driven by the rapid uptake of digital farming technologies using technologies such as sensors, precision GPS, robotics and automation, AI (machine learning, image processing), mobile applications and cloud technologies (see Table 2).

Digital image processing 'encompasses processes whose inputs and outputs are images and, in addition, encompasses processes that extract attributes from images, up to and including the recognition of individual objects' (Gonzalez et al., 2009). An image in the field of digital image processing refers to a two-dimension function of spatial coordinates, f (x,y), where f is the intensity or gray level at any pair of coordinates (x,y). Examples of image processing techniques in agriculture include the measurement of crop ground cover or the percentage of soil surface covered by plant foliage, which can be used to assess crop vigor (Mullan and Reynolds, 2010). Mullan and Reynolds (2010) studied the analysis of crop ground cover using digital ground cover (DGC) techniques. The technique used Photoshop as a tool to analyze digital images. This method has been utilized to study issues such as effects of ground cover on soil water evaporation (Mullan and Reynolds, 2010). Kakran and Mahajan (2012) have used digital image processing to determine wheat growth stages, which are determined by the percentage of green color. Three images of 6-,

Table 2 Smart technologies in Australian agriculture

Smart technologies	Examples
Artificial intelligence and machine learning technologies	Data visualizations and decision support systems
Image processing	3d imagery technologies Weed sensor technologies Precision pesticide
Sensor technologies	GPS controlled or tramline agriculture Various Ag machinery GPS systems Variable rate technologies RFID tags, fit bits and other livestock management Microclimate predictions
Drone technologies	Drones and sensor technologies Aerial imagery for scenario planning
Robotics	Autonomous ground vehicles for herding dairy cows Vertical farming and mechanization of horticulture Meat processing Robotic weeding AGBot2 at QUT Dairy Science group, University of Sydney, https://youtu.be/S4Dndp-Esd8 Robotic selective harvesting in horticulture
Big data/cloud computing/IoT	Big Data, access to farm machine and sensor data Supply chain integration Phenotyping
Mobile apps	Farm planning and decision making, diagnostics

14-, and 18-week-old wheat plants were used for the analysis. The system can be used to determine and predict wheat growth stages based on data from WSNs. Aerial pictures have also been used for crop yield prediction before harvesting (Thomas et al., 1997). Shahin et al. (2001) have used x-ray images of apples to monitor quality.

Leaf area index (LAI) is a variable used to assess crop growth, pests and disease. Zhang et al. (2012a,b) have studied the performance of LAI measurement methods for wheat. Two measurement methods, canopy spectral data acquisition and digital image processing, were compared for accuracy with traditional LAI measurement by using a length-width coefficient method. The authors concluded that canopy spectral data acquisition was more accurate than digital image processing. However, canopy spectral data acquisition requires a specialist spectroradiometer to measure spectral data. Although recent video sensor networks (SNs) provide some advantages in capturing data, there are still many challenges in improving digital image processing sufficiently to analyze data for use in DSS.

The normalized difference vegetation index (NDVI) is a technique used to measure canopy size and vegetation greenness, which can be used to estimate early cover, nitrogen content, water stress, post-anthesis stay-green and pre-anthesis biomass. The advantages of NDVI measurement are that it is quick, easy, of low cost and non-destructive (Govaerts and Verhulst, 2010). Pask et al. (2012) have described using a NDVI handheld sensor which has a high resolution suitable for the plot level measurements. Raun et al. (2001) have used a NDVI sensor to predict potential grain yield in winter wheat. NDVI techniques have been used to assess crop rotation, tillage and residue management (Govaerts and Verhulst, 2010), as well as to assess chlorophyll content (Lopes and Reynolds, 2012).

4 Core AI technologies

There are a number of computer science techniques that have been used to analyze agricultural data for decision making. Data mining and machine learning approaches have been used to predict crop yields. Recent studies have described the use of neural networks; (Das et al., 2018); random forest techniques, which have been used at global and/or regional scales (Folberth et al., 2019; Jeong et al., 2016); and genetic programming algorithms (Ali et al., 2018). A recent review of machine learning approaches for decision making in crop production has been provided by Chlingaryan et al. (2018).

4.1 Data mining in agriculture

Data mining has been described as 'a nontrivial extraction of previously unknown, potentially useful and reliable patterns from a set of data. It is the process of analysing data from different perspectives and summarizing it into useful information' (Mucherino et al., 2009). Data mining is a relatively new inter-disciplinary concept involving data analysis and knowledge discovery from databases (Rupnik et al., 2007). There are many examples of studies which have applied data mining techniques to determine the influence of crop and weather parameters and make predictions about yields (Guo and Xue, 2012, 2014; Medar and Rajpurohit, 2014; Bejo et al., 2014; Dahikar and Rode, 2014; Yengoh and Ardo, 2014; Ramesh and Vardhan, 2015).

Data mining is a multi-faceted approach which includes statistical analysis, data visualization, neural networks, knowledge discovery techniques, pattern recognition and database management (Feelders et al., 2000). Han et al. (2011) categorize data mining functionalities into six categories:

1 concept or class description: characterization and discrimination
2 mining frequent patterns, associations and correlations
3 classification and prediction
4 cluster analysis
5 outlier analysis
6 evolution analysis

The process of data mining consists of the following steps:

- data cleaning
- data integration
- data selection
- data transformation
- data mining
- pattern evaluation
- knowledge presentation

Data cleaning is a process to remove inconsistent and unwanted 'noise' data. Data integration is a process to combine data from multiple data sources. Data selection is a process to retrieve relevant data from the database for the analysis. Data transformation renders data into an appropriate format for the data mining process. It consists of processes such as data smoothing, aggregation, generalization, scale normalization and attribute construction. The steps from data cleaning to data transformation may be grouped as the pre-processing stage (Fig. 3).

Early data mining studies used the k-means approach. A study by Holmgren and Thuresson (1998) used a k-nearest neighbor approach to evaluate forest inventories and to estimate forest variables from the analysis of satellite imagery. A similar study by Rajagopalan and Lall (1999) applied the k-nearest

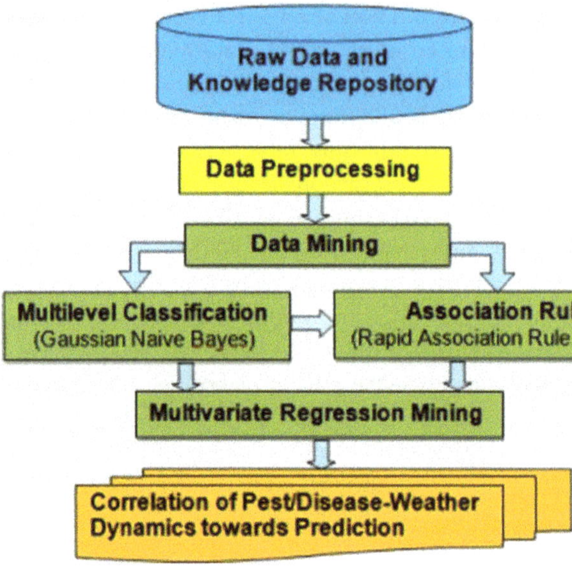

Figure 3 Example of data mining process in DSS.

neighbor approach to simulate daily rainfall and other weather variables. Jorquera et al. (2001) used the k-means method to assess pollution in the atmosphere. Verheyen et al. (2001) used the k-means approach for classifying soils in combination with GPS-based technologies. Meyer et al. (2004) also used a k-means approach to classify soils and crops such as apples. Leemans and Destain (2004) also used a k-means approach to analyze color images of fruits to assess their quality.

Ruß (2009) evaluated four different mining techniques to predict wheat yield: support vector regression (SVR), regression tree, multilayer perceptrons (MLPs) and radial basis function (RBF). Armstrong et al. (2007) have applied various clustering techniques on western Australian soil data. Jain and Arora (2012) have also proposed a method for data mining multiple patterns from clusters. The hierarchical cluster algorithm and hierarchical clustering explorer programs have been utilized to cluster data to help cotton farmers make decisions about sowing dates (Human Computer Interaction Lab, 2013).

Techniques such as logistic regression (LR) using SAS software, decision tree induction, rough sets (RSs) and hybridized rough set-based decision tree induction (RDT, J4.8) using Rosetta, Weka (Waikato Environment for Knowledge Analysis) and C++ programs, as well as Java implemented C4.5 (CJP), have been used to identify powdery mildew disease in mango (Jain et al., 2009). Romani et al. (2010) have used a novel unsupervised algorithm, called CLimate and rEmote sensing Association patteRns Miner (CLEARMiner), for mining association patterns in heterogeneous time series from climate and remote sensing data, with the aim of monitoring the growth of sugarcane. Rules generated by this new algorithm show association patterns in different periods of time in each time series, allowing accurate yield forecasting without having the burden of dealing with many data charts.

Abdullah et al. (2004) used data mining techniques on a range of agricultural, meteorological and pest data to assess patterns of pesticide use in Pakistan. Unsupervised clustering of data was performed using recursive noise removal (RNR). These clusters reveal patterns of pesticide use and potential reasons for overuse. Another interesting study for analyzing pesticide application practices using DSS is the research of Abdullah et al. (2012).

Divya et al. (2014) used data mining of data from a WSN to interpret interactions between crops, weather and pests to predict and manage thrips infestations. The data was validated using regression models. Ahamed et al. (2015) have used data mining techniques on environmental variables to predict annual yields of major crops in different districts in Bangladesh. A number of other recent studies have shown the potential of applying data mining techniques in agriculture (Raorane and Kulkarni, 2013; Mankar and Burange, 2014; Kalpana et al., 2014; Fathima and Geetha, 2014; Patel and Patel, 2014; Kaur et al., 2014; Geetha, 2015). Many researchers have successfully applied data mining techniques for crop yield prediction (Veenadhari et al., 2011; Raorane and Kulkarni, 2012; Ramesh and Vardhan, 2013, 2015; Medar and Rajpurohit, 2014). Data mining techniques have also been used in the prediction of rice crop yields in India (Gandhi and Armstrong, 2016a–d; Gandhi et al., 2016a,b, 2017).

4.2 Artificial neural networks (ANNs) in agriculture

Artificial neural networks (ANNs) are based on human brain biological processes. The neural network model comprises a number of layers including input, middle and output layers. ANN has been applied in various areas such as medical research (Ganesan et al., 2010; Udeshani et al., 2011; Elgader and Hamza, 2011; Mojarad et al., 2011; Singh et al., 2011; Halloran et al., 2011; Kharya, 2012; Narang et al., 2012; Saxena and Burse, 2012; Amato et al., 2013).

A number of studies have described the development of agronomic-based models using ANNs. Agronomic ANN applications include crop development modeling (Elizondo et al., 1994), pesticide and nutrient loss assessments (Yang et al., 1997; Starrett et al., 1997), soil water retention estimations (Schaap and Bouten, 1996), disease prediction (Batchelor et al., 1997) and fertility of hen eggs (Das and Evans, 1992). Kaul et al. (2005) showed that an ANN model can deliver more reliable and accurate crop yield forecasting compared to other methods such as regression models. Smith et al. (2009) have used Ward-style ANNs to develop air temperature prediction models for farmers. Ranjeet and Armstrong (2014) have also used neural networks to predict rice yields based on factors such as soil parameters. Dahikar and Rode (2014) have successfully applied ANNs using feed-forward back-propagation for agricultural crop yield prediction.

4.3 Bayesian networks (BNs) in agriculture

Bayesian networks (BNs) are considered to be ideal for situations enabling diagnostic reasoning on conditional dependencies to assess model structural and parameter uncertainty (Newlands and Townley-Smith, 2010). A BN is a method for representing beliefs and knowledge using probabilities, especially relevant for systems that are highly complex in their structure and functional interactions (Ascough et al., 2008). A BN uses the probabilistic components of a framework, as opposed to deterministic comparisons, to describe the connections among variables (Newlands and Townley-Smith, 2010). BNs are thus an increasingly popular method for modeling uncertain and complex domains.

Many studies have reported on the advantages and challenges using a BN for complex problems such as environmental modeling (Uusitalo, 2007). The technique provides a graphical model that encodes probabilistic relationships among variables of interest (Heckerman, 1997). This model has several advantages for data modeling when used in conjunction with statistical techniques. Some of these advantages were discussed in Heckerman (1997) and include the following:

1 The ability to handle situations where some data entries are missing by encoding dependencies among all variables
2 The ability to learn casual relationships which helps to understand a problem domain
3 Incorporation of casual and probabilistic semantics which allows it to combine prior knowledge and real-time data
4 An efficient way of avoiding overfitting of data

These advantages mean that BNs have the potential to deal with the complexities of agricultural environments.

There have been a few examples of the application of BNs in the agricultural domain. BNs have been used for predicting the yield response of winter wheat to fungicide programs (Tari, 1996). BNs have been used to assess the ease of calving in dairy cattle (Luo et al., 2002). The development of a DSS for growing malting barley without the use of pesticides utilized a BN approach (Kristensen and Rasmussen, 2002). A BN has been used for a DSS for the selection of irrigation systems for dairy farms in Australia (Robertson and Wang, 2004). BNs have also been used in water resource and land use management (Castelletti and

Soncini-Sessa, 2007; Aalders, 2008), weed infestation risk in corn (Bressan et al., 2009), pest control (Bi and Chen, 2010), crop disease (Perez-Ariza et al., 2012).

Sharma and Patil (2015) found that FIS and Bayesian techniques were more successful in forecasting rice production and demand than other techniques. Other studies (Quinn et al., 2013; Nearing et al., 2013; Maiti and Tiwari, 2014; Madadgar and Moradkhani, 2014; Sharma and Goyal, 2015; Shin et al., 2016) have also highlighted the application of a BN in an agricultural context. Gandhi and Armstrong (2016a,c) and Gandhi et al. (2016b) have compared a range of approaches to develop a DSS, including support vector machines, BNs and ANNs. They found that MLP and BayesNet achieved the highest accuracy, sensitivity and specificity.

4.4 Support vector machines (SVMs) in agriculture

Support vector machine (SVMs) are one of the newest supervised machine learning techniques. A number of studies have reported the application of SVMs in an agricultural context. For example, Camps-Valls et al. (2003) described the use of SVMs to classify crops, and Brudzewski et al. (2004) have highlighted its use in assessing milk quality. Another study by Du and Sun (2005) used SVMs to classify pizza sauce spreads, while Karimi et al. (2006) used SVMs for detecting weed and nitrogen stress in corn.

A number of other studies have shown the variety of uses for SVMS in agriculture (Tripathi et al., 2006; Fagerlund, 2007; Huang et al., 2010; Bharadwaj et al., 2012). A SVM has been applied for the estimation of crop biophysical parameters using aerial hyperspectral observations (Karimi et al., 2006). A SVM has also been used to analyze crop response patterns for different climate conditions for yield prediction (Tripathi et al., 2006; Brdar et al., 2011). Another study reported how a SVM was used to forecast demand and supply for pulp wood (Anandhi and Chezian, 2013).

4.5 Statistics and other techniques in agriculture

A number of other common statistical methods have been used to analyze agricultural data including various regression techniques (Tripathy et al., 2012). Linear regression is used to create a regression line that illustrates the relation between an independent variable and a dependent variable. Tripathy et al. (2012) have used the regression technique to identify hidden relationships between crop, weather and pest dynamics. Soltani and Sinclair (2012b) have developed a simple model using linear and nonlinear functions to simulate chickpea plant growth development, soil water balance, plant nitrogen balance, leaf development and senescence, as well as mass partitioning. These models depend on identifying crop growth stages, which can be measured in scales such as the Zadoks, Feekes and Haun scales (Zadoks et al., 1974; Department of Primary Industries, 2012).

Visualization techniques provide a means to understand and analyze complex data. Keim et al. (2008) provides an overview of visual analytics, its scope and concepts and details the most important technical research challenges in the field. Aigner et al. (2008) have provided a systematic review of the diversity of methods for visualizing time-oriented data. A number of other reviews have reviewed web-based solutions for data visualization (Edelson and Gordin, 1998; Nocke et al., 2008; Fuchs and Hauser, 2009; Shaw et al., 2009; Wei et al., 2009; Ladstädter et al., 2010; Sanga et al., 2013). Geospatial data mining has been suggested as a potential technique for the identification of relationships from spatial databases relevant for agricultural applications (Santos and Amaral, 2005; Kent, 2010; Ji and Cui, 2011).

Adeniyi (1993) has proposed the application of GIS visualization for agricultural resource planning and management. Oudemans et al. (2002) have reviewed the use of GISs, global positioning systems (GPSs) and RS for mapping and analyzing crop losses in crops such as cranberry. The analysis of spatial patterns using GISs and remote sensing imagery is an emerging field in ICT, which enables the user to visualize geographical entities both spatially and temporally (Karimipour et al., 2005). Sharma et al. (2010) has applied spatial data mining techniques for drought monitoring. Liu (2009) has described a GIS-based tool for modeling large-scale crop water relations with high spatial resolution, which can be used to support decision making in water management and crop production.

Armstrong et al. (2012) have used geospatial analysis in climate change studies in India and Australia. Kadiyala et al. (2015) have developed a similar model to assess the effect of climate change on groundnut cultivation, as have Mirschel et al. (2016). Memon et al. (2011) have reviewed the use of GPS and GIS technology in mapping wheat and cotton weeds. Dengiz (2013) has developed a spatial model of land suitability for rice cultivation using GIS techniques. Kihoro et al. (2013) have developed a map of areas suitable for rice cropping based on physical and climatic factors using a multi-criteria evaluation and GIS approach. Other studies have also reported using GIS in agriculture in areas such as sustainable bioenergy planning (Hiloidhari et al., 2017), assessment of heavy metal soil contamination (Hou et al., 2017), modeling soil carbon (Kaczynski et al., 2017), natural resource management under changing climate conditions (Kingra et al., 2016), peanut crop production (El-Sharkawy et al., 2016), land use classification, rice-based irrigated agriculture (Subramani and Raghu Prakash, 2016), assessing land suitability for agriculture (Shadeed, 2017), optimizing soil fertility management (Dembele et al., 2016), land use planning (Masoudi et al., 2017), sustainability of wheat production (Houshyar et al., 2018), site-specific fertilization requirements (Vekić et al., 2017) and evaluating ground water for irrigating crops (Kausar et al., 2016).

5 Case study 1: AgData DSS tool for western Australian broad acre cropping

There have been a number of initiatives in Australia for developing DSS-integrated AI techniques. AgData is a web-based decision support tool that has been developed by researchers in western Australia. The tool provides real-time access to seasonal weather data to predict crop yields for different varieties in various districts in western Australia. It is a cloud-based tool that leverages machine learning technologies to drive better crop decisions. AgData uses a context-aware interface with an AI prediction tool based on neural networks to predict crop yields, quality and cost-benefit scenarios. The tool integrates an AI prediction module using ANNs and data mining techniques. The web interface and various screenshots from the AgData DSS are shown in Figs. 4–6.

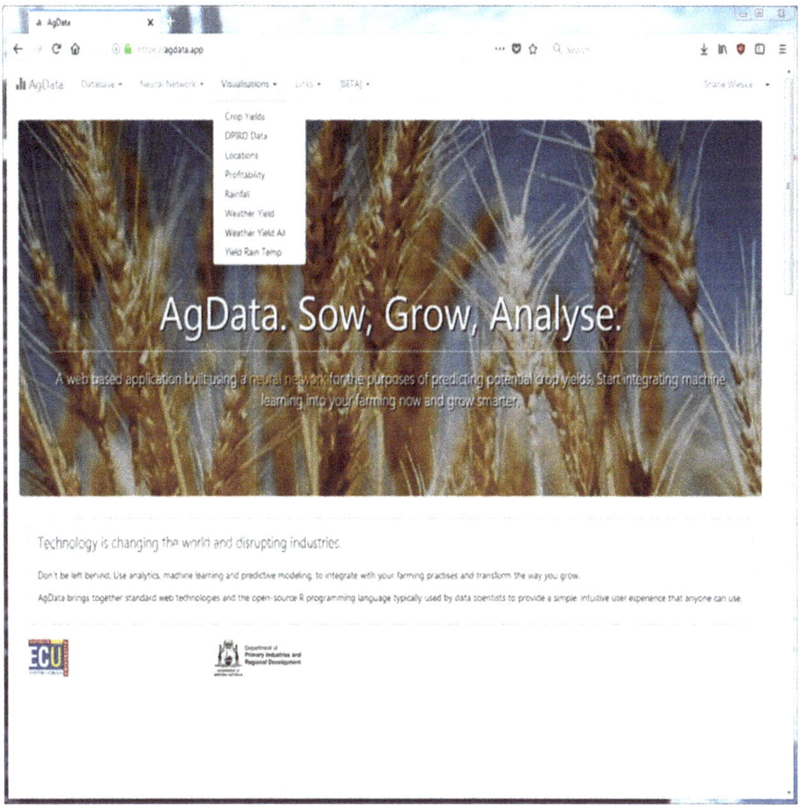

Figure 4 Web interface of AgData DSS.

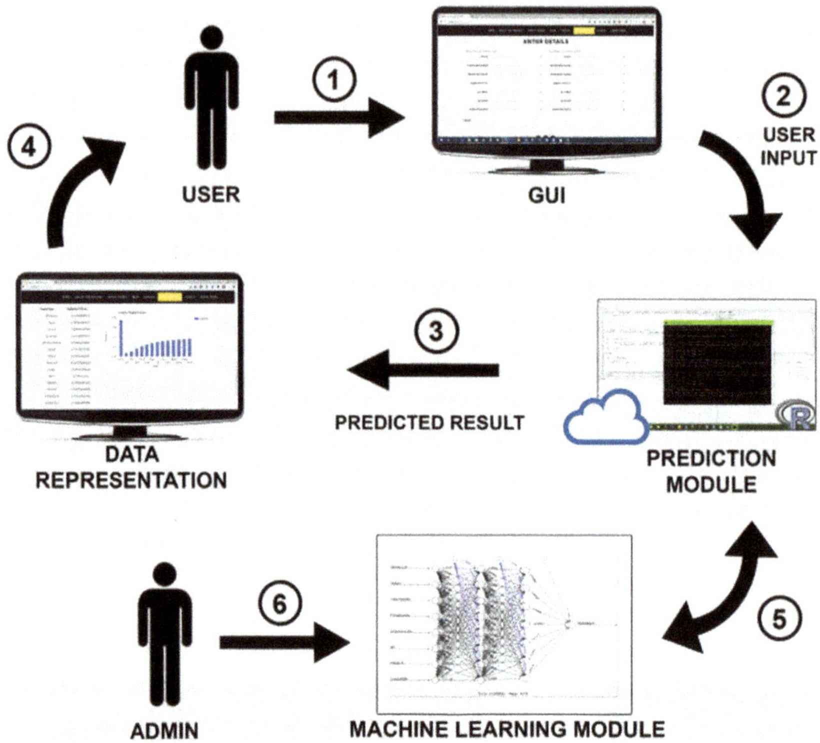

Figure 5 System design for AgData DSS.

6 Case study 2: GeoSense

GeoSense is a decision support tool developed by CSRE at IIT Bombay in collaboration with Japanese researchers. Geographical information and communication technologies (Geo-ICT), a WSN and cloud-computing technologies were used to develop a real-time DSS for crop water and nitrogen management, together with crop yield modeling/forecasting for precision farming. Further details of this DSS can be found in Adinaryana et al. (2015) and Sudharsan et al. (2012).

This system uses low-cost SN technology to obtain data on micro-climatic parameters in the field in real time. The design of the system consists of two types of wireless distributed sensing systems: FieldServer and Agrisens (from SPANN Lab) (Fig. 7). The distinctiveness of the Geosense system is its use of an integrated model with Geo-ICT and WSN for location-based information, combined with modeling and web-based GIS and short- and long-range communication/network systems. The system uses open-source tools to obtain and analyze dynamic real-time crop, weather and environmental data. The

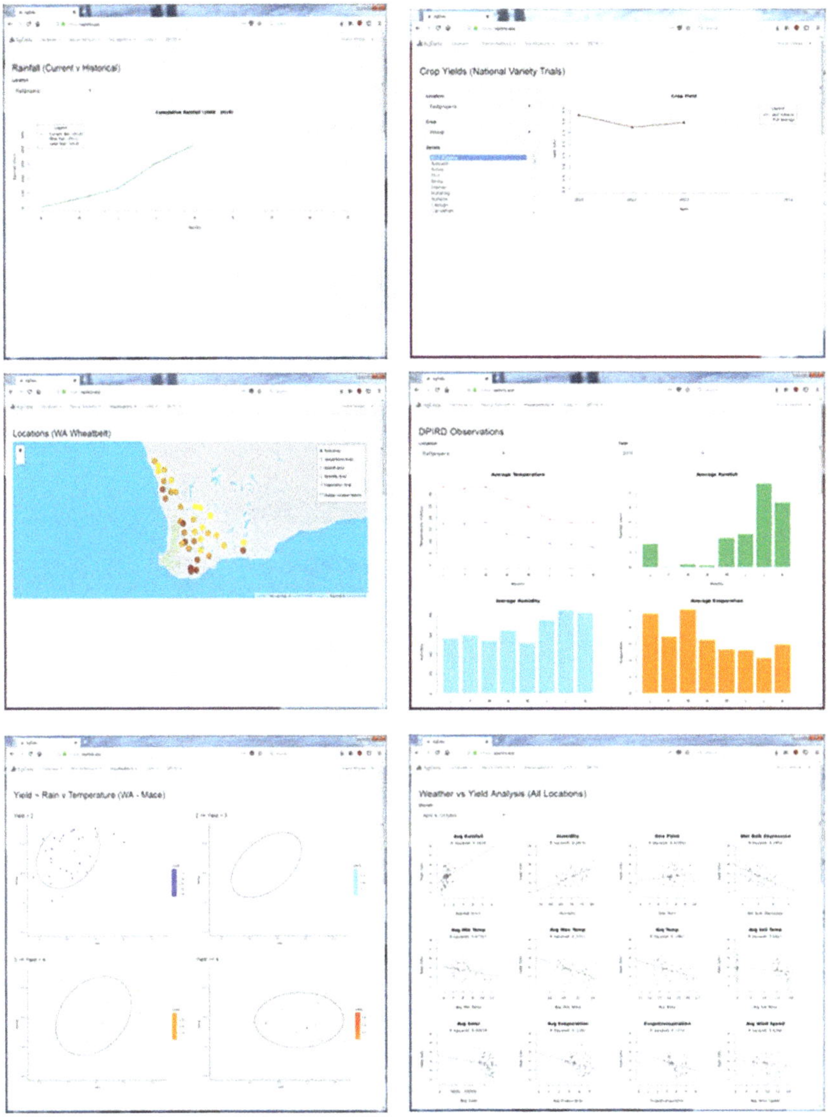

Figure 6 Screenshots from AgData DSS showing various rainfall, yield vs temperature predictions for different district and crop varieties.

web-based DSS contains unique features that are designed for novice users, running on http://geosense.dyndns-free.com:8091/server/with multi-mode communication and cloud-computing facilities (Fig. 8). The system is able to provide real-time, location-specific information on the basis of geospatial location/positioning technology (Figs. 9–11).

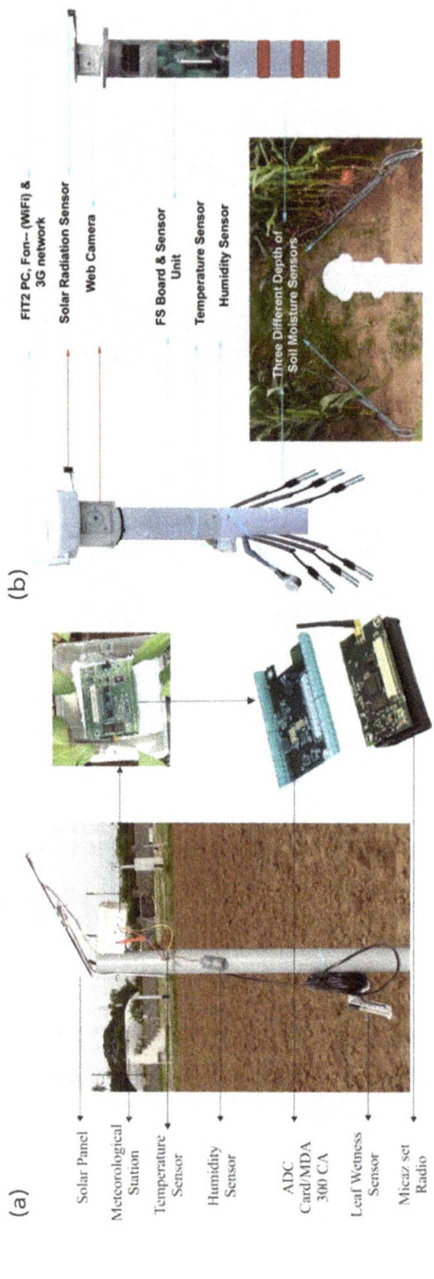

Figure 7 Sensing systems: (a) AgriSense and (b) FieldServer. *Source*: From Adinaryana et al. (2015).

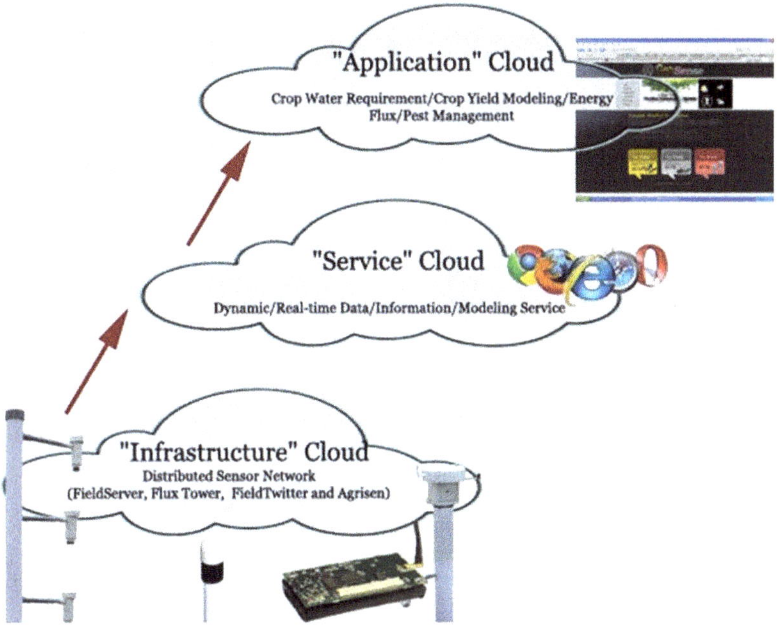

Figure 8 GeoSense cloud services. *Source*: Sudharshan et al. (2012).

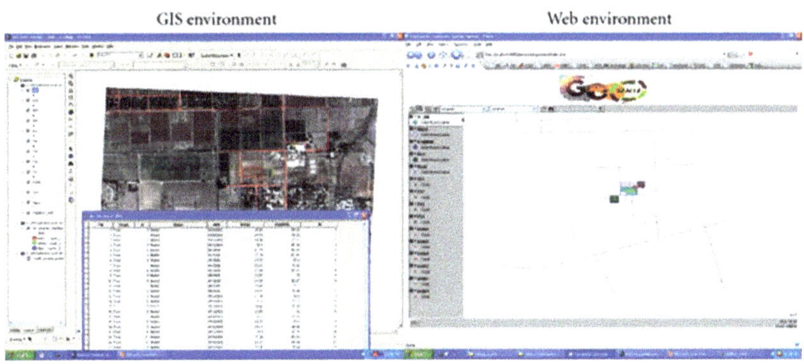

Figure 9 Location-based sensory information through open-source GIS. *Source*: Adinaryana et al. (2015) and Sudharsan et al. (2012).

7 Case study 3: Rice-based DSS

This DSS provides a decision support tool for rice production in India. It is summarized in Gandhi and Armstrong (2016a–d) and Gandhi et al. (2016a,b, 2017). The four main components of the DSS architecture are:

1 interface layer
2 input layer

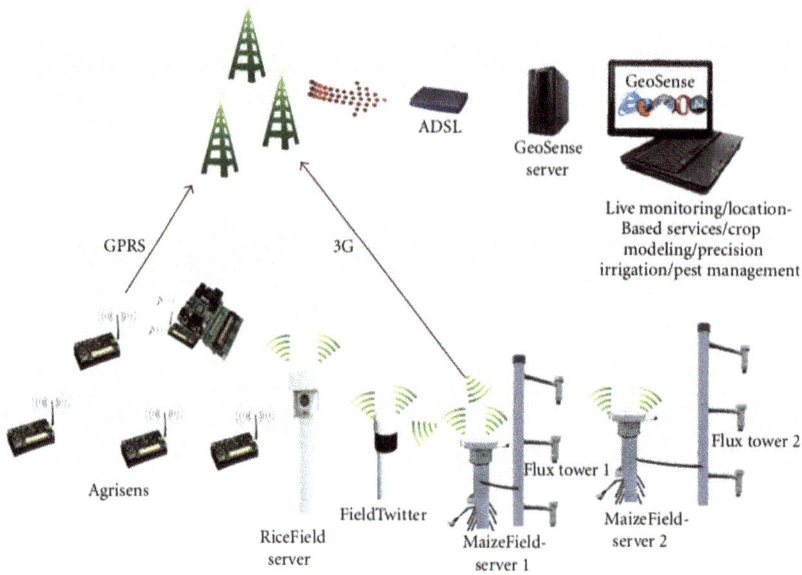

Figure 10 Integrated GeoSense architecture. *Source*: Adinaryana et al. (2015) and Sudharsan et al. (2012).

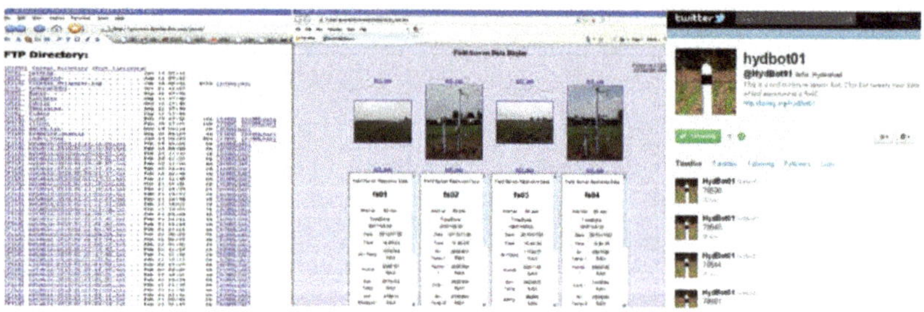

Figure 11 GeoSense near/real-time sensory data. *Source*: Adinaryana et al. (2015) and Sudharsan et al. (2012).

3 processing layer/model
4 database layer

The interface layer is the space where interactions between users and the system occur. The goal of this layer is to allow effective communication between the two. This layer offers five options for users to select from. The options are to use the DSS tool for Maharashtra state, data visualization for Maharashtra state, GIS visualization for Maharashtra state, a DSS tool for selected climatic zones and data visualization for selected climatic zones. Depending on the option selected in the interface layer, the input layer offers further options for

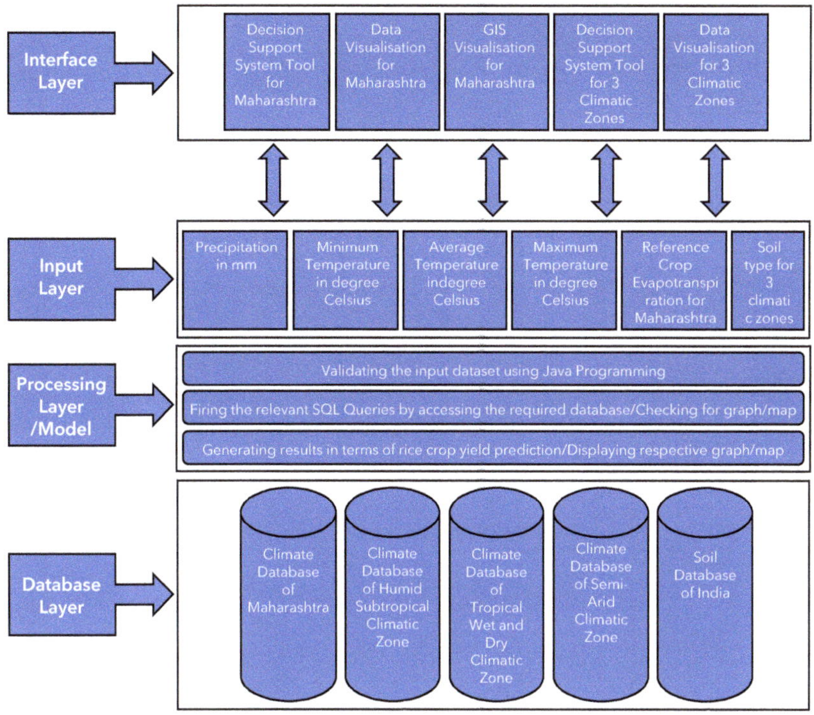

Figure 12 Architecture of the DSS.

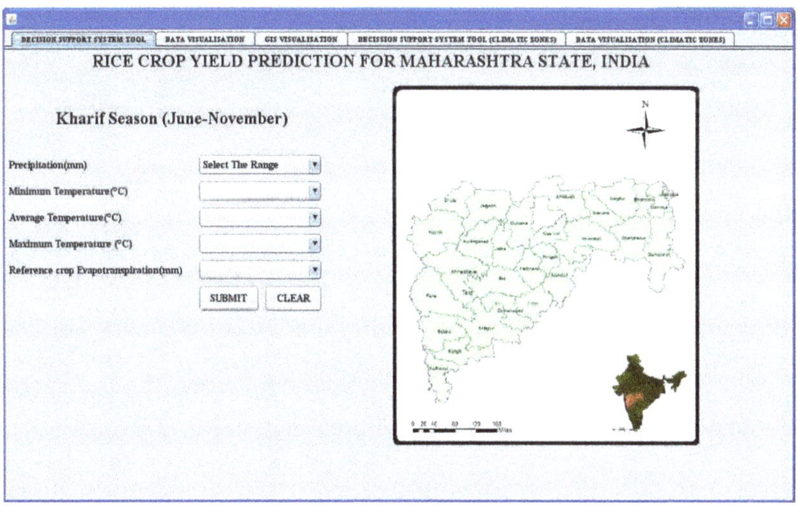

Figure 13 Graphic user interface of DSS prototype tool for the study area.

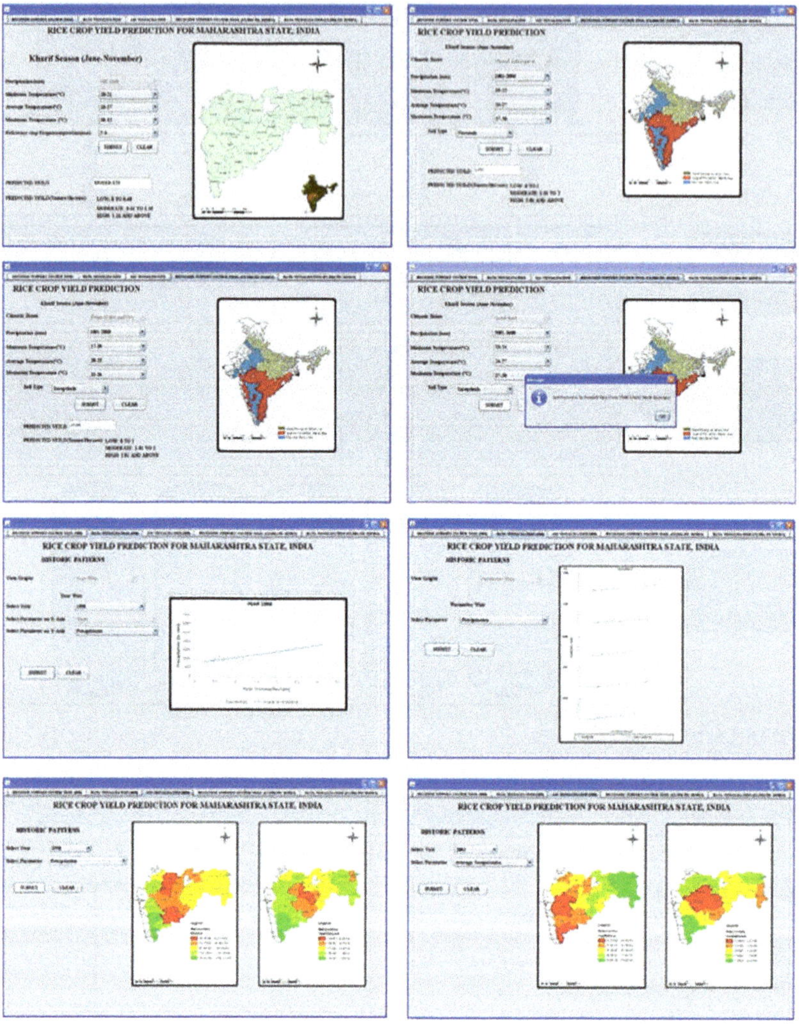

Figure 14 GIS visualizations and climate scenarios for various districts of Maharashtra.

processing or for visualization (see Figs. 12 and13). Based on the user input, the system formulates potential climate scenarios (based, in part, on historical data), which can be used to predict rice fields. The results generated from running these scenarios with the DSS tool are shown in Figs. 14–16.

8 Summary and future trends

The future of agricultural DSS is assured due to the support for these initiatives by the government, university and private enterprises. The adoption

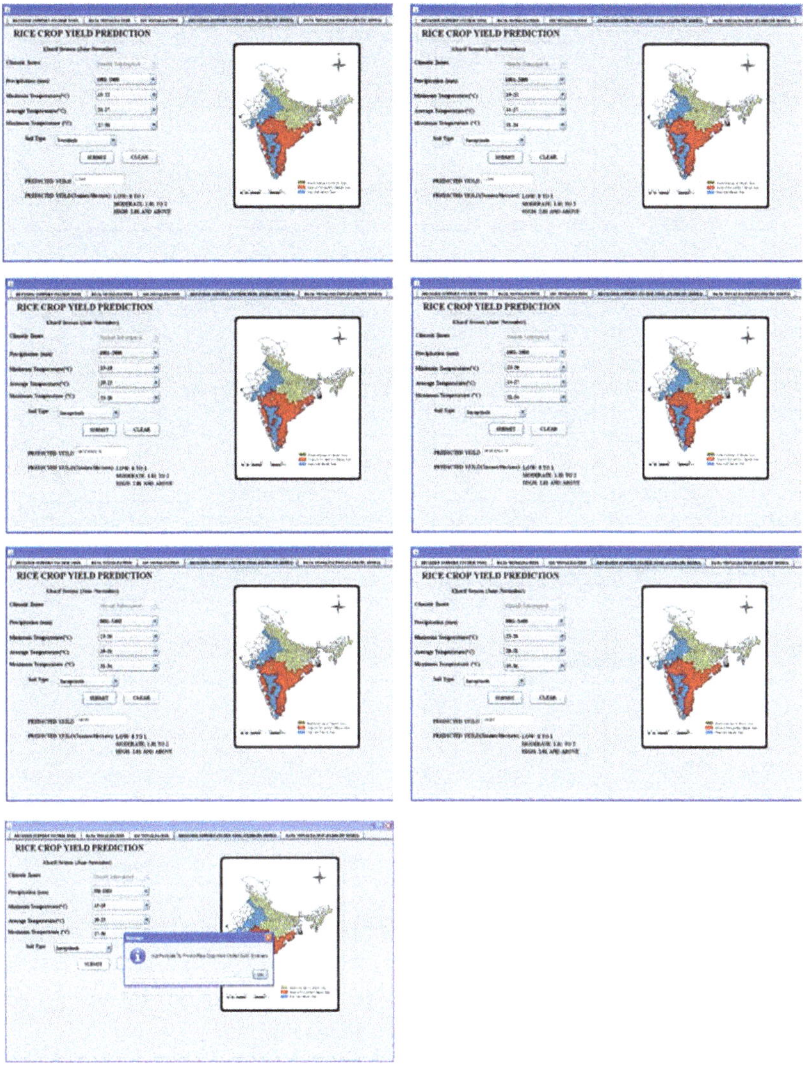

Figure 15 Climatic scenario for different agroclimatic zones.

of technologies such as AI, robotics and cloud computing has shown the possibilities for further improvements in the application of DSS in agriculture. The main drivers will be access to datasets that can be customized and better tools for farmer decision making.

9 Where to look for further information

Organizations focusing the use of ICT in agriculture and farmer decision support tools

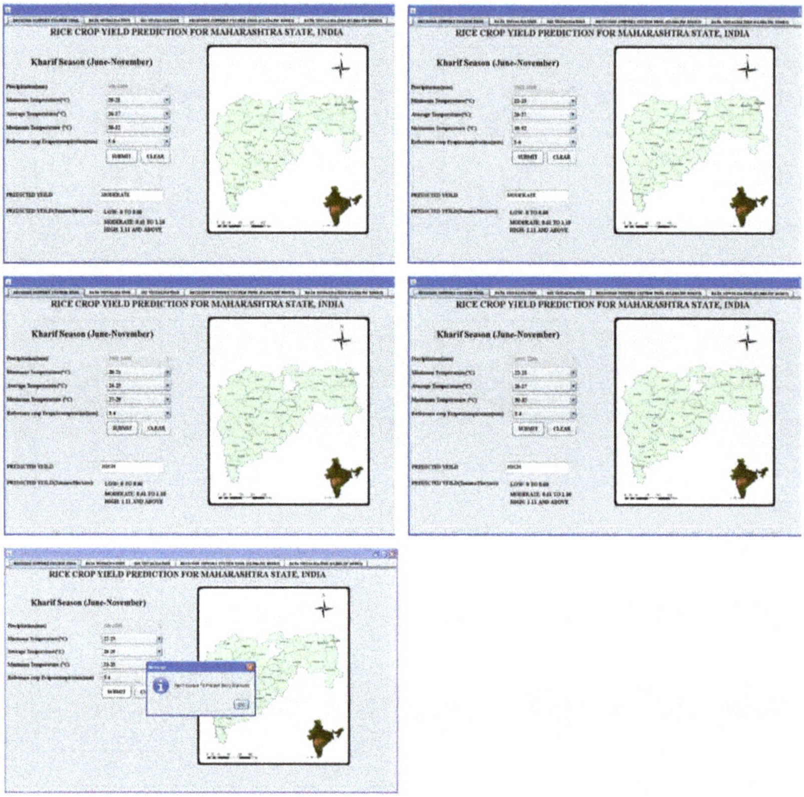

Figure 16 Climate scenarios for Maharashtra.

- eAgriculture.org
- CIGR CIGR International Commission of Agricultural and Biosystems Engineering http://www.cigr.org/
- ICRISAT https://www.icrisat.org/data-driven-solutions-to-support-smallholder-farmers-make-climate-smart-decisions/
- EFITA International Directory of Agriculture, Food and Environment https://www.efita.org/
- APFITA Asia-Pacific Federation for Information Technology in Agriculture http://www.apfita.org/
- ASICTA Australian Society of Information and Communication Technologies in Agriculture http://www.asicta.org/
- SPAA Society of the Precision Agriculture Australia https://spaa.com.au/

Agriculture decision tools

The tools are shown in Table 3.

Table 3 Agricultural decision support tools

	Description	Links
Government/Research		
CSIRO Data 61 initiative	Digital agriculture focuses on various areas including phenomics and sensing, software engineering, data analytics, precision agriculture, modeling, food innovation, farm systems management	https://www.csiro.au/en/Research/AF/Areas/Digital-agriculture
Iekbase service	Data-driven services using remote sensing and precision agriculture and machine learning	Iekbase.com
DPIRD eConnected Grainbelt	Provides a dashboard enables farmers, consultants and other users to capture and compare a range of weather data all on one page (or via APIs). This new feature has the ability to select a range of dates to graph, compare and analyze data	https://www.agric.wa.gov.au/r4r/econnected-grainbelt-project
eAgriculture RG ECU	Focuses on use of ICT to improve agricultural production using machine learning	http://www.ecu.edu.au/schools/computer-and-security-science/research-activity/eagriculture-research-group
UNE Smart Farms Initiative	Using precision agriculture and smart technologies to improve farm productivity	http://www.une.edu.au/research/research-centres-institutes/smart-farm
Australian Farm Institute P2D Project	Accelerating Precision Agriculture to Decision Agriculture (P2D) research program	http://farminstitute.org.au/p2dproject
eSagu	Agricultural expert advice is provided under eSagu framework for agriculture field crops (duration of the crop is at most one year) such as rice, cotton, maize, wheat, sugarcane, chillies	http://agriculture.iiit.ac.in/esagu2016/esagu.php
GeoSense	*GeoSense* – a real to near-real time DSS with Geo-ICT, wireless sensing network, data mining and cloud-computing systems for precision irrigation/protection/crop yield modeling	http://www.csre.iitb.ac.in/adi/agrolab.htm
Private Industry		
AgWorld	The cloud-based platform enables growers to share geosensed farm data.	https://agworld.com.au/
AgriWebb	Farm management software	https://www.agriwebb.com
CSBP NuLogic soil and plant analysis service	Customized fertilizer management based on soil and plant testing	https://csbp-fertilisers.com.au/agronomy/nulogic
Birchip cropping group	Agriculture decision support software platform	https://www.yieldprophet.com.au

10 References

Aalders, I. 2008. Modeling land-use decision behavior with Bayesian belief networks. *Ecology and Society* 13(1), 16. doi:10.5751/ES-02362-130116.

Abdullah, A., Brobst, S., Umer, M. and Khan, M. 2004. The case for an agri data warehouse: enabling analytical exploration of integrated agricultural data. In: *International Conference on Databases and Applications*. IASTED/ACTA Press, pp. 139-44.

Abdullah, A., Hussain, A. and Barnawi, A. 2012. Analysis of pesticide application practices using an intelligent agriculture decision support system (ADSS). In: Zhang, H., Hussain, A., Liu, D. and Wang, Z. (Eds), *Advances in Brain Inspired Cognitive Systems* (vol. 7366). Springer, Berlin, Heidelberg, pp. 382-91.

Adeniyi, P. 1993. Integration of remote sensing and GIS for agricultural resource management in Nigeria. *Advances in Remote Sensing* 2(3), 6-21.

Adinarayana, J., Sudharsan, D., Tripathy, A., Sawant, S., Merchant, S., Desai, U. and Kiura, T. 2012. GEOSENSE: an information, communication and dissemination system for decision support in precision farming. *Third National Conference on Agro-Informatics and Precision Agriculture*, India.

Adinarayana, Sudharsan, D., Ninomiya, S., Hirafuji, M. and Kiura, T. 2015. GeoSense: decision support system for precision faming in India. *Wulfania Journal* 2(2), 22-47.

Ahamed, A., et al. 2015. Applying data mining techniques to predict annual yield of major crops in different districts of Bangladesh. *16th IEEE/ACIS International Conference on Software Engineering, Artificial Intelligence, Networking and Parallel/Distributed Computing (SNPD)*, 1-3 June 2015, Takamatso, Japan.

Aigner, W., Miksch, S., Muller, W., Schumann, H. and Tominski, C. 2008. Visual methods for analyzing time-oriented data. *IEEE Transactions on Visualization and Computer Graphics* 14(1), 47-60. doi:10.1109/TVCG.2007.70415.

Ali, T. A. A., et al. 2018. Precision agriculture monitoring system using green internet of things (G-IoT). *2018 2nd International Conference on Trends in Electronics and Informatics (ICOEI)*. doi:10.1109/ICOEI.2018.8553866.

Amato, F., López, A., Peña-Méndez, E. M., Vaňhara, P., Hampl, A. and Havel, J. 2013. Artificial neural networks in medical diagnosis. *Journal of Applied Biomedicine* 11(2), 47-58. doi:10.2478/v10136-012-0031-x.

Anandhi, V. and Chezian, R. 2013. Support vector regression in forecasting. *International Journal of Advanced Research in Computer and Communication Engineering* 2(10), 4148-51.

Andrew, M., Grundy, M. and Harris, C. 2013. Decision support tools for agriculture. CSIRO, Australia.

Armstrong, L. J. and Nallan, S. A. 2016. Agricultural decision support framework for visualisation and prediction of Western Australian crop production. *2016 3rd International Conference on Computing for Sustainable Global Development (INDIACom)*, pp. 1907-12.

Armstrong, L. J., Diepeveen, D. and Maddern, R. 2007. The application of data mining techniques to characterize agricultural soil profiles. *Proceedings of the Sixth Australasian Conference on Data Mining and Analytics* 70, 85-100.

Armstrong, L. J., Diepeveen, D. A. and Tantisantisom, K. 2010. An eAgriculture-based decision support framework for information dissemination. *International Journal of*

Human Capital and Information Technology Professionals 1(4), 1–13. doi:10.4018/jhcitp.2010100101.

Armstrong, L., Adinarayana, J. and Dunstan, N. 2012. The use of ICT and geospatial techniques to improve Australian and Indian ground water and land resource management under climate change scenarios: a position paper. *Proceedings of the Geomatrix 12*, IIT Bombay, 26–29 February, Mumbai India.

Arshad, A. M., Armanto, M. E. and Zain, A. M. 2012. Expert system land evaluation for oil palm cultivation (ESLEOP). *Journal of Environmental Science and Engineering* 1(2B), 216–27.

Ascough, J. C., Maier, H. R., Ravalico, J. K. and Strudley, M. W. 2008. Future research challenges for incorporation of uncertainty in environmental and ecological decision-making. *Ecological Modelling* 219(3–4), 383–99. doi:10.1016/j.ecolmodel.2008.07.015.

Ashri, J. and Kaur, P. 2014. Performance evaluation of association rule mining on partitioned data. *International Journal of Enterprise Computing and Business Systems* 4(2), 1–7.

Batchelor, W., Yang, X. and Tshanz, A. 1997. Development of a neural network for soybean rust epidemics. *Transactions of the ASAE* 40, 247–52.

Bayu, S. and Aris, S. 2011. Application expert system of forward chaining and the rule based reasoning for simulation diagnose pest and disease red onion and chili plant. *Proceedings of the 1st International Conference on Information Systems for Business Competitiveness (ICISBC).*

Bejo, S., Mustaffha, S. and Ismail, W. 2014. Application of artificial neural network in predicting crop yield: a review. *Journal of Food Science and Engineering* 4, 1–9.

Bergsma, K., Sargent, S., Brecht, J. and Peart, R. 2013. An expert system for diagnosing chilling injury of vegetables. *HortScience* 26(6), 697C–697. doi:10.21273/HORTSCI.26.6.697C.

Bharadwaj, A., Dahiya, S. and Jain, R. 2012. Discretization based Support Vector Machine (D-SVM) for classification of agricultural datasets. *International Journal of Computer Applications* 40(1), 8–12. doi:10.5120/4918-7139.

Bhatia, J. and Gupta, A. 2014. Mining of quantitative association rules in agricultural data warehouse: a road map. *International Journal of Information Science and Intelligent System* 3(1), 187–98.

Bi, C. and Chen, G. 2010. Bayesian networks modeling for crop diseases. In: *Computer and Computing Technologies in Agriculture IV*. Springer, Berlin, Heidelberg, pp. 312–20.

Bisht, S. and Samantaray, S. D. 2015. Extracting spatial association rules in remotely sensed data of yellow rust disease in wheat crop at Udham Singh Nagar. *1st International Conference on Next Generation Computing Technologies*, Dehradun, India, pp. 847–51.

Brdar, S., Culibrk, D., Marinkovic, B., Crnobarac, J. and Crnojevic, V. 2011. Support vector machines with features contribution analysis for agricultural yield prediction. *Second International Workshop on Sensing Technologies in Agriculture, Forestry and Environment (EcoSense 2011)*, Belgrade, Serbia.

Bressan, G. M., Oliveira, V. A., Hruschka, E. R. and Nicoletti, M. C. 2009. Using Bayesian networks with rule extraction to infer the risk of weed infestation in a corn-crop. *Engineering Applications of Artificial Intelligence* 22(4–5), 579–92. doi:10.1016/j.engappai.2009.03.006.

Brudzewski, K., Osowski, S. and Markiewicz, T. 2004. Classification of milk by means of an electronic nose and SVM neural network. *Sensors and Actuators B* 98(2–3), 291–8. doi:10.1016/j.snb.2003.10.028.

Camps-Valls, G., Gomez-Chova, L., Calpe-Maravilla, J., Soria-Olivas, E., Martin-Guerrero, J. D. and Moreno, J. 2003. Support vector machines for crop classification using hyperspectral data. *Lecture Notes in Computer Science* 2652, 134-41. doi:10.1007/978-3-540-44871-6_16.

Castelletti, A. and Soncini-Sessa, R. 2007. Bayesian networks and participatory modelling in water resource management. *Environmental Modelling and Software* 22(8), 1075-88. doi:10.1016/j.envsoft.2006.06.003.

Chen, K., O'Leary, R. A. and Evans, F. H. 2019. A simple and parsimonious generalised additive model for predicting wheat yield in a decision support tool. *Agricultural Systems* 173(C), 140–50. doi:10.1016/j.agsy.2019.02.009.

Chlingaryan, A., Sukkarieh, S. and Whelan, B. 2018. Machine learning approaches for crop yield prediction and nitrogen status estimation in precision agriculture: a review. *Computers and Electronics in Agriculture* 151, 61–9. doi:10.1016/j.compag.2018.05.012.

Czimber, K. and Gálos, B. 2016. A new decision support system to analyse the impacts of climate change on the Hungarian forestry and agricultural sectors. *Scandinavian Journal of Forest Research* 31(7), 664–73. doi:10.1080/02827581.2016.1212088.

Dahikar, S. and Rode, S. 2014. Agricultural crop yield prediction using artificial neural network approach. *International Journal of Innovative Research in Electrical, Electronics, Instrumentation and Control Engineering* 2(1), 683–6.

Da Paz, R., Sehovic, A., Cook, D. M. and Armstrong, L. 2019. A novel approach to resource starvation attacks on message queuing telemetry transport brokers. *4th 2019 International Conference on Information Technology, Information Systems and Electrical Engineering (ICITISEE)*, JogJakarta, Indonesia.

Das, K. and Evans, M. D. 1992. Detecting fertility of hatching eggs using machine vision II: neural network classifiers. *Transactions of the ASAE* 35(6), 2035–41. doi:10.13031/2013.28832.

Das, B., Nair, B., Reddy, V. K. and Venkatesh, P. 2018. Evaluation of multiple linear, neural network and penalised regression models for prediction of rice yield based on weather parameters for west coast of India. *International Journal of Biometeorology* 62(10), 1809–22. doi:10.1007/s00484-018-1583-6.

Dath, A. and Balakrishnan, M. 2013. Development of an expert system for agricultural commodities. *The International Journal of Computer Science and Applications (TIJCSA)* 2(7), 74–90.

Dembele, D., Traore, K., Quansh, C., Osei, E. M., Bocar Dit Sire, B. and Ballo, M. 2016. Optimizing soil fertility management decision in Mali by remote sensing and GIS donnish. *Journal of Agricultural Research* 3(4), 22–34.

Dengiz, O. 2013. Land suitability assessment for rice cultivation based on GIS modeling. *Turkish Journal of Agriculture and Forestry* 37(3), 326–34. doi:10.3906/tar-1206-51.

Department of Primary Industries, Victoria, Australia. 2012. The decimal growth scale of cereals. Retrieved November 29, 2012, from http://www.dpi.vic.gov.au/agriculture/grain-crops/crop-production/decimal-growth-scale-cereals.

Derwin Suhartono, D., Aditya, W., Lestari, M. and Yasin, M. 2013. Expert system in detecting coffee plant diseases. *International Journal of Electrical Energy*, 156–62. doi:10.12720/ijoee.1.3.156-162.

Dhanya, C. T. and Nagesh Kumar, D. 2009. Data mining for evolution of association rules for droughts and floods in India using climate inputs. *Journal of Geophysical Research* 114(D2). doi:10.1029/2008JD010485.

Ding, Q., Ding, Q. and Perrizo, W. 2008. PARM–an efficient algorithm to mine association rules from spatial data. *IEEE Transactions on Systems, Man, and Cybernetics. Part B, Cybernetics* 38(6), 1513–24. doi:10.1109/TSMCB.2008.927730.

Divya, M., Manjunath, T. N. and Hegadi, R. S. 2014. A study on developing analytical model for groundnut pest management using data mining techniques. *2014 International Conference on Computational Intelligence and Communication Networks (CICN)*. IEEE, pp. 691–6.

Dong, J., Perrizo, W., Ding, Q. and Zhou, J. 2000. The application of association rule mining to remotely sensed data. *Proceedings of the 2000 ACM Symposium on Applied Computing* 1), 340–5.

Du, C.-J. and Sun, D.-W. 2005. Pizza sauce spread classification using colour vision and support vector machines. *Journal of Food Engineering* 66(2), 137–45. doi:10.1016/j.jfoodeng.2004.03.011.

Edelson, D. C. and Gordin, D. 1998. Visualization for learners: a framework for adapting scientists' tools. *Computers and Geosciences* 24(7), 607–16. doi:10.1016/S0098-3004(98)00363-X.

Elgader, H. A. A. and Hamza, M. H. 2011. Breast cancer diagnosis using artificial intelligence neural networks, Sudan University of Science and Technology. *Journal of Science and Technology* 12(1), 159–71.

Elizondo, D., McClendon, R. and Hoogenboom, G. 1994. Neural network models for predicting flowering and physiological maturity of soybean. *Transactions of the ASAE* 37(3), 981–8.

El-Sharkawy, M. M., Sheta, A. S., Abd El-Wahed, M. S., Arafat, S. M. and El Behiery, O. M. 2016. Precision agriculture using remote sensing and GIS for peanut crop production in arid land. *International Journal of Plant and Soil Science* 10, 1–9. doi:10.9734/IJPSS/2016/20539.

Fagerlund, S. 2007. Bird species recognition using support vector machines. *EURASIP Journal on Advances in Signal Processing* 2007(1), 8, Article ID 38637. doi:10.1155/2007/38637.

Fathima, G. and Geetha, R. 2014. Agriculture crop pattern using data mining techniques. *International Journal of Advanced Research in Computer Science and Software Engineering* 4(5), 781–6.

Feelders, A., Daniels, H. and Holseimer, M. 2000. Methodological and practical aspects of data mining. *Information and Management* 37(5), 271–81. doi:10.1016/S0378-7206(99)00051-8.

Folberth, C., Baklanov, A., Balkovic, J., Skalsky, R., Khabarov, N. and Obersteiner, M. 2019. Spatio-temporal downscaling of gridded crop model yield estimates based on machine learning. *Agricultural and Forest Meteorology* 264, 1–15. doi:10.1016/j.agrformet.2018.09.021.

Fuchs, R. and Hauser, H. 2009. Visualization of multi-variate scientific data. *Computer Graphics Forum* 28(6), 1670–90. doi:10.1111/j.1467-8659.2009.01429.x.

Gandhi, N. and Armstrong, L. 2016a. Applying data mining techniques to predict yield of rice in Humid Subtropical Climatic Zone of India. *Proceedings of the 10th INDIACom-2016, 3rd 2016 IEEE International Conference on Computing for Sustainable Global Development* (vol. 1), 16–18 March 2016. Institute of Electrical and Electronics Engineers, Inc, New Delhi, India, pp. 1901–6.

Gandhi, N. and Armstrong, L. 2016b. Assessing impact of seasonal rainfall on rice crop yield of Rajasthan, India using Association Rule Mining. *Proceedings of the*

2016 International Conference on Advances in Computing, Communications and Informatics. Institute of Electrical and Electronics Engineers, Inc., Piscataway, NJ, pp. 1021-4. doi:10.1109/ICACCI.2016.7732178.

Gandhi, N. and Armstrong, L. 2016c. A review of the application of data mining techniques for decision making in agriculture. *Proceedings of IEEE 2nd International Conference on Contemporary Computing and Informatics* (vol. 1), 14-17 December 2016. Institute of Electrical and Electronics Engineers, Inc., Piscataway, NJ, pp. 1-6. doi:10.1109/IC3I.2016.7917925.

Gandhi, N. and Armstrong, L. 2016d. Rice crop yield forecasting of Tropical Wet and Dry climatic zone of India using data mining techniques. *2016 IEEE International Conference on Advances in Computer Applications*. Institute of Electrical and Electronics Engineers, Inc., Piscataway, NJ, pp. 357-63. doi:10.1109/ICACA.2016.7887981.

Gandhi, N., Armstrong, L. and Petkar, O. 2016a. Proposed Decision Support System (DSS) for Indian rice crop yield prediction. *Proceedings of the International Conference on 2016 IEEE Technological Innovations in ICT for Agriculture and Rural Development (TIAR)*. Institute of Electrical and Electronics Engineers, Inc., Piscataway, pp. 13-8. doi:10.1109/TIAR.2016.7801205.

Gandhi, N., Armstrong, L. and Petkar, O. 2016b. Predicting rice crop yield using bayesian networks. *Proceedings of the 2016 International Conference on Advances in Computing, Communications and Informatics* (vol. 1), 21-24 Sepember 2016. Institute of Electrical and Electronics Engineers, Inc., Piscataway, NJ, pp. 795-9, doi:10.1109/ICACCI.2016.7732143.

Gandhi, N., Armstrong, L. and Nandawadekar, M. 2017. Application of data mining techniques for predicting rice crop yield in semi-Arid Climate Zone of India. *Proceedings of 2017 IEEE Conference on Technological Innovations in ICT for Agriculture and Rural Development (TIAR 2017)*. IEEE, Piscataway, NJ, pp. 116-20.

Ganesan, D. N., Venkatesh, D. K., Rama, D. M. A. and Palani, A. M. 2010. Application of neural networks in diagnosing cancer disease using demographic data. *International Journal of Computer Applications* 1(26), 81-97. doi:10.5120/476-783.

Garg, K., Wani, S. and Patil, M. 2016. A simple and farmer-friendly decision support system for enhancing water use efficiency in agriculture: tool development, testing and validation. *Soil and Water Management* 110, 1716-29.

Geetha, M. 2015. A survey on data mining techniques in Agriculture. *International Journal of Innovative Research in Computer and Communication Engineering* 3(2), 887-92.

Giusti, E. and Marsili-Libelli, S. 2015. A Fuzzy Decision Support System for irrigation and water conservation in agriculture. *Environmental Modelling and Software* 63, 73-86. doi:10.1016/j.envsoft.2014.09.020.

Gonzalez, R. C., Woods, R. E. and Eddins, S. L. 2009. *Digital Image Processing Using MATLAB* (2nd edn.). Gatesmark Pub. S.I., Knoxville, TN.

Govaerts, B. and Verhulst, N. 2010. The normalize difference vegetation Index (NDVI) GreenSeeker™ handheld sensor: toward the integrated evaluation of crop management. Part A: concepts and case studies. Available at: http://www.plantstress.com/methods/Greenseeker.PDF.

Guerrero, J. M., Guijarro, M., Montalvo, M., Romeo, J., Emmi, L., Ribeiro, A. and Pajares, G. 2013. Automatic expert system based on images for accuracy crop row detection in maize fields. *Expert Systems with Applications* 40(2), 656-64. doi:10.1016/j.eswa.2012.07.073.

Guo, W. and Xue, H. 2012. An incorporative statistic and neural approach for crop yield modelling and forecasting. *Neural Commuting and Applications* 21, 109–17.

Guo, W. W. and Xue, H. 2014. Crop yield forecasting using artificial neural networks: a comparison between spatial and temporal models. *Mathematical Problems in Engineering* 2014, 1-7. doi:10.1155/2014/857865.

Halloran, M., McGinley, B., Conceicao, R. C., Morgan, F., Jones, E. and Glavin, M. 2011. Spiking neural networks for breast cancer classification in a dielectrically Heteroge-Neous breast. *Progress in Electromagnetics Research* 113, 413–28. doi:10.2528/PIER10122203.

Han, J., Pei, J. and Kambe, M. 2011. *Data Mining: Concepts and Techniques*. Elsevier.

Hardjomidjojo, H., Imantho, H. and Suyamto, D. 2014. Decision support system on spatial planning for food security policy analysis (case study: Jawa Barat Province). *Proceedings of the 9th Conference of the Asian Federation for Information Technology in Agriculture "ICT's for Future Economic and Sustainable Agricultural Systems"*, Perth, Australia, pp. 190–9.

Hassan, M., Abu Samah, B., Mohamed Shaffril, H. A. and D'Silva, J. L. 2011. Perceived usefulness of ICT usage among JKKK members in peninsular Malaysia. *Journal of Asian Social Science* 7(10), 255–66. doi:10.5539/ass.v7n10p255.

Haykin, S. 1999. *Neural Networks: a Comprehensive Foundation* (2nd edn.). Pearson Prentice Hall.

Heckerman, D. 1997. Bayesian networks for data mining. *Data Mining and Knowledge Discovery* 1(1), 79–119. doi:10.1023/A:1009730122752.

Hiloidhari, M., Baruah, D. C., Singh, A., Kataki, S., Medhi, K., Kumari, S., Ramachandra, T. V., Jenkins, B. M. and Thakur, I. S. 2017. Emerging role of geographical information systems (GIS) in sustainable bioenergy planning. *Bioresource Technology* 242, 218–26. doi:10.1016/j.biortech.2017.03.079.

Hitoshi, Y., Yuichi, Y. and Hokoku, C. 2012. Development of the diagnostic expert system for tea processing. *Tea Research Journal* 109, 37–55.

Hochman, Z. and Carberry, P. S. 2011. Emerging consensus on desirable characteristics of tools to support farmers' management of climate risk in Australia. *Agricultural Systems* 104(6), 441–50. doi:10.1016/j.agsy.2011.03.001.

Holmgren, P. and Thuresson, T. 1998. Satellite remote sensing for forestry planning: a review. *Scandinavian Journal of Forest Research* 13(1–4), 90–110. doi:10.1080/02827589809382966.

Hou, D., O'Connor, D., Nathanail, P., Tian, L. and Ma, Y. 2017. Integrated GIS and multivariate statistical analysis for regional scale assessment of heavy metal soil contamination: a critical review. *Environmental Pollution* 231(1), 1188–200. doi:10.1016/j.envpol.2017.07.021.

Houshyar, E., Wu, X. F. and Chen, G. Q. 2018. Sustainability of wheat and maize production in the warm climate of southwestern Iran: an emergy analysis. *Journal of Cleaner Production* 172, 2246–55. doi:10.1016/j.jclepro.2017.11.187.

Hu, Y., Wang, J. and Li, P. 2010. Expert system for greenhouse production management. In: Vizureanu, P. (Ed.), *Expert Systems*. InTechOpen. ISBN: 978-953-307-032-2.

Huang, B., Xie, C. and Tay, R. 2010. Support vector machines for urban growth modelling. *Geoinformatica* 14(1), 83–99. doi:10.1007/s10707-009-0077-4.

Human Computer Interaction Lab. 2013. Hierarchical clustering explorer. Retrieved 1 July 2013, from http://www.cs.umd.edu/hcil/multi-cluster/.

Jabjone, S. and Wannasang, S. 2014. Decision support system using artificial neural network to predict rice production in Phimai District, Thailand. *International Journal of Computer and Electrical Engineering* 6(2), 162–6. doi:10.7763/IJCEE.2014.V6.814.

Jadhav, S., Yelapure, S. and Babar, V. 2011. Rule based expert system in the use of inorganic fertilizers for sugarcane crop. *International Journal of Computer and Applications* 36(4).

Jain, R. and Arora, A. 2012. Aproach for mining multiple patterns from clusters. *Journal of the Indian Society of Agricultural Statistics* 67(1).

Jain, R., Sonajharia, M. and Ramasubramaniam, V. 2009. Machine learning for forewarning crop diseases. *Journal of Indian Society of Agricultural Statistics* 63(1), 11.

Jaiswal, R. K., Ghosh, N. C., Lohani, A. K. and Thomas, T. 2015. Fuzzy AHP-based multi-criteria support for watershed prioritization. *Water Resource Management* 29(12), 4205–27. doi:10.1007/s11269-015-1054-3.

Jeong, J. H., Resop, J. P., Mueller, N. D., Fleisher, D. H., Yun, K., Butler, E. E., Timlin, D. J., Shim, K. M., Gerber, J. S., Reddy, V. R. and Kim, S. H. 2016. Random forests for global and regional crop yield predictions. *PLoS ONE* 11(6), e0156571. doi:10.1371/journal.pone.0156571.

Ji, W. and Cui, J. 2011. Application of Geographical Information System (GIS) in agricultural land classification and grading. *2011 International Conference on Agricultural and Natural Resources and Engineering Advances in Biomedical Engineering* 3(5), pp. 201–5.

Ji, B., Sun, Y., Yang, S. and Wan, J. 2007. Artificial neural networks for rice yield prediction in mountainous regions. *Journal of Agricultural Science* 145(3), 249–61. doi:10.1017/S0021859606006691.

Jim, R. and Mary, M. 2003. On the construction and training of reformulated radial basis function neural networks. *IEEE Transactions on Neural Networks* 14.

Jorquera, H., Perez, R., Cipriano, A. and Acuna, G. 2001. Short term forecasting of air pollution episodes. In: Zannetti, P. (Ed.), *Environmental Modeling 4*. WIT Press, UK.

Kaczynski, R., Siebielec, G., Hanegraaf, M. C. and Korevaar, H. 2017. Modelling soil carbon trends for agriculture development scenarios at regional level. *Geoderma* 286(104), 104–15. doi:10.1016/j.geoderma.2016.10.026.

Kadiyala, M., Nedumaran, S., Singh, P., Chukka, S., Irshad, A. and Bantilan, S. 2015. An integrated crop model and GIS decision support system for assisting agronomic decision making under climate change. *Science of the Total Environment* 521–522, 123–34.

Kakran, A. and Mahajan, R. 2012. Monitoring growth of wheat crop using digital image processing. *International Journal of Computer Applications* 50(10), 18–22. doi:10.5120/7807-0940.

Kalpana, R., Shanthi, N. and Arumugam, S. 2014. A survey on data mining techniques in agriculture. *International Journal of Advances in Computer Science and Technology* 3(8), 426–31.

Karimi, Y., Prasher, S. O., Patel, R. M. and Kim, S. H. 2006. Application of support vector machine technology for Weed and nitrogen stress detection in corn. *Computers and Electronics in Agriculture* 51(1–2), 99–109. doi:10.1016/j.compag.2005.12.001.

Karimipour, F., Delavar, M. R. and Kinaie, M. 2005. Water quality management using GIS data mining. *Journal of Environmental Informatics* 5(2), 61–71. doi:10.3808/jei.200500047.

Kaul, M., Hill, R. L. and Walthall, C. 2005. Artificial neural network for corn and soybean prediction. *Agricultural Systems* 85(1), 1–18. doi:10.1016/j.agsy.2004.07.009.

Kaur, M., Gulati, H. and Kundra, H. 2014. Data mining in agriculture on crop price prediction: techniques and applications. *International Journal of Computer Applications* 99(12), 1–3. doi:10.5120/17422-8273.

Kaura, R., Dina, S. and Pannub, S. 2013. Expert system to detect and diagnose the leaf diseases of cereals. *International Journal of Current Engineering and Technology* 3(4), 1480–3.

Kausar, R., Hussain, A., Hussain, N., Ahmad, F., et al. 2016. Evaluating groundwater suitability for irrigating crops using GIS technique in tehsil Noorpur, district Khushab. *International Journal of Agriculture and Applied Sciences (Pakistan)* 5, 15–21.

Keim, D., Mansmann, F., Schneidewind, J., Ziegler, H. and Thomas, J. 2008. Visual analytics: scope and challenges. In: *Visual Data Mining: Theory, Techniques and Tools for Visual Analytics*. Lecture Notes in Computer Science, No. 4404. Springer, pp. 76–90.

Kent, S. 2010. Integrating geospatial analysis. *Business Intelligence Journal* 15(3), 329.

Khan, F. and Singh, D. 2014. Association rule mining in the field of agriculture: a survey. *International Journal of Scientific and Research Publications* 4(7).

Kharya, S. 2012. Using data mining techniques for diagnosis and prognosis of cancer disease. *International Journal of Computer Science, Engineering and Information Technology* 2(2), 55–66. doi:10.5121/ijcseit.2012.2206.

Kihoro, J., Bosco, N. J. and Murage, H. 2013. Suitability analysis for rice growing sites using a multicriteria evaluation and GIS approach in great Mwea region, Kenya. *SpringerPlus* 2(1), 265.

Kingra, P. K., Majumder, D. and Singh, S. P. 2016. Application of remote sensing and GIS in agriculture and natural resource management under changing climatic conditions. *Agricultural Research Journal* 53(3), 295–302. doi:10.5958/2395-146X.2016.00058.2.

Kristensen, K. and Rasmussen, I. A. 2002. The use of a Bayesian network in the design of a decision support system for growing malting barley without use of pesticides. *Computers and Electronics in Agriculture* 33(3), 197–217. doi:10.1016/S0168-1699(02)00007-8.

Ladstädter, F., Steiner, A. K., Lackner, B. C., Pirscher, B., Kirchengast, G., Kehrer, J., Hauser, H., Muigg, P. and Doleisch, H. 2010. Exploration of climate data using interactive visualization. *Journal of Atmospheric and Oceanic Technology* 27(4), 667–79. doi:10.1175/2009JTECHA1374.1.

Leemans, V. and Destain, M.-F. 2004. A real time grading method of apples based on features extracted from defects. *Journal of Food Engineering* 61(1), 83–9. doi:10.1016/S0260-8774(03)00189-4.

Liu, J. 2009. A GIS-based tool for modelling large-scale crop-water relations. *Environmental Modelling and Software* 24(3), 411–22 doi:10.1016/j.envsoft.2008.08.004.

Liu, J., Goering, C. and Tian, L. 2001. A neural network for setting target corn yields. *Transaction of the ASAE* 44(3), 705–13.

Long, B. and Parton, K. 2012. Decision Support Systems (DSS) – where success is failure of continued use. *Proceedings of the 16th Australian Agronomy Conference*, NSW, pp. 14–8.

Lopes, M. S. and Reynolds, M. P. 2012. Stay-green in spring wheat can be determined by spectral reflectance measurements (normalized difference vegetation index) independently from phenology. *Journal of Experimental Botany* 63(10), 3789–98. doi:10.1093/jxb/ers071.

Luo, M. F., Boettcher, P. J., Schaeffer, L. R. and Dekkers, J. C. M. 2002. Estimation of genetic parameters of calving ease in first and second parities of Canadian Holsteins using Bayesian methods. *Livestock Production Science* 74(2), 175–84. doi:10.1016/S0301-6226(01)00294-9.

Madadgar, S. and Moradkhani, H. 2014. Spatio-temporal drought forecasting within Bayesian networks. *Journal of Hydrology* 512, 134–46. doi:10.1016/j.jhydrol.2014.02.039.

Mahmoud, S. H., Mohammad, F. S. and Alazba, A. A. 2015. Delineation of potential sites for rainwater harvesting structures using a geographic information system-based decision support system. *Hydrology Research* 46(4), 591–606. doi:10.2166/nh.2014.054.

Maiti, S. and Tiwari, R. K. 2014. A comparative study of artificial neural networks, Bayesian neural networks and adaptive neuro-fuzzy inference system in groundwater level prediction. *Environmental Earth Sciences* 71(7), 3147–60. doi:10.1007/s12665-013-2702-7.

Manimaran, J. and Velmurugan, T. 2015. Analysing the quality of association rules by computing an interestingness measures. *Indian Journal of Science and Technology* 8(15). doi:10.17485/ijst/2015/v8i15/76693.

Mankar, A. and Burange, M. 2014. Data mining - an evolutionary view of Agriculture. *International Journal of Application or Innovation in Engineering and Management* 3(3), 102–5.

Masoudi, M., Jokar, P. and Sadeghi, M. 2017. Land use planning using a quantitative model and geographic information system (GIS) in Darab County, Iran. *Journal of Materials and Environmental Science* 8(8), 2975–85.

Medar, R. and Rajpurohit, V. 2014. A survey of data mining techniques for crop yield prediction. *International Journal of Advance Research in Computer Science and Management Studies* 2(9), 1–8.

Memon, R. A., et al. 2011. Use of GPS and GIS technology in surveying and mapping of wheat and cotton weeds in Khairpur district, Sindh, Pakistan. *Pakistan Journal of Botany* 43, 1873–8.

Meyer, G. E., Neto, J. C., Jones, D. D. and Hindman, T. W. 2004. Intensified fuzzy clusters for classifying plant, soil, and residue regions of interest from color images. *Computers and Electronics in Agriculture* 42(3), 161–80. doi:10.1016/j.compag.2003.08.002.

Mirschel, W., Wenkel, K.-O., Berg, M., Wieland, R., Nendel, C., Köstner, B., Topazh, A. G., Terleev, V. V. and Badenko, V. L. 2016. A spatial model-based decision support system for evaluating agricultural landscapes under the aspect of climate change. In: Mueller, L., Sheudshen, A. and Eulenstein, F. (Eds), *Novel Methods for Monitoring and Managing Land and Water Resources in Siberia*. Springer Water. Springer, Cham, Switzerland. doi:10.1007/978-3-319-24409-9_23.

Mojarad, S., Dlay, S., Woo, W. and Sherbet, G. 2011. Cross validation evaluation for breast cancer prediction using multilayer perceptron neural networks. *American Journal of Engineering and Applied Sciences* 4(4), 576–85. doi:10.3844/ajeassp.2011.576.585.

Montalvo, M., Guerrero, J. M., Romeo, J., Emmi, L., Guijarro, M. and Pajares, G. 2013. Automatic expert system for weeds/crops identification in images from maize fields. *Expert Systems with Applications* 40(1), 75–82. doi:10.1016/j.eswa.2012.07.034.

Mucherino, A., Papajorgji, P. and Pardalos, P. M. 2009. A survey of data mining techniques applied to agriculture. *Springer, Operational Research International Journal* 9(2), 121–40. doi:10.1007/s12351-009-0054-6.

Mullan, D. J. and Reynolds, M. P. 2010. Quantifying genetic effects of ground cover on soil water evaporation using digital imaging. *Functional Plant Biology* 37(8), 703–12. doi:10.1071/FP09277.

Naderloo, L., Alimardani, R., Omid, M., Sarmadian, F., Javadikia, P., Torabi, M. Y. and Alimardani, F. 2012. Application of ANFIS to predict crop yield based on different energy inputs. *Measurement* 45(6), 1406–13. doi:10.1016/j.measurement.2012.03.025.

Naidu, V., Ravisankar, H., Dhagat, S., Kamalvanshi, V. and Sharma, A. 2013. Expert system for identification of weed seedlings. *Indian Journal of Weed Science* 45(4), 278–81.

Nakornphanom, K., Lursinsap, C. and Rugchatjaroen, A. 2002. Fault immunization model for elliptic radial basis function neuron. *Proceedings of the 9th International Conference on Neural Information Processing*, pp.1027–31.

Narang, S., Verma, H. and Sachdev, U. 2012. A review of breast cancer detection using ART model of neural networks. *International Journal of Advanced Research in Computer Science and Software Engineering* 2(10), 311–8.

Nearing, G. S., Gupta, H. V. and Crow, W. T. 2013. Information loss in approximately Bayesian estimation techniques: a comparison of generative and discriminative approaches to estimating agricultural productivity. *Journal of Hydrology* 507, 163–73. doi:10.1016/j.jhydrol.2013.10.029.

Negid, N. 2014. Expert System for Wheat Yields Protection in Egypt (ESWYP). *International Journal of Innovative Technology and Exploring Engineering (IJITEE)* 3(11), 1–4.

Neuhaus, A. and Bowden, W. 2014. A unique interactive Decision Support System (NUlogic®) for advisors to guide sustainable and profitable fertiliser and soil amendment applications in Western Australia, part 1: past, present and future development of NUlogic® agronomy models. *Proceedings of the 9th Conference of the Asian Federation for Information Technology in Agriculture "ICT's for Future Economic and Sustainable Agricultural Systems"*, Perth, Australia, pp. 342–52.

Newlands, N. K. and Townley-Smith, L. 2010. Predicting energy crop yield using Bayesian networks. In: *Proceedings of the Fifth IASTED International Conference* (vol. 711), pp. 14–106.

Ni, G. 2015. Architecture research on citrus planting expert system. In: *2015 International Conference on Education Technology, Management and Humanities Science (ETMHS 2015).*

Nocke, T., Sterzel, T., Böttinger, M. and Wrobel, M. 2008. Visualization of climate and climate change data: an overview. In: *Digital Earth Summit on Geoinformatics*, pp. 226–32.

O'Neal, M. R., Engel, B. A., Ess, D. R. and Frankenberger, J. R. 2002. Neural network prediction of maize yield using alternative data coding algorithms. *Biosystems Engineering* 83(1), 31–45. doi:10.1006/bioe.2002.0098.

Oudemans, P. V., Pozdnyakova, L., Hughes, M. G. and Rahman, F. 2002. GIS and remote sensing for detecting yield loss in cranberry culture. *Journal of Nematology* 34(3), 207–12.

Pachepsky, Y. A., Timlin, D. and Várallyay, G. 1996. Artificial neural networks to estimate soil water retention from easily measurable data. *Soil Science Society of America Journal* 60(3), 727–33. doi:10.2136/sssaj1996.03615995006000030007x.

Parker, C. and Campion, S. 1997. Improving the uptake of decision support systems in agriculture. *First European Conference for Information Technology in Agriculture*, Copenhagen.

Pask, A., Pietragalla, J., Mullan, D. and Reynolds, M. P. (Eds.). 2012. *Physiological Breeding II: a Field Guide to Wheat Phenotyping*. The International Maize and Wheat Improvement Center, Mexico.

Patel, H. and Patel, D. 2014. A brief survey of data mining techniques applied to agricultural data. *International Journal of Computer Applications* 95(9), 6–8. doi:10.5120/16620-6472.

Perez-Ariza, C., Nicholson, A. and Flores, M. 2012. Prediction of coffee rust disease using Bayesian networks. *Proceedings of the Sixth European Workshop on Probabilistic Graphical Models*, pp. 259–66.

Prasad, G. and Babu, A. 2006. A study on various expert systems in agriculture. *Georgian Electronic Scientific Journal: Computer Science and Telecommunications* 4(11), 81–6.

Puteh, S., Rizon, M., Juhari, M., Nor Khairah, J., Siti Kamarudin, S., Aryati, B. and Nursalasawati, R. 2004. Backpropagation algorithm for rice yield prediction. *Proceedings of the 9th of the Ninth International Symposium on Artificial Life and Robotics*, Beppu, Japan, Oita, pp. 586–9.

Quinn, J. M., Monaghan, R. M., Bidwell, V. J. and Harris, S. R. 2013. A Bayesian Belief Network approach to evaluating complex effects of irrigation-driven agricultural intensification scenarios on future aquatic environmental and economic values in a New Zealand catchment. *Marine and Freshwater Research* 64(5), 460–74. doi:10.1071/MF12141.

Rajagopalan, B. and Lall, U. 1999. A k-nearest-neighbor simulator for daily precipitation and other weather variables. *Water Resources Research* 35(10), 3089–101. doi:10.1029/1999WR900028.

Ramesh, D. and Vardhan, B. 2013. Data mining techniques and applications to agricultural yield data. *International Journal of Advanced Research in Computer and Communication Engineering* 2(9), 3477–80.

Ramesh, D. and Vardhan, B. 2015. Analysis of crop yield prediction using data mining techniques. *International Journal of Research in Engineering and Technology* 4(1), 47–473.

Ramteke, M. A. and Dhande, S. S. 2013. Improving decision support systems with data mining, data warehousing, OLAP and OLTP technologies. *International Journal of Engineering and Computer Science* 2(4).

Ranjeet, T. and Armstrong, L. 2014. An artificial neural network for predicting crops yield in Nepal. *Proceedings of the 9th Conference of the Asian Federation for Information Technology in Agriculture 'ICT's for Future Economic and Sustainable Agricultural Systems'*, Perth, Australia, pp. 376–86.

Raorane, A. and Kulkarni, R. 2012. Data Mining: an effective tool for yield estimation in the agricultural sector. *International Journal of Emerging Trends and Technology in Computer Science* 1(2), 75–9.

Raorane, A. and Kulkarni, R. 2013. Review-role of data mining in agriculture. *International Journal of Computer Science and Information Technologies* 4(2), 270–2.

Raun, W. R., Solie, J. B., Johnson, G. V., Stone, M. L., Lukina, E. V., Thomason, W. E. and Schepers, J. S. 2001. In-season prediction of potential grain yield in winter wheat using canopy reflectance. *Agronomy Journal* 93(1), 131–8. doi:10.2134/agronj2001.931131x.

Ravisankar, H., Sivaraju, K., Krishnamurthy, V. and Raju, C. 2010. Expert system for identification and management of abiotic stresses in tobacco (*Nicotiana tabacum*). *Indian Journal of Agricultural Sciences* 80(2).

Reddy, P. and Ankaiah, R. 2005. A framework of information technology-based agriculture information dissemination system to improve crop productivity. *Current Science* 88(12), 1905–13.

Riul, A. Jr., de Sousa, H. C., Malmegrim, R. R., dos Santos Jr., D. S., Carvalho, A. C. P. L. F., Fonseca, F. J., Oliveira Jr., O. N.and Mattoso, L. H. C. 2004. Wine classification by taste sensors made from ultra-thin films and using neural networks. *Sensors and Actuators B* 98(1), 77–82. doi:10.1016/j.snb.2003.09.025.

Robertson, D. and Wang, Q. J. 2004. Bayesian networks for decision analyses-an application to irrigation system selection. *Australian Journal of Experimental Agriculture* 44(2), 145–50. doi:10.1071/EA02178.

Romani, L. A., de Avila, A. M. H., Zullo Jr., J., Chbeir, R., Traina Jr., C. and Traina, A. J. 2010. Clearminer: a new algorithm for mining association patterns on heterogeneous time series from climate data. In: *Proceedings of the 2010 ACM Symposium on Applied Computing*, pp. 900–5.

Romeo, J., Pajares, G., Montalvo, M., Guerrero, J. M., Guijarro, M. and De La Cruz, J. M. 2013. A new Expert System for greenness identification in agricultural images. *Expert Systems with Applications* 40(6), 2275–86. doi:10.1016/j.eswa.2012.10.033.

Roseline, P., Tauro, C. and Ganesan, N. 2012. Design and development of fuzzy expert system for integrated disease management in finger millets. *International Journal of Computer Applications* 56(1), 31–6. doi:10.5120/8857-2815.

Ruß, G. 2009. Data mining of agricultural yield data: a comparison of regression models. In: Perner, P. (Ed.), *Advances in Data Mining Applications and Theoretical Aspects* (vol. 5633). Springer-Verlag, Berlin Heidelberg, pp. 24–37.

Rupnik, R., Kukar, M. and Krisper, M. 2007. Integrating data mining and decision support through data mining based decision support system. *The Journal of Computer Information Systems* 47(3), 89.

Sanga, C., Sumari, N. and Tumbo, S. 2013. On the development of climate data visualization tool for interpretation of empirical results from climate model: does it add value to different stakeholders? *Proceedings of the 6th UbuntuNet Alliance Annual Conference*, pp. 223–35.

Santos, M. Y. and Amaral, L. A. 2005. Geo-spatial data mining in the analysis of a demographic database. *Soft Computing* 9(5), 374–84. doi:10.1007/s00500-004-0417-0.

Saxena, S. and Burse, K. 2012. A survey on neural network techniques for classification of breast cancer data. *International Journal of Engineering and Advanced Technology* 2(1), 234–7.

Schaap, M. G. and Bouten, W. 1996. Modeling water retention curves of sandy soils using neural networks. *Water Resources Research* 32(10), 3033–40. doi:10.1029/96WR02278.

Seflek, A. and Carman, K. 2010. A design of an expert system for selecting pumps used in agricultural irrigation. *Mathematical and Computational Applications* 15(1), 108–16. doi:10.3390/mca15010108.

Selvakumar, A., Arul, N. and Mohammed, G. 2011. An implementation of expert system in garlic using (ABC) Algorithm, Electronics Computer Technology (ICECT). *3rd International Conference* (vol. 1), pp. 45–8.

Shadeed, M. 2017. Developing a GIS-based agro-land suitability map for the Faria agricultural catchment, Palestine. *International Journal of Global Environmental Issues* 16(1-3), 190-204.

Shafinah, K., Sahari, N., Sulaiman, R., Soyapi, M., Yusoff, M. and Ikram, M. 2013. A framework of an expert system for crop pest and disease management. *Journal of Theoretical and Applied Information Technology* 58(1), 182-90.

Shahin, M. A., Tollner, E. W. and McClendon, R. W. 2001. Artificial intelligence classifiers for sorting apples based on watercore. *Journal of Agricultural Engineering Research* 79(3), 265-74. doi:10.1006/jaer.2001.0705.

Sharma, A. and Goyal, M. 2015. Bayesian network model for monthly rainfall forecast. *2015 IEEE International Conference on Research in Computational Intelligence and Communication Networks (ICRCICN)*, pp. 241-6.

Sharma, S. and Patil, S. V. 2015. Key indicators of rice production and consumption, correlation between them and supply-demand prediction. *International Journal of Productivity and Performance Management* 64(8), November, 1113-37. doi:10.1108/IJPPM-06-2014-0088.

Sharma, S., Singh, K. and Singh, A. 2010. An Expert System for diagnosis of diseases in rice plant. *International Journal of Artificial Intelligence* 1(1), 26-31.

Shaw, A., Sheppard, S., Burch, S., Flanders, D., Wiek, A., Carmichael, J., Robinson, J. and Cohen, S. 2009. Making local futures tangible–synthesizing, downscaling, and visualizing climate change scenarios for participatory capacity building. *Global Environmental Change* 19(4), 447-63. doi:10.1016/j.gloenvcha.2009.04.002.

Shin, J. Y., Ajmal, M., Yoo, J. and Kim, T. W. 2016. A bayesian network-based probabilistic framework for drought forecasting and outlook. *Advances in Meteorology* 2016, 1-10. doi:10.1155/2016/9472605.

Singh, S., Gupta, P. and Sharma, M. 2011. Breast cancer detection and classification of histopathological images. *International Journal of Engineering, Science and Technology* 3(5), 4228-32.

Smith, B. A., Hoogenboom, G. and McClendon, R. W. 2009. Artificial neural networks for automated year round temperature prediction. *Computers and Electronics in Agriculture* 68(1), 52-61. doi:10.1016/j.compag.2009.04.003.

Soltani, A. and Sinclair, T. R. 2012a. *Modeling Physiology of Crop Development, Growth and Yield*. CABI Publishing, Wallingford, UK and Cambridge, MA.

Soltani, A. and Sinclair, T. R. 2012b. SSM-models (SSM_ppm). Retrieved 10 September 2012, from https://sites.google.com/site/cropmodeling/ssm-models.

Sridhar, R., Babu, P., Parimala, M. and Thirupathi Rao, N. 2010. Implementation of Web Based chilli expert advisory system using ABC optimization algorithm. *International Journal on Computer Science and Engineering(IJCSE)* 2(6), 2141-4.

Starrett, S. K., Starrett, S. K. and Adams, G. L. 1997. Using artificial neural networks and regression to predict percentage of applied nitrogen leached under turfgrass. *Communications in Soil Science and Plant Analysis* 28(6-8), 497-507. doi:10.1080/00103629709369806.

Subramani, T. and Raghu Prakash, G. 2016. Rice based irrigated agriculture using GIS. *International Journal of Emerging Trends and Technology in Computer Science (IJETTCS)* 5(3), 114-24.

Sudharsan, D., Adinarayana, J.,Tripathy, A. K.,Ninomiya, S.,Hirafuji, M.,Kiura, T.,Desai, U. B., Merchant, S. N., Raji Reddy, D. R.and Sreenivas, G. 2012. GeoSense: a multimode information and communication system. *ISRN Sensor Networks* 2012, Article ID 215103 doi:10.5402/2012/215103.

Tanaka, K. and Kiura, T. 2014. Crop yield prediction systems for rainfed areas and mountainous areas in Thailand. *Proceedings of the 9th Conference of the Asian Federation for Information Technology in Agriculture 'ICT's for Future Economic and Sustainable Agricultural Systems'*, Perth, Australia.

Tanna, P. and Ghodasara, D. Y. 2014. Using apriori with WEKA for frequent pattern mining. *International Journal of Engineering Trends and Technology* 12(3), 127-31. doi:10.14445/22315381/IJETT-V12P223.

Tari, F. 1996. A Bayesian Network for predicting yield response of winter wheat to fungicide programmes. *Computers and Electronics in Agriculture* 15(2), 111-21. doi:10.1016/0168-1699(96)00011-7.

Thakkar, R., Kayasth, M. and Desai, H. 2014. Rule based and association rule mining on agriculture dataset. *International Journal of Innovative Research in Computer and Communication Engineering* 2(11), 6381-4.

Theodosiou, T., Valsamidis, S., Hatziliadis, G. and Nikolaidis, M. 2012. Measuring, archetyping and mining *Olea europaea* production data. *Journal of Systems and Information Technology* 14(4), 318-35. doi:10.1108/13287261211279062.

Thomas, G., Taylor, G. and Wood, J. 1997. Mapping yield potential with remote sensing. *Precision Agriculture* 1, 713-20.

Tripathi, S., Srinivas, V. V. and Nanjundiah, R. S. 2006. Downscaling of precipitation for climate change scenarios: a support vector machine approach. *Journal of Hydrology* 330(3-4), 621-40. doi:10.1016/j.jhydrol.2006.04.030.

Tripathy, A. K., Adinarayana, J. and Sudharsan, D. 2009. Geospatial data mining for agriculture pest management - a framework. *2009 17th International Conference on Geoinformatics*.

Tripathy, A. K., Adinarayana, J., Merchant, S. N., Desai, U. B., Vijayalakshmi, K., RajiReddy, D. and Kiura, T. 2012. Data mining and wireless sensor network for 50 groundnut pest thrips dynamics and predictions. *Journal of Emerging Trends in Computing and Information Sciences* 3(6), 913-29.

Udeshani, K. A. G., Meegama, R. G. N. and Fernando, T. G. I. 2011. Statistical feature based Neural Network Approach for the detection of lung cancer in chest x-ray images. *International Journal of Image Processing* 5(4), 425-34.

Uusitalo, L. 2007. Advantages and challenges of Bayesian networks in environmental modelling. *Ecological Modelling* 203(3-4), 312-8. doi:10.1016/j.ecolmodel.2006.11.033.

Veenadhari, S., Misra, B. and Singh, C. 2011. Data mining techniques for predicting crop productivity - a review article. *International Journal of Computer Science and Technology* 2(1), 98-100.

Vekić, T. T., Ivezić, V., Popović, B., Semialjac, Z., Kerovec, D. and Lončarić, Z. 2017. GIS modeling of site-specific fertilization requirements. *Environmental Science* doi:10.19044/esj.2017.v13n12p%p.

Verheyen, K., Adriaens, D., Hermy, M. and Deckers, S. 2001. High-resolution continuous soil classification using morphological soil profile descriptions. *Geoderma* 101(3-4), 31-48. doi:10.1016/S0016-7061(00)00088-4.

Wei, Y., Santhana-Vannan, S. and Cook, R. 2009. Discover, visualize, and deliver geospatial data through OGC standards-based WebGIS system. *2009 17th International Conference on Geoinformatics*, pp. 1-6.

Wolfert, S., Ge, L., Verdouw, C. and Bogaardt, M.-J. 2017. Big data in smart farming - a review. *Agricultural Systems* 153, 69-80. doi:10.1016/j.agsy.2017.01.023.

Xiao'an, B., Na, Z., Junmin, S., Lingling, Z. and Jun, G. 2011. Application of association rule mining algorithm based on interestingness to agricultural pest control. *International Agricultural Engineering Journal* 20(4), 52-7.

Yadav, S., Singhal, N. and Yadav, V. 2014. RICEsmart: an expert system to enhance rice yield. *International Journal of Computer Science and Information Technologies* 5(5), 6877-9.

Yang, C., Prasher, S., Sreekanth, S., Patni, N. and Masse, L. 1997. An artificial neural network model for simulating pesticide concentrations in soil. *Transactions of the ASAE* 40, 1285-94.

Yengoh, G. T. and Ardo, J. 2014. Crop yield gaps in Cameroon. *Ambio* 43(2), 175-90. doi:10.1007/s13280-013-0428-0.

Zadoks, J. C., Chang, T. T. and Konzak, C. F. 1974. A decimal code for the growth stages of cereals. *Weed Research* 14(6), 415-21. doi:10.1111/j.1365-3180.1974.tb01084.x.

Zhang, H. T., Yang, Q., Lai, C. S. and Lai, L. L. 2012a. New trends for decision support systems. Paper presented at the *2012 IEEE International Conference on Systems, Man, and Cybernetics (SMC)*, 14-17 October.

Zhang, R., Ba, J., Ma, Y., Wang, S., Zhang, J. and Li, W. 2012b. A comparative study on wheat leaf area index by different measurement methods. Paper presented at the *The First International Conference on Agro-Geoinformatics*, 2-4 August.

Chapter 6

Improving data management and decision-making in precision agriculture

Soumyashree Kar, Rohit Nandan, Rahul Raj, Saurabh Suradhaniwar and J. Adinarayana, Indian Institute of Technology Bombay (IIT Bombay), India

1 Introduction

Increasing productivity is a key objective of any crop improvement or breeding program. This has led to an increased use of fertilizers, pesticides, and machinery to help improve yield. However, these practices can have negative environmental impacts such as a loss of soil fertility and biodiversity (Bender et al., 2016). Precision agriculture (PA) has been adopted to improve crop production while optimizing resource allocation. There has been an increasing use of PA systems integrated with unmanned aerial vehicle (UAV), wireless sensor network (WSN), and radio frequency identification (RFID) technologies. While UAVs scan crops and livestock, WSN connects and synchronizes data collection and processing, and RFID technology allows tracking across the supply chain (Muralidhara and Geethanjali, 2018).

Site-specific precision farming has three main components which are (Sahoo et al., 2007):

1 identify each field location (based, for example, on site-specific nutrient or crop protection requirements),

http://dx.doi.org/10.19103/AS.2020.0069.07

2 capture, interpret, and analyze agronomic data at an appropriate scale and frequency, and
3 use this data to optimize resource allocation for the maximum benefit of each field location.

PA is based on the concept of 'differential' treatment based on field variations, as contrasted with traditional uniform field management. Precision farming is, therefore, also sometimes referred to as site-specific farming (Swinton and Lowenberg-DeBoer, 1998; Li et al., 2019), prescription farming (Rawlins, 1996; Griffin, 2018), or variable rate technology (VRT) farming (Sawyer, 1994; Rahim et al., 2018). The following sections discuss the tools and technologies used in PA for effective decision-making, the methods for identifying variability in crop production systems, and recent developments in PA.

2 Remote sensing technologies

Remote sensing (RS) techniques have been primarily used in PA for identifying variability across larger areas. There are a range of optical, thermal, hyperspectral, and synthetic aperture radar (SAR) satellite imaging techniques as well as laser-based light detection and ranging (LIDAR) techniques. There are trade-offs in using each RS method which determine their usefulness in particular situations. Suboptimal temporal and spatial resolution, cloud cover, and high data acquisition costs limit the use of optical and hyperspectral RS data. Poor spatial resolution and expensive data processing costs also limit the use of SAR imaging for field-scale applications.

Mulla (2013) has conducted an extensive review of satellite RS platforms, their spectral, spatial resolution, revisit frequency, and suitability for PA, and has summarized the main multispectral vegetation indices used in data processing. LAI, NDVI, PRI, CWSI, and other vegetation indices are used to measure crop health, for example, by measuring canopy structure and color over time. Xue and Su (2017) have also documented both multispectral (i.e. broad-band) and narrow-band vegetation indices along with the procedures for validating the estimates of crop status derived from these indices.

Recent developments in image processing techniques include image fusion and transformation, feature extraction and application of deep learning to obtain super-resolution images. Zhang et al. (2016) have reviewed the deep learning algorithms used for RS data analysis, which include image restoration and de-noising, image sharpening, spectral feature extraction and classification, spatial feature extraction and mapping, and target recognition. Kamilaris and Prenafeta-Boldú (2018) have also explored deep learning methods primarily used for PA applications besides providing a set of open-source agricultural datasets. The authors have reviewed over 15 PA applications, the corresponding deep learning methods and the data used thereof.

UAV also play an important role in collecting high spatial and temporal resolution farm data in near real time. Cameras mounted on a drone/UAV/UAS platform can take images at a higher resolution than satellite cameras. These high-resolution images contain information about crop physical/chemical properties. The data can be processed with various image interpretation, photogrammetry, and/or machine learning techniques to identify targets for PA applications. Drones also carry sensors as well as cameras, as shown in Fig. 1.

Drones are helpful in every stage of farming, whether before sowing or during crop growth. A LIDAR-installed drone can, for example, find how well a field is leveled so that the farmer can level it uniformly before sowing seeds. Soil nitrogen and water can also be estimated using drone-based hyperspectral sensors (Zhang et al., 2019; Kim et al., 2019) which can help target fertilizer application precisely. Drone-based technologies can be used to analyze the spatial distribution of soil nitrogen present on the farm.

When the crop is growing, thermal sensors can be used to estimate leaf water content which identifies areas of water deficit for targeted irrigation. Thermal cameras have also shown the capability for presymptomatic detection of biotic stress (Ben-Gal et al., 2009; Bellvert et al., 2015). Hyperspectral images can also be used to estimate leaf water content (Gago et al., 2015). RGB cameras and LIDAR sensors can be used to estimate plant height and other crop biophysical parameters which help in understanding crop development over a period of time and indicate the need for targeted nutrition to improve growth (Holman et al., 2016; De Souza et al., 2017; Spoorthi et al., 2017). Wi-Fi

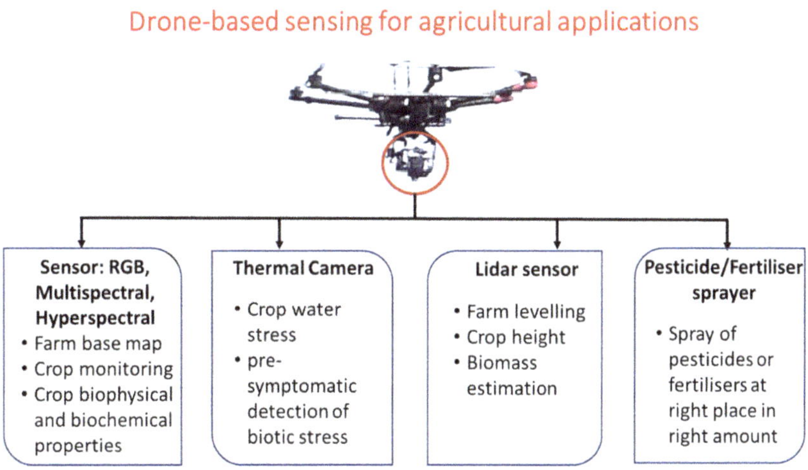

Figure 1 Sensors used on drones for various agricultural applications.

technology has been recently integrated in drones that enable First Person View (FPV) functionality by associating the UAV with HD cameras such as GoPro, DJI, and Parrot. This allows streaming of real-time flight video to a smartphone or tablet.

Drones can be classified by weight and payload capacity. As an example, the Indian Directorate General of Civil Aviation (DGCA) classifies drones into five weight-based categories: (a) Nano (<250 gm), (b) Micro (>250 gm and <2 kg), (c) Small (>2 kg but <25 kg), (d) Medium (>25 kg but <150 kg), and (e) Large: (>150 kg; DGCA, 2019). In each of these classes, drones can also be categorized into fixed-wing airplanes and rotary motor helicopters. While the fixed-wing drones can fly at higher speeds ranging from 25 to 45 mph and can cover a range of 500–750 acres per hour, depending on battery size, rotary motor drones can hover and focus and can fly at constant speed, enabling them to focus on a particular location. Rotary motor drones (with limited battery life) are increasingly being used in more confined locations which require safe takeoff and landing in small areas (Puri et al, 2017). Examples of common drone types are:

1 Honeycomb AgDrone System: Fixed-wing drone, with a maximum of 858 acres coverage at a maximum height of 400 feet, suitable for mapping and monitoring of agricultural plots.
2 DJI Matrice 100: Fixed-wing quadcopter with dual battery support (which increases almost 40 min of flight times), Global Position System (GPS), flight controller, and so forth, which enables flight operation in all environmental conditions; regarded as the best for agricultural applications with operating temperature ranging from −10 to 40°c.
3 The DJI T600 Inspire Quadcopter: preferred for its fast charging, video recording, and easy navigational capabilities.
4 DJI MG-1S: mostly preferred for its precision, effectiveness, intuitiveness, ease of use, safety, build, and sophisticated field monitoring capabilities.
5 Agras MG-1 - DJI octocopter: designed to assist farmers in spraying pesticides, insecticides or fertilizers over large agricultural areas, due to its liquid payload capacity upto 10 kg and coverage of 4000–6000 m^2 area in just 10 min (i.e. 70 times faster than manual spraying).
6 EBEE SQ-SenseFly: the high-performance agriculture drone (integrated with multispectral and RGB cameras, besides the automatic 3D flight planning and seamless compatibility with the Pix4D drone image processing software) is designed for an end-to-end agricultural monitoring which extends from planting to harvesting.
7 Lancaster 5 Precision Hawk: an autonomous drone enabled with features to optimize flight plan during data collection.

8 SOLO AGCO: provides an optimal solution for better farm management, with its intelligent mission planning and cloud-based high-resolution mapping software to increase flight efficiency.
9 Albatross UAV by Applied Aeronautics: equipped with 5 h of flight time in a single mission and a multispectral sensor, which allows monitoring of crop health, soil fertility, and pest spread at reduced operating costs (thereby leading to improved yields).
10 AgEagle RX-60: an easy-to-operate and durable fixed-wing drone, capable of capturing NIR/NDVI aerial maps on-the-fly.

3 Geographic information system (GIS) technologies

Geographic information system (GIS) enables the association of data with location information, accessed using GPS systems. The use of geographic information in PA enables a geostatistical analysis of soil, weather, and crop conditions for PA applications such as mapping, soil, and crop analysis decision-making and variable-rate applications. Key geostatistical methods include interpolation, kriging, variography, as well as simulation techniques which generate interpolated surfaces to replicate the spatial characteristics found in the sample data. Geostatistical simulations are especially used in development and analysis of region-specific crop growth models (Priya and Shibasaki, 2001; Ines et al., 2002; Tan and Shibasaki, 2003; Sailaja et al., 2019). The importance of spatial scale in sampling is also an important characteristic in field planning and data collection which is carried out using GIS simulations (Scudiero et al., 2015; Antle et al., 2017). Córdoba et al. (2016) have demonstrated a procedure to delineate multivariate homogeneous zones for site-specific farming based on target crop characteristics. Kadiyala et al. (2015) have developed an integrated crop model and GIS-based decision support system for assisting agronomic decision-making.

The use of GIS is enabled by software packages such as ARCVIEW (Minami et al., 2000), IDRISI (Eastman, 2003), and MeteoInfo (Wang, 2014). Most of these more sophisticated and expensive programs are used in research and by larger companies rather than by individual farmers. GRASS GIS (Neteler et al., 2012) is an open-source GIS software widely used for geospatial analysis.

The use of geospatial methods in PA are explored by Chen et al. (2015). They propose a cyber-physical infrastructure for PA which provides interoperable access for sensors, data, and algorithms within an integrated information system. Similarly, Guo et al. (2018) have developed a cyber-physical system (CPS) framework for managing greenhouses. The system includes a collaborative network of sensors and robots, integrated using collaborative control theory (CCT) for monitoring, detecting, and responding (MDR) to variations in greenhouse conditions. Intelligent geospatial systems are also being developed with fog computing capacities at the network edge for increased automation of machine-to-machine (M2M) communication without human intervention. Chen (2017) has designed a hierarchy-based food traceability concept mapping framework, which integrates each entity in the stakeholder layer (farmer, producer, processor, distributor, retailer, and customer) and attempts to minimize quality losses in food supply chain.

Delgado et al. (2019) have proposed Sustainable Precision Agriculture and Environment (SPAE), a geospatial cloud framework which could integrate existing technologies with big data analysis. The authors suggest a WebGIS framework able to scale PA from site level to regional level. A big data analysis framework could integrate multiple geospatial databases into an automated system and combine artificial intelligence (AI), the internet of things (IoT), drones, robots, and big data to form the 'Digital Twin', that is, precision agricultable at a global scale.

Zhang et al. (2018) have described the Geospatial Sensor Web (GSW), a cyber-physical infrastructure designed to:

1 achieve integrated and sharable management of diverse sensing resources,
2 obtain real-time or near real-time and spatiotemporal continuous data,
3 conduct interoperable and online geoscience data processing and analysis, and
4 provide focusing services with web-based geoscience information and knowledge.

Rezník et al. (2017) have used geospatial techniques to combine satellite imagery with in-situ sensor data analysis to identify and manage disaster risks in agriculture.

4 Sensors and sensor networks

Sensors and WSN are an integral component of precision farming. They provide on-field proximal sensing solutions. Real-time measurements required during field operations include identification and analysis of spatial variabilities in field topology, soil type, soil moisture and temperature, relative humidity and temperature, crop growth and nutrition status, canopy (or leaf) water status, leaf area index, leaf angle, and so forth, which might need immediate remedial actions. Sensors can be broadly categorized into (Zhang et al., 2002):

- yield sensors,
- soil sensors, and
- crop sensors.

These are discussed in the following paragraphs.

Yield sensors are used for measuring crop yields, and include impact or mass flow sensors, weight-based sensors, optical yield sensors, or g-ray sensors. These sensors are used to generate yield maps for applications such as variable-rate fertilization (Pelletier and Upadhyaya, 1999; Tamayo et al., 2010). These sensors use differential GPS (DGPS) to facilitate real-time spatially variable yield data collection.

Soil sensors typically use spectrophotometric methods to capture the soil spectral reflectance in the visible (400–700 nm) and NIR wavebands (1600–2600 nm) to predict the content of organic matter and moisture present in surface and subsurface soils (Hummel et al., 2001). Some soil sensors measure soil electrical conductivity (EC) using electrodes to detect soil properties (Lund et al., 2005). Drummond et al. (2000) have mapped the soil subsurface by combining an EC-based soil probe and a penetrometer. Although EC technologies are widely used, accurately calibrating such soil sensors is generally a limiting factor. Sudduth et al. (2001) have reviewed accuracy issues in electromagnetic induction sensing of soil EC for PA. Adamchuk et al. (2004) have surveyed the different technologies (which include electrical and electromagnetic, optical and radiometric, mechanical, acoustic, pneumatic, and electrochemical measurement methods) used for real-time mapping of soil properties. Recent developments in soil sensing include impedance spectroscopy at low radio frequency (RF; Kumar et al., 2018) and nanostructured biosensors (Antonacci et al., 2018).

Crop sensors use vegetation indices to measure characteristics such as leaf temperature, water, chlorophyll, nitrogen, and carbon content (Bauer et al., 2016; Yin et al., 2019), as well as assessing plant photosynthetic rate (Ji et al., 2016; Jian et al., 2018). Spectrophotometric sensors use spectral reflectance properties to map these properties (Stamenković et al., 2018). These techniques have been used to detect weeds by identifying their spectral characteristics (Wang et al., 2001; Navulur and Prasad, 2017). Detection of aphid infestation has also been performed by monitoring changes in plant temperature through an infrared plant-temperature transducer (Michels et al., 2000).

The significance of WSN has increased in importance with advances in information communication and dissemination technologies (ICDTs) which enable optimized allocation of resources in agricultural production (Lee et al., 2010). In order to standardize communication between service-provider and service-user in a WSN system, the Open Geospatial Consortium (OGC) has developed the sensor web enablement (SWE) architecture (Botts et al., 2006). This facilitates interoperability through service-based architectures for sensor data discovery by identifying the exact lineage of collected data (Botts and Robin, 2007).

Sensors can also be mounted on equipment such as crop sprayers. Optical sensors have been developed to measure flow rates of granular fertilizers, providing feedback control for VRT sprayers (Swisher et al., 1999). While DGPS-enabled field-scouts are used for enhanced spatial precision, autonomous vehicles with multi-angle NIR and multispectral cameras onboard also enable high spectral precision in VRT farming. Although the use of robotic harvesting systems is well established, the establishment of an IoT platform by using WSNs, in combination with the robotic harvesters, provides a complete on-field management system. Robotic platforms for performing multiple functions in agricultural systems are already available in many countries, such as in the United States (Slaughter et al., 1995; Pangels et al., 2002), Japan (Umeda et al., 1999), France (Cordesses et al., 2000), Italy (Foglia and Reina, 2006), Israel (Bechar and Vigneault, 2017), and India (Megalingam et al., 2017). Blackmore et al. (2005) have concluded that the future of agricultural mechanization is 'robotic agriculture'.

5 Statistical and crop simulation models

In PA applications, it is imperative to identify the relationships between variables such as soil, topography, crop characteristics, and weather conditions, since the interplay between these factors determines changes in crop yield and other variables. Statistical models quantify changes in crop performance such as crop yield (the response variable) due to the changes in environmental factors (the explanatory variables). Mixed model regression analysis is predominantly used to estimate the effect of each variable on the response variable, that is, identifying the proportion of variation in the response variable contributed by each covariate. These models are primarily predictive and either parametric or nonparametric. Table 1 highlights examples of common models used.

Another key tool in PA are crop models. Crop models are computer-based representation of a soil-water-atmosphere-plant system which can be used to inform decisions such as selection of crop variety, sowing dates, and crop management. Crop models date back to the 1960s (de Wit, 1965). This has led to crop models such as DSSAT, Agricultural Production Systems Simulator (APSIM), Hybrid maize, and AquaCrop (Jones et al., 2003; Keating et al., 2003; Yang et al., 2004; Raes et al., 2009).

The general framework of crop models is shown in Fig. 2. These models are categorized into empirical and process-based models. Empirical models use regression equations between variables such as weather data and crop yield. These types of models need to be calibrated before use and the calibrated model for a specific location is not suitable for other locations and climatic conditions. Process-based models are computationally more complex and

Table 1 Examples of statistical models used for precision agriculture

Serial#	Model	Application	Source
Parametric models			
1	Logistic regression	Identification and mapping of site-specific soil properties, and the probability of soil erosion; improving the spatial resolution of soil surveys	Mueller et al. (2005)
2	Linear Discriminant Analysis (LDA)	Improved identification of plants from weeds, from multispectral images	López-Granados et al. (2008)
3	Naïve Bayes	Development of agriculture cyber-physical systems based on classification of agricultural land soils	Airlangga and Liu (2019)
4	Bayesian Penalized Regression	Analysis of WSN data for soil-plant interaction and variability	Pollice et al. (2019)
5	Mixed Models	Development of a seed drill depth control system for precision furrowing and seeding	Nielsen et al. (2018)
Nonparametric models			
1	K–Nearest Neighbors	Paddy crop and weed classification using color features	Kamath et al. (2018)
2	Support Vector Machines (SVM)	Digital soil mapping	Ji et al. (2017)
3	Random Forests	Detection and classification of postharvest growth using LIDAR data	Koenig et al. (2015)
4	Radial Basis Function Network (RBFN)	Classification of diseased plant leaves	Muthukannan et al. (2015)
5	Nonparametric Bayesian Networks	Development of a farm management information system: ArgoDSS	Kukar et al. (2019)
6	Perceptron	Optimization of fertilizer application, and yield estimation	Ruß (2009)
7	Simple Neural Networks	Detection of weeds in agricultural fields using multispectral images onboard an Unmanned Ground Vehicle (UGV)	Potena et al. (2017)

more flexible than empirical models. Crop models have been frequently used for the following types of decision-making:

1. Understanding soil-water-climate-plant interactions,
2. Planning site-specific experiments,
3. Impact assessment of climate variability on crop performance,
4. Investigating optimal crop management practices, and
5. Yield prediction.

Examples of climate impact assessment include Grace (1988), Adams et al. (1990), and Ainsworth and Long (2005).

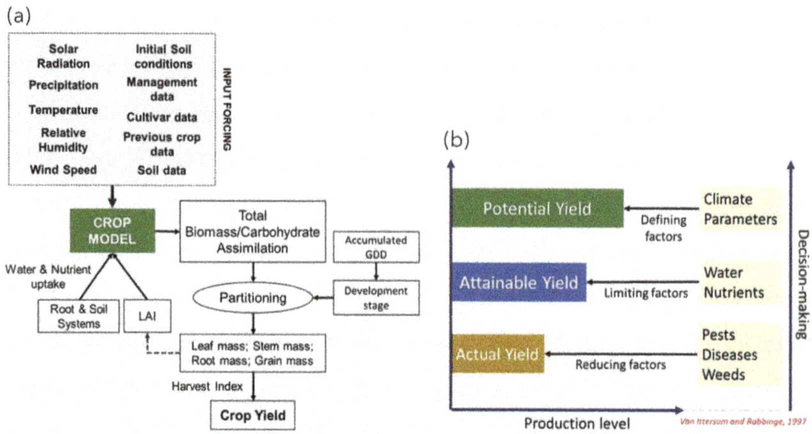

Figure 2 (a) General framework of crop models; (b) principles of decision-making and crop yields levels (van Ittersum and Rabbinge, 1997).

6 Identifying variability in crop production systems

Types of variability include:

- spatial variability,
- temporal variability, and
- biotic and abiotic variability.

Spatial variability (which includes variations in both soil and environmental conditions) is probably the most significant source of variability in crop production systems. Spatial models are used to capture the effects of spatial heterogeneity in individual locations (Duarte and Vencovsky, 2005; Rodríguez-Álvarez et al., 2016). There have been a number of reviews on optimizing experimental designs (Tisne et al., 2013; Junker et al., 2015). Alpha Lattice designs have been found to be particularly effective in managing variability in field conditions (Peternelli and de Resende, 2015). However, it remains challenging to take full account of spatial heterogeneity such as micro-agro-climatic field conditions which can be very difficult to predict. Both satellite-based (Khanal et al., 2017) and UAV RS (Katsigiannis et al., 2016) technologies have been used to understand spatiotemporal variability in agricultural fields.

One use of spatial models is to assess the proportion of phenotypic variance due to genotypic or spatial effects. The use of such models, either based on field trials (Tisne et al., 2013) or data generated by high-throughput plant phenotyping (HTPP) systems (Cabrera-Bosquet et al., 2012), has significantly contributed to crop improvement programs. Tisne et al. (2013) have suggested minimizing spatial heterogeneity by continuous rotation of individual plants being analyzed in HTPP systems to homogenize the microenvironmental conditions experienced by plants. Leiser et al. (2012) have suggested improved breeding designs to alleviate this problem.

Modeling interrow competition effects and spatial variability has been assessed by Sarker and Singh (2015) using lentil cultivars as an example. Spatial variability is also evaluated in field trials of several other crops, like chickpea, sorghum and barley. Use of spatial models has resulted in changing the ranking of genotypes (based on mean yield), subsequently resulting in a different set of high-yield genotypes. In the case of field trials, model comparisons are largely based on the goodness of fit of models inferred using the Akaike Information Criterion based on the Deviance (AICD) estimator (Akaike, 1974).

Temporal variability occurs due to factors such as changes in temperature and other environmental conditions, changing phenophases of plants, and the corresponding changes in genotype x environment (GxE) interactions (McBratney et al., 2005; Lake et al., 2008). Time-series data are primarily used to capture such differences in patterns of phenotypic development among different cultivars. Linear time-series forecasting methods have been widely used owing to their simplicity. However, deep learning neural networks are now being increasingly used to decipher the complex plant interactions involving multiple variables (Wang and Xiong, 2014; Ndikumana et al., 2018). Some studies have used combined spatiotemporal analysis to examine crop variability with respect to yield and nutrient management (Batchelor et al., 2002; Starr, 2005; Hedley, 2015; Dong et al., 2017).

Biotic variability occurs due to differences in soil properties and environmental conditions. A deficit in requirements such as soil moisture, soil pH, or nutrients can lead to biotic stress. Diseases and insect pests result in abiotic stress in plants. Early detection of biotic and abiotic stress is critical for effective crop management (Falkenberg et al., 2007; Fang and Ramasamy, 2015). Imaging techniques (Chaerle and Van Der Straeten, 2000), especially fluorescence imaging (Behmann et al., 2015) and hyperspectral RS (Carter and Knapp, 2001; Prabhakar et al., 2011; Mahlein et al., 2013; Pantazi et al., 2017), have been used in identification of stressed plants, though fluorescence imaging is complex to operate in field conditions. There has been an increasing use of sensor networks to monitor potential stress (Moshou et al., 2011; Mahlein, 2016). A key challenge is to distinguish individual biotic and abiotic factors contributing to variable rates of stress within a field (Yadeta and Thomma, 2013).

It remains the case that many analytical methods still tend to understate the complexity of environment and plant growth interactions (Pauli et al., 2016). Deciphering the variability caused by plant, soil, and weather interaction requires 'envirotyping', followed by Genotype-by-Environment-by-Management interaction (GxExM) modeling of phenotypes (Xu, 2016). Envirotyping requires an adequate amount of high spatial resolution field-level environmental data, which could be best acquired using affordable open-source sensor networks. Sawant et al. (2017) have provided a detailed architecture of SWE standards for agro-meteorological data collection and analysis, and have provided a comparative analysis of existing SWE-based architectures. Notable open-source platforms for the analysis of spatiotemporal interaction characteristics of plants are Yet Another Time Series Model (YATSM) and Time Series Tools (TSTools) in Quantum GIS software (Holden, 2015a,b). Di Paola et al. (2016) have reviewed the applications, advantages, and limitations of more than 70 crop models.

7 Summary and future trends

The application of PA is limited by several factors, including:

- the costs involved in installation of proximal sensors,
- site-specific data collection,
- technical complexity in high-resolution data analysis,
- implementing effective statistical methods to identify spatiotemporal variability in fields, and
- knowledge gaps in understanding GxE interactions.

Some of these challenges are being addressed by utilizing big data in smart farming applications, as reviewed by Wolfert et al. (2017). Recent innovations include the following:

- SIMAGRI is an agro-climate DSS developed by Han et al. (2018) which integrates seasonal variations into crop models. It enables identification of the most suitable management practices and agricultural inputs for particular climatic conditions for three crops: wheat, maize, and soybean.
- AgroDSS is a cloud-based time-series analysis toolbox, developed by Kukar et al. (2019) for identifying and examining pest population dynamics in plants. Farmers can directly upload the images of their infected crops and obtain diagnostic information about identifying and managing pests.
- CropGIS (Machwitz et al., 2019) is another web-based DSS. It incorporates high spatial resolution crop biomass measurements by integrating optical satellite RS data with the APSIM crop model to measure and predict crop biomass development.
- AgriPrediction, an IoT-based DSS developed by dos Santos et al. (2019), integrates LoRa technology and time-series prediction modeling in order to predict stress in crops. Pre-visual stress in crops is detected

by identifying anomalies from very high-resolution time-series data. The authors also provide a comparative chart of previously developed DSS, along with their applications, data analytical, and communication technology used.

A key challenge remains the limited number of real-world applications of these technologies. The potential is for every farmer to benefit from DSS to improve decision-making for greater efficiency and sustainability.

8 Where to look for further information

There are several research projects carried out by the Agro-Informatics Lab, Centre of Studies In Resources Engineering (CSRE), Indian Institute of Technology Bombay (IIT Bombay) in collaboration with The University of Tokyo and National Agricultural Research Center/Tsukuba, Japan, where the main objective is to provide innovative solutions (using disruptive technologies) for precision agriculture and the farming community as a whole. The details can be found at: http://www.csre.iitb.ac.in/adi/agrolab.htm. The site also includes details on the projects:

- DSFS: Data Sciences for Farming Support Systems for Sustainable Crop Production Under Climate Change, and
- GrIDSense: ICT in Water and Pest/Disease Management for Yield Improvement in Horticulture (Citrus).

An IoT platform (SenseQube) for Smart Agriculture and decision making is developed by the Geo-Computational Systems and IoT Group, IIT Bombay. It includes functionalities like sensor-based field mapping, real time insights, water-use efficiency and crop water requirement estimation. The following link includes detailed information:

- http://www.senseqube.com/.

Other organizations and centres which expertise in DSS development for PA are:

- Research and Development Division, National Agricultural Statistics Service, US Department of Agriculture, USA (https://www.usda.gov/).
- Center for Spatial Information Science and Systems, George Mason University, USA (http://csiss.gmu.edu/index.htm).
- UR Technologies and Information Systems for Agricultural System (TSCF), France (https://www.irstea.fr/en/research/research-units/tscf).

Further examples on the use of disruptive technologies for PA by CGIAR can be found via the following links:

- https://www.cgiar.org/how-we-work/hero/high-tech-farming-with-franceli no-rodrigues-cimmyt/https://www.rezatec.com/resources/projects/me xican-compass/.
- https://wle.cgiar.org/thrive/2017/03/20/precision-agriculture-bigger-y ields-smaller-farms.

9 References

Adamchuk, V. I., Hummel, J. W., Morgan, M. T. and Upadhyaya, S. K. 2004. On-the-go soil sensors for precision agriculture. *Computers and Electronics in Agriculture* 44(1), 71–91. doi:10.1016/j.compag.2004.03.002.

Adams, R. M., Rosenzweig, C., Peart, R. M., Ritchie, J. T., McCarl, B. A., Glyer, J. D., Curry, R. B., Jones, J. W., Boote, K. J. and Allen, L. H. 1990. Global climate change and US agriculture. *Nature* 345(6272), 219–24. doi:10.1038/345219a0.

Ainsworth, E. A. and Long, S. P. 2005. What have we learned from 15 years of free-air CO_2 enrichment (FACE)? A meta-analytic review of the responses of photosynthesis, canopy properties and plant production to rising CO_2. *The New Phytologist* 165(2), 351–71. doi:10.1111/j.1469-8137.2004.01224.x.

Airlangga, G. and Liu, A. 2019. Initial machine learning framework development of agriculture cyber physical systems. In: Journal of Physics: Conference Series (vol. 1196, no. 1, p. 012065). IOP Publishing. doi:10.1088/1742-6596/1196/1/012065.

Akaike, H. 1974. A new look at the statistical model identification. In: *Selected Papers of Hirotugu Akaike*. Springer, New York, NY, pp. 215–22.

Antle, J. M., Basso, B., Conant, R. T., Godfray, H. C. J., Jones, J. W., Herrero, M., Howitt, R. E., Keating, B. A., Munoz-Carpena, R., Rosenzweig, C., Tittonell, P. and Wheeler, T. R. 2017. Towards a new generation of agricultural system data, models and knowledge products: design and improvement. *Agricultural Systems* 155, 255–68. doi:10.1016/j.agsy.2016.10.002.

Antonacci, A., Arduini, F., Moscone, D., Palleschi, G. and Scognamiglio, V. 2018. Nanostructured (Bio) sensors for smart agriculture. *TrAC Trends in Analytical Chemistry* 98, 95–103. doi:10.1016/j.trac.2017.10.022.

Batchelor, W. D., Basso, B. and Paz, J. O. 2002. Examples of strategies to analyze spatial and temporal yield variability using crop models. *European Journal of Agronomy* 18(1–2), 141–58. doi:10.1016/S1161-0301(02)00101-6.

Bauer, J., Siegmann, B., Jarmer, T. and Aschenbruck, N. 2016. On the potential of wireless sensor networks for the in-situ assessment of crop leaf area index. *Computers and Electronics in Agriculture* 128, 149–59. doi:10.1016/j.compag.2016.08.019.

Bechar, A. and Vigneault, C. 2017. Agricultural robots for field operations. Part 2: operations and systems. *Biosystems Engineering* 153, 110–28. doi:10.1016/j.biosystemseng.2016.11.004.

Behmann, J., Mahlein, A. K., Rumpf, T., Römer, C. and Plümer, L. 2015. A review of advanced machine learning methods for the detection of biotic stress in precision crop protection. *Precision Agriculture* 16(3), 239–60. doi:10.1007/s11119-014-9372-7.

Bellvert, J., Marsal, J., Girona, J. and Zarco-Tejada, P. J. 2015. Seasonal evolution of crop water stress index in grapevine varieties determined with high-resolution remote sensing thermal imagery. *Irrigation Science* 33(2), 81–93. doi:10.1007/s00271-014-0456-y.

Bender, S. F., Wagg, C. and van der Heijden, M. G. A. 2016. An underground revolution: biodiversity and soil ecological engineering for agricultural sustainability. *Trends in Ecology and Evolution* 31(6), 440–52. doi:10.1016/j.tree.2016.02.016.

Ben-Gal, A., Agam, N., Alchanatis, V., Cohen, Y., Yermiyahu, U., Zipori, I., Presnov, E., Sprintsin, M. and Dag, A. 2009. Evaluating water stress in irrigated olives: correlation of soil water status, tree water status, and thermal imagery. *Irrigation Science* 27(5), 367–76. doi:10.1007/s00271-009-0150-7.

Blackmore, S., Stout, B., Wang, M. and Runov, B. 2005. Robotic agriculture–the future of agricultural mechanisation. In: *Proceedings of the 5th European Conference on Precision Agriculture*, pp. 621-8.

Botts, M. and Robin, A. 2007. OpenGIS Sensor Model Language (SensorML) implementation specification. Open Geospatial Consortium. Available at: http://www.opengeospatial.org/standards (accessed on 26 May 2014).

Botts, M., Robin, A., Davidson, J. and Simonis, I. 2006. OpenGIS Sensor Web Enablement (SWE) architecture document. Open Geospatial Consortium. Available at: http://www.opengeospatial.org/standards (accessed on 26 May 2014).

Cabrera-Bosquet, L., Crossa, J., von Zitzewitz, J., Serret, M. D. and Luis Araus, J. 2012. High-throughput phenotyping and genomic selection: the frontiers of crop breeding converge F. *Journal of Integrative Plant Biology* 54(5), 312–20.

Carter, G. A. and Knapp, A. K. 2001. Leaf optical properties in higher plants: linking spectral characteristics to stress and chlorophyll concentration. *American Journal of Botany* 88(4), 677–84. doi:10.2307/2657068.

Chaerle, L. and Van Der Straeten, D. 2000. Imaging techniques and the early detection of plant stress. *Trends in Plant Science* 5(11), 495–501. doi:10.1016/s1360-1385(00)01781-7.

Chen, R. Y. 2017. An intelligent value stream-based approach to collaboration of food traceability cyber physical system by fog computing. *Food Control* 71, 124–36. doi:10.1016/j.foodcont.2016.06.042.

Chen, N., Zhang, X. and Wang, C. 2015. Integrated open geospatial web service enabled cyber-physical information infrastructure for precision agriculture monitoring. *Computers and Electronics in Agriculture* 111, 78–91. doi:10.1016/j.compag.2014.12.009.

Cordesses, L., Cariou, C. and Berducat, M. 2000. Combine harvester control using real time kinematic GPS. *Precision Agriculture* 2(2), 147–61. doi:10.1023/A:1011473630247.

Córdoba, M. A., Bruno, C. I., Costa, J. L., Peralta, N. R. and Balzarini, M. G. 2016. Protocol for multivariate homogeneous zone delineation in precision agriculture. *Biosystems Engineering* 143, 95–107. doi:10.1016/j.biosystemseng.2015.12.008.

De Souza, C. H. W., Lamparelli, R. A. C., Rocha, J. V. and Magalhães, P. S. G. 2017. Height estimation of sugarcane using an unmanned aerial system (UAS) based on structure from motion (SfM) point clouds. *International Journal of Remote Sensing* 38(8–10), 2218–30. doi:10.1080/01431161.2017.1285082.

de Wit, C. T. 1965. Photosynthesis of leaf canopies (No. 663). Pudoc.

Delgado, J. A., Short, N. M., Roberts, D. P. and Vandenberg, B. 2019. Big data analysis for sustainable agriculture on a Geospatial Cloud Framework. *Frontiers in Sustainable Food Systems* 3, 54. doi:10.3389/fsufs.2019.00054.

DGCA. 2019. Requirements for operation of civil remotely piloted aircraft system (RPAS). Available at: http://www.dgca.nic.in/misc/draft%20cars/CAR%20-%20UAS%20(Draft_Nov2017).pdf (last accessed on 2 September 2019).

Di Paola, A., Valentini, R. and Santini, M. 2016. An overview of available crop growth and yield models for studies and assessments in agriculture. *Journal of the Science of Food and Agriculture* 96(3), 709–14. doi:10.1002/jsfa.7359.

Dong, J., Burnham, J. G., Boots, B., Rains, G. and Dellaert, F. 2017. 4D crop monitoring: spatio-temporal reconstruction for agriculture. In: *2017 IEEE International Conference on Robotics and Automation (ICRA)*. IEEE, pp. 3878–85.

dos Santos, U. J. L., Pessin, G., da Costa, C. A. and da Rosa Righi, R. 2019. AgriPrediction: a proactive internet of things model to anticipate problems and improve production in agricultural crops. *Computers and Electronics in Agriculture* 161, 202–13. doi:10.1016/j.compag.2018.10.010.

Drummond, P. E., Christy, C. D. and Lund, E. D. 2000. Using an automated penetrometer and soil EC probe to characterize the rooting zone. In: *Proceedings of the 5th International Conference on Precision Agriculture*, Bloomington, Minnesota, USA, 16–19 July 2000. American Society of Agronomy, pp. 1–9.

Duarte, J. B. and Vencovsky, R. 2005. Spatial statistical analysis and selection of genotypes in plant breeding. *Pesquisa Agropecuária Brasileira* 40(2), 107–14.

Eastman, J. R. 2003. *IDRISI Kilimanjaro: Guide to GIS and Image Processing*. Clark Labs, Clark University, Worcester, MA.

Falkenberg, N. R., Piccinni, G., Cothren, J. T., Leskovar, D. I. and Rush, C. M. 2007. Remote sensing of biotic and abiotic stress for irrigation management of cotton. *Agricultural Water Management* 87(1), 23–31. doi:10.1016/j.agwat.2006.05.021.

Fang, Y. and Ramasamy, R. P. 2015. Current and prospective methods for plant disease detection. *Biosensors* 5(3), 537–61. doi:10.3390/bios5030537.

Foglia, M. M. and Reina, G. 2006. Agricultural robot for radicchio harvesting. *Journal of Field Robotics* 23(6–7), 363–77. doi:10.1002/rob.20131.

Gago, J., Douthe, C., Coopman, R. E., Gallego, P. P., Ribas-Carbo, M., Flexas, J., Escalona, J. and Medrano, H. 2015. UAVs challenge to assess water stress for sustainable agriculture. *Agricultural Water Management* 153, 9–19. doi:10.1016/j.agwat.2015.01.020.

Grace, J. 1988. 3. Plant response to wind. *Agriculture, Ecosystems and Environment* 22–23, 71–88. doi:10.1016/0167-8809(88)90008-4.

Griffin, W. 2018. Prescription farming It's just what the doctor ordered. *Australian Canegrower*, 04 June 2018, p. 16.

Guo, P., Dusadeerungsikul, P. O. and Nof, S. Y. 2018. Agricultural cyber physical system collaboration for greenhouse stress management. *Computers and Electronics in Agriculture* 150, 439–54. doi:10.1016/j.compag.2018.05.022.

Han, E., Baethgen, W. E., Ines, A. V., Mer, F., Souza, J. S., Berterretche, M., Atunez, G. and Barreira, C. 2018. SIMAGRI: an agro-climate decision support tool. *Computers and Electronics in Agriculture* 161, 241–51.

Hedley, C. 2015. The role of precision agriculture for improved nutrient management on farms. *Journal of the Science of Food and Agriculture* 95(1), 12–9. doi:10.1002/jsfa.6734.

Holden, C. E. 2015a. TSTools: linking time and space visualization for remotely sensed time series. *Zenodo*. doi:10.5281/zenodo.34182.

Holden, C. E. 2015b. Yet another time series model (YATSM). *Zenodo*. doi:10.5281/zenodo.17129.

Holman, F., Riche, A., Michalski, A., Castle, M., Wooster, M. and Hawkesford, M. 2016. High throughput field phenotyping of wheat plant height and growth rate in field plot trials using UAV based remote sensing. *Remote Sensing* 8(12), 1031. doi:10.3390/rs8121031.

Hummel, J. W., Sudduth, K. A. and Hollinger, S. E. 2001. Soil moisture and organic matter prediction of surface and subsurface soils using an NIR soil sensor. *Computers and Electronics in Agriculture* 32(2), 149–65.

Ines, A. V. M., Gupta, A. D. and Loof, R. 2002. Application of GIS and crop growth models in estimating water productivity. *Agricultural Water Management* 54(3), 205–25. doi:10.1016/S0378-3774(01)00173-1.

Ji, Y., Jiang, Y., Li, T., Zhang, M., Sha, S. and Li, M. 2016. An improved method for prediction of tomato photosynthetic rate based on WSN in greenhouse. *International Journal of Agricultural and Biological Engineering* 9(1), 146–52.

Ji, W., Adamchuk, V., Chen, S., Biswas, A., Leclerc, M. and Rossel, R. V. 2017. The use of proximal soil sensor data fusion and digital soil mapping for precision agriculture. In: *Pedometrics 2017*, 298p.

Jian, Y., Xinying, L., Man, Z. and Han, L. 2018. Photosynthetic rate prediction of tomato plant population based on PSO and GA. *IFAC-PapersOnLine* 51(17), 61–6. doi:10.1016/j.ifacol.2018.08.061.

Jones, J. W., Hoogenboom, G., Porter, C. H., Boote, K. J., Batchelor, W. D., Hunt, L. A., Wilkens, P. W., Singh, U., Gijsman, A. J. and Ritchie, J. T. 2003. The DSSAT cropping system model. *European Journal of Agronomy* 18(3–4), 235–65. doi:10.1016/S1161-0301(02)00107-7.

Junker, A., Muraya, M. M, Weigelt-Fischer, K., Arana-Ceballos, F., Klukas, C., Melchinger, A. E., Meyer, R. C., Riewe, D. and Altmann, T. 2015. Optimizing experimental procedures for quantitative evaluation of crop plant performance in high throughput phenotyping systems. *Frontiers in plant science* 5, 770.

Kadiyala, M. D., Nedumaran, S., Singh, P., Srinivas, C., Irshad, M. A. and Bantilan, M. C. 2015. An integrated crop model and GIS decision support system for assisting agronomic decision making under climate change. *Science of the Total Environment* 521–522, 123–34. doi:10.1016/j.scitotenv.2015.03.097.

Kamath, R., Balachanra, M. and Prabhu, S. 2018. Paddy crop and weed classification using color features for computer vision based precision agriculture. *International Journal of Engineering and Technology (UAE)* 7(4), 2909–16.

Kamilaris, A. and Prenafeta-Boldú, F. X. 2018. Deep learning in agriculture: a survey. *Computers and Electronics in Agriculture* 147, 70–90. doi:10.1016/j.compag.2018.02.016.

Katsigiannis, P., Misopolinos, L., Liakopoulos, V., Alexandridis, T. K. and Zalidis, G. 2016. An autonomous multi-sensor UAV system for reduced-input precision agriculture applications. In: *2016 24th Mediterranean Conference on Control and Automation (MED)*. IEEE, pp. 60–4.

Keating, B. A., Carberry, P. S., Hammer, G. L., Probert, M. E., Robertson, M. J., Holzworth, D., Huth, N. I., Hargreaves, J. N. G., Meinke, H., Hochman, Z., McLean, G., Verburg, K., Snow, V., Dimes, J. P., Silburn, M., Wang, E., Brown, S., Bristow, K. L., Asseng, S., Chapman, S., McCown, R. L., Freebairn, D. M. and Smith, C. J. 2003. An overview of APSIM, a model designed for farming systems simulation. *European Journal of Agronomy* 18(3–4), 267–88. doi:10.1016/S1161-0301(02)00108-9.

Khanal, S., Fulton, J. and Shearer, S. 2017. An overview of current and potential applications of thermal remote sensing in precision agriculture. *Computers and Electronics in Agriculture* 139, 22–32. doi:10.1016/j.compag.2017.05.001.

Kim, D., Son, Y., Park, J., Kim, T. and Jeon, J. 2019. Evaluation of calibration method for field application of UAV-based soil water content prediction equation. *Advances in Civil Engineering* 2019, 1–10. doi:10.1155/2019/2486216.

Koenig, K., Höfle, B., Hämmerle, M., Jarmer, T., Siegmann, B. and Lilienthal, H. 2015. Comparative classification analysis of post-harvest growth detection from terrestrial LiDAR point clouds in precision agriculture. *ISPRS Journal of Photogrammetry and Remote Sensing* 104, 112–25. doi:10.1016/j.isprsjprs.2015.03.003.

Kukar, M., Vračar, P., Košir, D., Pevec, D. and Bosnić, Z. 2019. AgroDSS: A decision support system for agriculture and farming. *Computers and Electronics in Agriculture* 161, 260–71. doi:10.1016/j.compag.2018.04.001.

Kumar, R., Weber, R. J., Pandey, G. and Iowa State University Research Foundation (ISURF). 2018. Low RF-band impedance spectroscopy based sensor for in-situ, wireless soil sensing. U.S. Patent Application 10/073,074.

Lake, J. V., Bock, G. R. and Goode, J. A. (Eds). 2008. *Precision Agriculture: Spatial and Temporal Variability of Environmental Quality* (vol. 210). John Wiley & Sons.

Lee, W. S., Alchanatis, V., Yang, C., Hirafuji, M., Moshou, D. and Li, C. 2010. Sensing technologies for precision specialty crop production. *Computers and Electronics in Agriculture* 74(1), 2–33. doi:10.1016/j.compag.2010.08.005.

Leiser, W. L., Rattunde, H. F., Piepho, H. P. and Parzies, H. K. 2012. Getting the most out of sorghum low-input field trials in West Africa using spatial adjustment. *Journal of Agronomy and Crop Science* 198(5), 349–59.

Li, Z., Taylor, J., Frewer, L., Zhao, C., Yang, G., Li, Z., Liu, Z., Gaulton, R., Wicks, D., Mortimer, H., Cheng, X., Yu, C. and Sun, Z. 2019. A comparative review on the state and advancement of Site-Specific Crop Management in the UK and China. *Frontiers of Agricultural Science and Engineering* 6(2). doi:10.15302/J-FASE-2018240.

López-Granados, F., Peña-Barragán, J. M., Jurado-Expósito, M., Francisco-Fernández, M., Cao, R., Alonso-Betanzos, A. and Fontenla-Romero, O. 2008. Multispectral classification of grass weeds and wheat (*Triticum durum*) using linear and nonparametric functional discriminant analysis and neural networks. *Weed Research* 48(1), 28–37. doi:10.1111/j.1365-3180.2008.00598.x.

Lund, E. D., Adamchuk, V. I., Collings, K. L., Drummond, P. E. and Christy, C. D. 2005. Development of soil pH and lime requirement maps using on-the-go soil sensors. *Precision Agriculture* 5, 457–64.

Machwitz, M., Hass, E., Junk, J., Udelhoven, T. and Schlerf, M. 2019. CropGIS–a web application for the spatial and temporal visualization of past, present and future crop biomass development. *Computers and Electronics in Agriculture* 161, 185–93. doi:10.1016/j.compag.2018.04.026.

Mahlein, A.-K. 2016. Plant disease detection by imaging sensors-parallels and specific demands for precision agriculture and plant phenotyping. *Plant Disease* 100(2), 241-51. doi:10.1094/PDIS-03-15-0340-FE.

Mahlein, A.-K., Rumpf, T., Welke, P., Dehne, H.-W., Plümer, L., Steiner, U. and Oerke, E.-C. 2013. Development of spectral indices for detecting and identifying plant diseases. *Remote Sensing of Environment* 128, 21–30. doi:10.1016/j.rse.2012.09.019.

McBratney, A., Whelan, B., Ancev, T. and Bouma, J. 2005. Future directions of precision agriculture. *Precision Agriculture* 6(1), 7–23. doi:10.1007/s11119-005-0681-8.

Megalingam, R. K., Bandhyopadhyay, S., Vivek, G. V. and Rahi, M. J. 2017. Hand gesture based wireless robotic arm control for agricultural applications. In: *IOP Conference Series: Materials Science and Engineering* (vol. 225, no. 1, p. 012204). IOP Publishing.

Michels, G. J., Piccinni, G., Rush, C. M. and Fritts, D. A. 2000. Using infrared transducers to sense greenbug infestation in winter wheat. In: *5th International Conference on Precision Agriculture*. CD-ROM, American Society of Agronomy, Precision Agriculture Center, University of Minnesota, St Paul, MN.

Minami, M., Sakala, M. and Wrightsell, J. 2000. *Using ArcMap*. ESRI, Redlands, CA, p. 528.

Moshou, D., Bravo, C., Oberti, R., West, J. S., Ramon, H., Vougioukas, S. and Bochtis, D. 2011. Intelligent multi-sensor system for the detection and treatment of fungal diseases in arable crops. *Biosystems Engineering* 108(4), 311–21. doi:10.1016/j.biosystemseng.2011.01.003.

Mueller, T. G., Cetin, H., Fleming, R. A., Dillon, C. R., Karathanasis, A. D. and Shearer, S. A. 2005. Erosion probability maps: calibrating precision agriculture data with soil surveys using logistic regression. *Journal of Soil and Water Conservation* 60(6), 462–8.

Mulla, D. J. 2013. Twenty five years of remote sensing in precision agriculture: key advances and remaining knowledge gaps. *Biosystems Engineering* 114(4), 358–71. doi:10.1016/j.biosystemseng.2012.08.009.

Muralidhara, B. L. and Geethanjali, B. 2018. A review on different technologies used in Agriculture. *International Journal of Pure and Applied Mathematics* 119(16), 4117–34.

Muthukannan, K., Latha, P., Selvi, R. P. and Nisha, P. 2015. Classification of diseased plant leaves using neural network algorithms. *ARPN Journal of Engineering and Applied Sciences* 10(4), 1913–9.

Navulur, S. and Prasad, M. G. 2017. Agricultural management through wireless sensors and internet of things. *International Journal of Electrical and Computer Engineering* 7(6), 3492.

Ndikumana, E., Ho Tong Minh, D., Baghdadi, N., Courault, D. and Hossard, L. 2018. Deep recurrent neural network for agricultural classification using multitemporal SAR Sentinel-1 for Camargue, France. *Remote Sensing* 10(8), 1217. doi:10.3390/rs10081217.

Neteler, M., Bowman, M. H., Landa, M. and Metz, M. 2012. GRASS GIS: a multi-purpose open source GIS. *Environmental Modelling and Software* 31, 124–30. doi:10.1016/j.envsoft.2011.11.014.

Nielsen, S. K., Munkholm, L. J., Lamandé, M., Nørremark, M., Edwards, G. T. C. and Green, O. 2018. Seed drill depth control system for precision seeding. *Computers and Electronics in Agriculture* 144, 174–80. doi:10.1016/j.compag.2017.12.008.

Pangels, H., Pilarski, T., Fitzpatrick, K., Happold, M., Ollis, M., Whittaker, W., Stentz, A. and Carnegie-Mellon University. 2002. Agricultural harvester with robotic control. U.S. Patent 6,336,051.

Pantazi, X. E., Moshou, D., Oberti, R., West, J., Mouazen, A. M. and Bochtis, D. 2017. Detection of biotic and abiotic stresses in crops by using hierarchical self organizing classifiers. *Precision Agriculture* 18(3), 383–93. doi:10.1007/s11119-017-9507-8.

Pauli, D., Chapman, S. C., Bart, R., Topp, C. N., Lawrence-Dill, C. J., Poland, J. and Gore, M. A. 2016. The quest for understanding phenotypic variation via integrated approaches in the field environment. *Plant Physiology* 172(2), 622–34.

Pelletier, G. and Upadhyaya, S. K. 1999. Development of a tomato load/yield monitor. *Computers and Electronics in Agriculture* 23(2), 103–17. doi:10.1016/S0168-1699(99)00025-3.

Peternelli, L. A. and de Resende, M. D. V. 2015. Experimental designs for next generation phenotyping. In: *Phenomics*. Springer, Cham, pp. 15–32.

Pollice, A., Jona Lasinio, G., Rossi, R., Amato, M., Kneib, T. and Lang, S. 2019. Bayesian measurement error correction in structured additive distributional regression with an application to the analysis of sensor data on soil-plant variability. *Stochastic Environmental Research and Risk Assessment* 33(3), 747-63. doi:10.1007/s00477-019-01667-1.

Potena, C., Nardi, D. and Pretto, A. 2017. Fast and accurate crop and weed identification with summarized train sets for precision agriculture. In: *International Conference on Intelligent Autonomous Systems*. Springer, Cham, pp. 105–21. doi:10.1007/978-3-319-48036-7_9.

Prabhakar, M., Prasad, Y. G., Thirupathi, M., Sreedevi, G., Dharajothi, B. and Venkateswarlu, B. 2011. Use of ground based hyperspectral remote sensing for detection of stress in cotton caused by leafhopper (Hemiptera: Cicadellidae). *Computers and Electronics in Agriculture* 79(2), 189–98. doi:10.1016/j.compag.2011.09.012.

Priya, S. and Shibasaki, R. 2001. National spatial crop yield simulation using GIS-based crop production model. *Ecological Modelling* 136(2–3), 113–29. doi:10.1016/S0304-3800(00)00364-1.

Puri, V., Nayyar, A. and Raja, L. 2017. Agriculture drones: a modern breakthrough in precision agriculture. *Journal of Statistics and Management Systems* 20(4), 507–18. doi:10.1080/09720510.2017.1395171.

Raes, D., Steduto, P., Hsiao, T. C. and Fereres, E. 2009. AquaCrop the FAO crop model to simulate yield response to water: II. Main algorithms and software description. *Agronomy Journal* 101(3), 438–47. doi:10.2134/agronj2008.0140s.

Rahim, H., Wahab, M. A. M. A., Amin, M. Z. M., Harun, A. and Haimid, M. T. 2018. Perception and acceptance of farmers on precision farming technology in selected granary area. *Economic and Technology Management Review* 13, 97–110.

Rawlins, S. L. 1996. Moving from precision to prescription farming: *The Next Plateau*. In: Robert, P. C., Rust, R. H. and Larson, W. E. (Eds), *Precision Agriculture*, (precisionagricu3), pp. 283–94.

Řezník, T., Lukas, V., Charvát, K., Charvát, K., Křivánek, Z., Kepka, M., Herman, L. and Řezníková, H. 2017. Disaster risk reduction in agriculture through geospatial (Big) data processing. *ISPRS International Journal of Geo-Information* 6(8), 238. doi:10.3390/ijgi6080238.

Rodríguez-Álvarez, M. X., Boer, M. P., van Eeuwijk, F. A. and Eilers, P. H. 2016. Spatial models for field trials. arXiv preprint arXiv:1607.08255.

Ruß, G. 2009. Data mining of agricultural yield data: a comparison of regression models. In: *Industrial Conference on Data Mining*. Springer, Berlin, Heidelberg, pp. 24–37. doi:10.1007/978-3-642-03067-3_3.

Sahoo, R. N., Tomar, R. K., Pandey, S., Sahoo, P. M., Chakraborty, D. and Kalra, N. 2007. Precision farming: concept and application in Indian context. *Indian Journal of Crop Science* 2(1), 25–7.

Sailaja, B., Voleti, S. R., Subrahmanyam, D., Nathawat, M. S. and Rao, N. H. 2019. Regional rice yield estimation by integration of spatial technologies and crop model. *Journal of Remote Sensing and GIS* 4(2), 57–66.

Sarker, A. and Singh, M. 2015. Improving breeding efficiency through application of appropriate experimental designs and analysis models: a case of lentil (*Lens culinaris* Medikus subsp. *culinaris*) yield trials. *Field Crops Research* 179, 26-34.

Sawant, S., Durbha, S. S. and Jagarlapudi, A. 2017. Interoperable agro-meteorological observation and analysis platform for precision agriculture: a case study in citrus crop water requirement estimation. *Computers and Electronics in Agriculture* 138, 175-87.

Sawyer, J. E. 1994. Concepts of variable rate technology with considerations for fertilizer application. *Journal of Production Agriculture* 7(2), 195-201. doi:10.2134/jpa1994.0195.

Scudiero, E., Skaggs, T. H. and Corwin, D. L. 2015. Regional-scale soil salinity assessment using Landsat ETM+ canopy reflectance. *Remote Sensing of Environment* 169, 335-43. doi:10.1016/j.rse.2015.08.026.

Slaughter, D. C., Curley, R. G., Chen, P. and Giles, D. K. 1995. Robotic cultivator. U.S. Patent 5,442,552.

Spoorthi, S., Shadaksharappa, B., Suraj, S. and Manasa, V. K. 2017, February. Freyr drone: Pesticide/fertilizers spraying drone-an agricultural approach. In: *2017 2nd International Conference on Computing and Communications Technologies (ICCCT)*. IEEE, pp. 252-5.

Stamenković, Z., Randjić, S., Santamaria, I., Marković, D., Van Vaerenbergh, S. and Pešović, U. 2018. Decision support system for plant and crop treatment and protection based on wireless sensor networks. In: *2018 41st International Spring Seminar on Electronics Technology (ISSE)*, pp. 1-5). IEEE.

Starr, G. C. 2005. Assessing temporal stability and spatial variability of soil water patterns with implications for precision water management. *Agricultural Water Management* 72(3), 223-43. doi:10.1016/j.agwat.2004.09.020.

Sudduth, K. A., Drummond, S. T. and Kitchen, N. R. 2001. Accuracy issues in electromagnetic induction sensing of soil electrical conductivity for precision agriculture. *Computers and Electronics in Agriculture* 31(3), 239-64. doi:10.1016/S0168-1699(00)00185-X.

Swinton, S. M. and Lowenberg-DeBoer, J. 1998. Evaluating the profitability of site-specific farming. *Journal of Production Agriculture* 11(4), 439-46. doi:10.2134/jpa1998.0439.

Swisher, D. W., Sudduth, K. A. and Borgelt, S. C. 1999. Optical measurement of granular fertilizer flow rates for precision agriculture. ASAE Proc. Paper, (99-3111).

Tamayo, R. A. C., Ibarra, M. L. and Macías, J. A. G. 2010. Better crop management with decision support systems based on wireless sensor networks. In: *2010 7th International Conference on Electrical Engineering Computing Science and Automatic Control*. IEEE, pp. 412-7.

Tan, G. and Shibasaki, R. 2003. Global estimation of crop productivity and the impacts of global warming by GIS and EPIC integration. *Ecological Modelling* 168(3), 357-70. doi:10.1016/S0304-3800(03)00146-7.

Tisne, S., Serrand, Y., Bach, L., Gilbault, E., Ben Ameur, R., Balasse, H., Voisin, R., Bouchez, D., Durand-Tardif, M., Guerche, P., Chareyron, G., Da Rugna, J., Camilleri, C. and Loudet, O. 2013. Phenoscope: an automated large-scale phenotyping platform offering high spatial homogeneity. *The Plant Journal: for Cell and Molecular Biology* 74(3), 534-44. doi:10.1111/tpj.12131.

Umeda, M., Kubota, S. and Iida, M. 1999. Development of "STORK", a watermelon-harvesting robot. *Artificial Life and Robotics* 3(3), 143-7.

van Ittersum, M. K. and Rabbinge, R. 1997. Concepts in production ecology for analysis and quantification of agricultural input-output combinations. *Field Crops Research* 52(3), 197–208.

Wang, Y. Q. 2014. MeteoInfo: GIS software for meteorological data visualization and analysis. *Meteorological Applications* 21(2), 360–8. doi:10.1002/met.1345.

Wang, J. and Xiong, S. 2014. A hybrid forecasting model based on outlier detection and fuzzy time series–a case study on Hainan wind farm of China. *Energy* 76, 526–41. doi:10.1016/j.energy.2014.08.064.

Wang, N., Zhang, N., Dowell, F. E., Sun, Y. and Peterson, D. E. 2001. Design of an optical weed sensor usingplant spectral characteristics. *Transactions of the ASAE* 44(2), 409. doi:10.13031/2013.4673.

Wolfert, S., Ge, L., Verdouw, C. and Bogaardt, M. J. 2017. Big data in smart farming–a review. *Agricultural Systems* 153, 69–80. doi:10.1016/j.agsy.2017.01.023.

Xu, Y. 2016. Envirotyping for deciphering environmental impacts on crop plants. *Theoretical and Applied Genetics* 129(4), 653–73. doi:10.1007/s00122-016-2691-5.

Xue, J. and Su, B., 2017. Significant remote sensing vegetation indices: a review of developments and applications. *Journal of Sensors* 2017, Article ID 1353691, 17p.

Yadeta, K. A. and Thomma, B. P. 2013. The xylem as battleground for plant hosts and vascular wilt pathogens. *Frontiers in Plant Science* 4, 97. doi:10.3389/fpls.2013.00097.

Yang, H. S., Dobermann, A., Lindquist, J. L., Walters, D. T., Arkebauer, T. J. and Cassman, K. G. 2004. Hybrid-maize–a maize simulation model that combines two crop modeling approaches. *Field Crops Research* 87(2–3), 131–54. doi:10.1016/j.fcr.2003.10.003.

Yin, G., Verger, A., Qu, Y., Zhao, W., Xu, B., Zeng, Y., Liu, K., Li, J. and Liu, Q. 2019. Retrieval of high spatiotemporal resolution leaf area index with Gaussian processes, wireless sensor network, and satellite data fusion. *Remote Sensing* 11(3), 244. doi:10.3390/rs11030244.

Zhang, N., Wang, M. and Wang, N. 2002. Precision agriculture–a worldwide overview. *Computers and Electronics in Agriculture* 36(2–3), 113–32. doi:10.1016/S0168-1699(02)00096-0.

Zhang, L., Zhang, L. and Du, B. 2016. Deep learning for remote sensing data: a technical tutorial on the state of the art. *IEEE Geoscience and Remote Sensing Magazine* 4(2), 22–40. doi:10.1109/MGRS.2016.2540798.

Zhang, X., Chen, N., Chen, Z., Wu, L., Li, X., Zhang, L., Di, L., Gong, J. and Li, D. 2018. Geospatial sensor web: a cyber-physical infrastructure for geoscience research and application. *Earth-Science Reviews* 185, 684–703. doi:10.1016/j.earscirev.2018.07.006.

Zhang, Y., Sui, B., Shen, H. and Ouyang, L. 2019. Mapping stocks of soil total nitrogen using remote sensing data: a comparison of random forest models with different predictors. *I* 160, 23–30. doi:10.1016/j.compag.2019.03.015.

Part 2

Case studies

Chapter 7

Decision support systems (DSS) for better fertiliser management

Dhahi Al-Shammari, Patrick Filippi, James P. Moloney, Niranjan S. Wimalathunge, Brett M. Whelan and Thomas F. A. Bishop, The University of Sydney, Australia

1 Introduction

World population growth has almost doubled since the year 2000, and the projections suggest that the number will reach 9.5 billion by 2050 (Henchion et al., 2017). This places more pressure on agricultural systems to meet the increasing demand for food with constant improvements in agricultural productivity and maintenance of soil fertility (Rasool et al., 2007). Soil fertility can be maintained or enhanced by the use of fertilisers which have been responsible for increasing productivity by 40–60% worldwide (Roberts, 2009). Fertilisers contain nutrients that promote the growth of plants and increase productivity (McDonald, 1989; Haynes and Naidu, 1998). Plants consume large amounts

http://dx.doi.org/10.19103/AS.2020.0069.13

of macronutrients such as nitrogen (essential for plant growth generally), phosphorus (essential for respiration and photosynthesis) and potassium (assists formation of protein and to move carbohydrates, water and nutrients within plant tissues) (Hodges, 2010; Hazelton and Murphy, 2016). A deficiency of nutrients can adversely affect plants. Examples include a reduction in plant size and growth for P deficiency (Sharma et al., 2013) and poor development of root systems and stunted structure in the case of K deficiency (Hodges, 2010). N deficiency can cause an exponential decline in leaf formation and expansion rates, and eventually lead to a reduction in crop productivity (Broadley et al., 2000; Živčák et al., 2014). In contrast, the excessive use of fertilisers has adverse consequences economically and environmentally (Oliver et al., 2013). For instance, excess amounts of P and K can lead to leaching and runoff (Gourley et al., 2012), and excess N may encourage plants to continue in their vegetative stages longer, which in turn leads to delays in fruiting and flowering, causing a lowering of potential yields (Hodges, 2010).

Today, nitrogen consumption has risen to be about 2.5 times more than P fertiliser and about 4 times higher than K worldwide (Gellings and Parmenter, 2016). In Australia, it has been reported that N use in particular recorded the highest among the other fertilisers to reach about 159 million tonnes between 2010 and 2014 (Angus and Grace, 2017). However, N can be lost easily from the soil system by leaching, runoff, volatilisation, denitrification or by plant removal (Hodges, 2010), often resulting in a high volume of inputs with low potential returns (Angus, 2001). In New South Wales, Australia, the cotton industry recorded losses of about 10 million dollars annually in cotton fields due to over-applied fertilisers, which can be as much as 40 kg of N/ha (Rochester, 2008). With the rise of fertiliser prices, there is a need to review previous and current tools and continue to examine new approaches for better nitrogen-use efficiency (NUE), and subsequently achieve greater outputs using fewer inputs. NUE for the purpose of this chapter is defined as the calculated ratio between the applied nitrogen and that removed from the cropping system at harvest (Brentrup and Pallière, 2010).

Variability of soil nutrients within a field is common, with some fields being able to produce high yields in some parts without any nutrient supplements, whereas other areas might not be able to meet their yield potential without added fertilisers (Maleki et al., 2008). Therefore, applications of fertilisers should be matched with the yield potential to minimise their inputs while maximising the yield (Hoang and Coelli, 2011). This can be achieved through what is known in precision agriculture (PA) as variable rate applications of fertilisers. PA technologies offer promising tools for improved agricultural management which considers the variability within the field, and not treating it as a uniform entity (Cook and Bramley, 1998; Whelan, 2007; Chlingaryan et al., 2018). For example, variable rate technologies (VRT), which are tools for PA, are usually used to apply the right amount of nutrients at the right place based on spatial variability determined by decision support system (DSS) tools. The definition of DSS varies in the literature, but it is generally agreed that the objective of DSS is to provide the required support to improve decision-making (Arnott and Pervan, 2014). Many DSS approaches have been presented to guide variable rate applications, including the use of crop models and soil tests. This could simultaneously reduce the impact of fertilisers on the environment and help to cost-effectively manage inputs for better production results (Oliver et al., 2013).

These DSS differ in their complexity and input data required to produce a decision. Some of the approaches used by DSS only use a single predictor to forecast potential yield, which is then used to determine fertiliser applications. For example, a simple linear relationship between wheat and water was modelled by French and Schultz (1984a,b). This empirical model only provides the 'best attainable yield' benchmark based on the amount of rain and expected evaporation. The French and Schultz (F&S) (1984a,b) approach does not account for stored soil water which may contribute to crop water use. This model was modified to account for stored soil water by Robertson and Kirkegaard (2005). With recent advancements in modelling and geospatial technologies, it is possible to predict soil-stored water in near real time for the whole profile (Wimalathunge and Bishop, 2019). More complex approaches for forecasting potential yield involve using machine-learning-based approaches which exploit a data cube of soil, weather and remote sensing for predictive purposes (Whitbread and Hancock, 2008; Filippi et al., 2019).

The objective of this chapter is to review some of the approaches used by DSS to determine fertiliser-application decisions. Furthermore, this chapter includes two case studies to estimate season-specific nitrogen requirements of wheat crops at a within-field scale in Australia. These models forecast yield in two key periods of the season in which farmers make decisions for fertiliser applications - pre-sowing and mid-season. The first approach is based on collating various spatio-temporal variables that affect plant growth to forecast the expected yield using the same approach as Filippi et al. (2019). The second model is based on forecasting yield using a soil water balance model (Li et al., 2019). Both of these approaches subsequently use the forecasted yield to calculate the amount of N required for each spatial location with an N-replacement formula, which is based on the forecasted yield and protein content.

2 Direct methods for determining crop nitrogen requirements for decision support

Potential crop yield (expected yield) is the amount of yield that is expected to be achieved per unit of land at the end of the growing season. Crop yield can be forecasted based on the interaction between plants and environmental and soil factors as well as management practices with the absence of all or some biotic and abiotic stresses (Long et al., 2015). As the season progresses, potential yield can vary due to the effect of limiting factors such as temperature and soil moisture and other stresses. Therefore, management decisions need to be updated within the season, particularly those decisions related to fertiliser applications to maximise nutrient use efficiency (NUE). NUE is crucial in modern agriculture to reduce environmental risk and increase profitability (Fageria, 2016). Farmers and advisors use a range of tools to estimate N requirements to reach a target yield based on the limiting factors available such as soil moisture and nutrients (Dunsford et al., 2015).

In this section we will discuss some methods and approaches which offer potential for delineating and monitoring within-field variation of nutrient (directly) as well as approaches based on yield potential. The following two sections discuss simple methods for direct decision-making on fertiliser application. The chapter then discusses indirect approaches which are based on forecasting crop yield and which can be used for fertiliser applications.

2.1 Traditional approach – expert knowledge aided by soil tests

Soil testing has a long history in agriculture to guide nutrient applications (Khan et al., 2001). Soil testing provides information about levels of soil fertility, type and amount of fertiliser required, soil constraints (e.g. acidity, salinity, sodicity), and assessment of land capabilities. Specifically, fertiliser recommendations based on soil testing depends on the accuracy of collected samples (Cope et al., 1981). Appropriate sampling schemes are required when collecting soil samples due to the high spatial variability of nutrients across a field or paddock. Soil nutrients can vary considerably in space and time, meaning that multiple samples must be taken on a regular basis (Schwenke et al., 2019). Moreover, in terms of nitrogen the amount of soil N usually changes over time due to mineralisation, immobilisation, denitrification and leaching. Therefore, soil testing is generally required at the beginning of the season.

The most traditional, and simplest method is the rule-of-thumb approach (Creelman et al., 2015), which has been adapted across Australia to help farmers in their decisions for fertiliser applications (Angus, 2008; Lawrence et al., 2000). These rule-of-thumb approaches are based on the experience of local growers and advisors to give a rough estimation for N mineralisation (Schwenke et al., 2019). For instance, farmers who want to rely on their experience use soil testing results to estimate the N available in the soil. This soil testing is basically performed by taking representative samples from a paddock/field during a certain period. Pre-sowing and in-season mineralisation are considered by most advisors when making fertiliser recommendations. Based on these tests and the targeted yield, advisors can calculate the required N. Farmers and advisors can then update their decisions with the progress of the season depending on the climate conditions. However, these rules-of-thumb provide limited explanation or understanding without taking into account other factors that might affect N mineralisation (Lawrence et al., 2000; Hayman and Alston, 1999). The above-mentioned methods are based on soil samples which are typically bulked over a paddock and this basically provides a required N-rate per paddock. This approach typically does not account for variability within a paddock, which is essential for ensuring the efficiency and profitability of applied fertiliser. Therefore, more sophisticated tools and information are needed for better decision-making for fertiliser applications that meet both the profitability and environmental sustainability targets (Gourley et al., 2007).

2.2 Decisions direct from previous yield maps (nutrient replacement)

The maintenance or increase in soil fertility is necessary to achieve optimum production through better fertiliser decisions. Yield maps are often used to provide fertiliser recommendations based on the amount of nutrients removed by crops. Yield maps often delineate areas within a field which vary in their yield potential; thus, different fertiliser applications could be applied depending on the yield variability (Dey, 2015). For example, a formula based on the maintenance of nutrients was used to calculate the amount of nitrogen removed with every kilogram of wheat crop using a yield map (Whelan and Taylor, 2013). The amount of nitrogen removed by the previous crop was calculated, and an average protein content of 14% was assumed to produce the nitrogen prescription map (Eq. 1).

$$N^{removed}(kg\,N/ha) = yield(kg/ha) * protein(\%) * 0.00175 \quad (1)$$

The resulting N-removed maps can be used to calculate the N required across a field to apply different levels of fertilisers. The drawback of relying on this approach is that it assumes that the attainable yield is not changing in the subsequent seasons and this is not practical, especially when considering the variation in weather conditions from season to season and from one region to another.

3 Indirect methods for determining crop nitrogen requirements for decision support: simulation models

Simulation models are computer tools that often use different inputs to predict/forecast yield (Dintwa et al., 2004). Inputs such as soil characteristics, weather data, sowing dates and so on are used to simulate plant growth which is ultimately used to predict/forecast the final yield (Landau et al., 1998). These models can simulate nutrient dynamics in the soil, as well as the effect of climatic conditions on plant growth (Nendel and Kersebaum, 2004). Many crop-simulation models have been developed to simulate plant growth that take into account the importance of nutrients for plant growth (Van Duivenbooden et al., 1995). For example, The Agricultural Production Systems sIMulator (APSIM) and the maize calculator are discussed to provide examples of simulation models for management strategies.

APSIM consists of soil and plant management modules which simulate the dynamics of carbon, water and nitrogen within the soil system (Probert et al., 1998). More specifically, the APSIM-SOILN module simulates N mineralisation to predict the N supply from the previous crop roots and residues, and soil (Keating et al., 2003). APSIM has been tested in many farming systems (Robertson et al., 2000). It has been reported that APSIM scored high satisfaction among the agricultural community in parts of Australia; however, there are calls for more validation in other areas (Carberry et al., 2002). Even though APSIM has helped producers to adjust their decisions about N-management practices to meet their target yield, it showed lack of capabilities and performance in dealing with issues such as pests, frost and waterlogging (Carberry et al., 2002). Whelan and Taylor (2013) reported that APSIM is time-consuming and requires many inputs to be provided for decision-making which is not suitable for most users.

Yield Prophet (YP) is a user-friendly tool adapted from APSIM and assists growers with their management decisions as the season progresses. YP provides reports for the potential yield anytime during the growing season based on the simple inputs provided by farmers (Hunt et al., 2006). However, similar to APSIM, YP does not provide any direct advice or recommendation, and does not take into account disease, pest, weed competition, extreme events such as extreme weather, flood or soil constraints (Donatelli et al., 2017). Generally, the APSIM model provides yield forecast at the plot or field-scale which can be used for uniform management. In this case, within-field variability cannot be captured and treated separately. However, in a study conducted by Wong et al. (2005), the APSIM was used in conjunction with soil property maps to simulate yield and nitrate leaching within field at 5 m resolution (Wong et al., 2005). The conclusion of this study was that within-field variability can be determined separately to decrease nitrate leaching by splitting the N applications. This gives an indication that simulation models can be used as inputs for PA to mitigate the risk of nutrient leaching under different climate and soil variation.

The maize calculator is a DSS tool that was developed to optimise the use of fertiliser applications for maize crops in New Zealand (Stone and Hochman, 2004). It was developed in response to the producers' demand for an effective method which allows them to optimise fertiliser applications for maize crops (Stone and Hochman, 2004). This DSS utilises the PARJIB model, which simulates the response of the crop, and therefore yield, to the nutrient supply. The PARJIB model uses soil chemical tests as relative indicators, not absolute indicators, as well as the weather conditions (Reid et al., 2002). This calculator was used intensively by farmers in its first three years with a root-mean-square deviation (RMSD) of 0.7 t/ha (Stone and Hochman, 2004). However, it ultimately failed as a fertiliser DSS despite being supported by a user-friendly interface. There was no evidence that farmers changed their practices for fertiliser applications using the maize calculator and this was reported by the annual surveys of fertiliser use by maize farmers. The surveys showed that the applications of phosphorus and potassium on maize crop did not change. The failure of the Maize Calculator to change farmers' practices was attributed partly to the fact that the decision of fertiliser applications was affected by pre-knowledge of the requirements of subsequent pasture or crop (Stone and Hochman, 2004).

4 Indirect methods for determining crop nitrogen requirements for decision support: yield forecasts using data-driven approaches

Agricultural data comes in many different forms and types - such as remotely sensed imagery, proximal sensing data, weather data, and experimental data - and the quantity of this data is constantly increasing. This data comes in different formats including text, raster imagery, multimedia, geospatial data and so on (Ngo et al., 2018).

Yield forecasting using a data-driven approach is not a new concept in agriculture, but the data used to forecast and predict yield have developed and expanded as a result of the continuous development of modern agriculture and the emergence of digital tools which have opened new opportunities in extracting valuable information (Ma et al., 2014). Approaches are now combining large amounts of data from different spatial and temporal datasets from many fields (Kamilaris et al., 2017), into one model, and machine learning can then be used to forecast yield (Filippi et al., 2019).

An example of forecasting crop yield is a data-driven model which used a data cube of covariates in space and time (space-time cube) which was developed by Filippi et al. (2019). This approach is a data-driven model where the input data can vary in spatial and temporal resolutions. This data-driven approach can forecast yield at field and sub-field (pixel-level) resolution and at different time points in the season.

Filippi et al. (2019) tested the model on three crop types (wheat, canola and barley) from three successive seasons located in the wheatbelt of Western Australia (WA). Using grower collected data in combination with the publicly available data, they could forecast crop yield in pre-, within- and late-season with Lin's concordance correlation coefficient (LCCC) ranging from 0.89 to 0.92

at the field resolution. Crop yield forecasts can help farmers and advisors to forecast the outputs which can be the basis of their management practices (Basso et al., 2013). Filippi et al. (2019) suggested that the tested model is a promising tool to support decision-making related to fertilisers in terms of variable rate application which is usually influenced by in-season weather. The flexibility of this model in terms of number of data inputs and forecasting yield at early stages of crop development make it interesting to test this model for PA applications to provide nutrient prescription maps. This model can be used in conjunction with the nitrogen replacement equation (Section 2.2) to calculate areas with different potential yield and from this helps the growers to update their decisions for fertiliser applications.

5 Indirect methods for determining crop nitrogen requirements for decision support: yield forecasts based on water supply

Many studies have been conducted to explain the relationship between yield and water use or rainfall, which is usually a limiting factor (Ali and Talukder, 2008). French and Schultz (1984a) have modelled the relationship between yield potential and water. This method is widely used to forecast grain yield in Australia due to its simplicity. In their method, they proposed WUE as a benchmark for wheat performance in South Australia (SA). In this sense, WUE is defined as the ability of plants to produce plant grain or biomass from the stored water in the soil and rainfall during the growing season. They used 61 sites to draw a linear relationship between potential yield and each water unit (French and Schultz, 1984a; Whitbread and Hancock, 2008). They concluded that soil-water content at sowing and sowing time were the most important factors that affect yield (French and Schultz, 1984a). In their second study, they emphasised that including evaporation and transpiration is needed to indicate the evaporation lost directly to the atmosphere and thus the available water for transpiration is used to calculate the potential yield (French and Schultz, 1984b). Robertson and Kirkegaard (2005) modified F&S equation by including the stored soil water to the equation which contributes to the crop water use (Eq. 2).

$$\text{Yield(kg/ha)} = \text{crop water supply} - \text{evaporation} \times \text{best yield benchmark*} \quad (2)$$

Where crop water supply is an estimation of water available to the crop, evaporation is the amount of evaporation from soil which depends on the amount of in-crop rainfall and soil type, and best yield benchmark is crop specific. For example, best yield benchmark for wheat is about 20 kg of wheat per hectare for every millimetre of water used by the plant.

As long as the correlation between soil moisture and yield is strong, this correlation can be very valuable for in-season decision-making. In-season soil moisture nowcasts and forecast can be done using more developed methods which recently emerged as a result of development in remote sensing and modelling. In this chapter, the modified French and Schultz equation was used in Case study 2, but with novel soil-moisture nowcasting model that predicts soil moisture in space and time (Wimalathunge and Bishop, 2019). In this model, soil moisture can be nowcasted for any stage of plant growth which can be related to forecasted yield. The forecasted yield can then be used in the nutrient replacement equation (Eq. 6) to provide a nutrient prescription map.

6 Decision support in action: case studies

Nutrient requirement maps can be generated through the use of various yield-forecasting approaches. These models can help indicate areas where to apply the appropriate amount of fertiliser to match nutrients with the targeted yield. These models vary in their complexity, and the data used to predict or forecast crop yield can vary considerably. Here, two approaches that can forecast crop yields at a within-field scale are discussed as tools to help in nitrogen decision-making for a specific crop (wheat). The first case study has been conducted using a data-driven model based on Filippi et al. (2019) to generate N-prescription maps. This model uses a modest amount of data which rely on publicly available and private data. The data-driven model requires access to the yield data for multiple years and paddocks to train it on previous yield maps to produce predicted yield maps. These maps can be updated within the season as the weather and other factors change, and thus fertiliser maps can be updated accordingly. On the other hand, the second model is a simple yield prediction model based on soil moisture data only. This model based on the work of Wimalathunge and Bishop (2019) requires only the modelling of soil moisture to be used to generate yield potential maps. These maps also can be updated within the season as the conditions change. Both of these produces can generate N-prescription maps at the within-field scale.

The study site is located in the wheatbelt of Western Australia (WA) (Fig. 1). Grain crops and livestock are the major agricultural industries in this region. The area is largely composed of sandy soil with notable amounts of gravel, and one of the most widespread soil types are Sodosols (Filippi et al., 2019). The majority of broad-acre systems rely on rain in their production; therefore, farm businesses are very sensitive to the year to year seasonal variation (Ellis and Albrecht, 2017). The area receives an average annual rainfall of 476 mm, and the study site received an average rainfall of 455 mm for the study year (2014). The paddock has a very gentle slope ranging from 0.19 to 4.7% and an elevation ranging from 94 to 134 m.

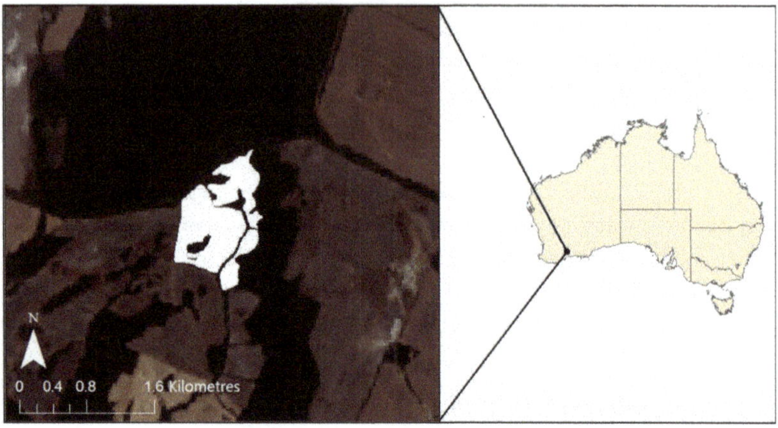

Figure 1 Location of study site in Western Australia.

A wheat crop (season 2014) was selected to be the experimental crop because it is the major crop in WA, comprising of 65% of annual grain production (Llewellyn and Powles, 2001). It is reported that only 40% of N is recovered by crop in dryland wheat farms, and so more research is encouraged to be conducted in these areas (Angus and Grace, 2017). In WA, the annual losses of N fertiliser are reported to be about 29% due to ammonia volatilisation and 14–72 kg N/ha due to leaching during the growing season (Barton et al., 2016).

7 Case study 1: nitrogen fertiliser applications using a data-driven approach

7.1 Datasets

This case study used freely available data as well as on-farm datasets (Table 1). These spatio-temporal data were assembled into a space-time cube (STC) in preparation for modelling. Two separate STCs were created for the generation of two forecasted yield maps for April (pre-sowing) and July (mid-season).

Forecasted wheat yield maps covering 85.92 ha at 10-metre resolution were generated for April and July time periods using a random forest model fitted to the STC. These two time periods were selected on the fact that some important management decisions are made during these periods related to fertiliser applications (Fig. 2).

7.2 Calculation of nitrogen requirements

The forecasted yield map in April was used to calculate the N requirements. This map was then updated for N requirements using the yield forecasted

Table 1 Data description used in this study

Data description	Resolution (m)	Source	April model	July model	Availability
Yield	10	On farm	✓	✓	Private
Soil EC_a (Shallow at 50 cm and Deep at 1.5 m) from dual EM	10	On farm	✓	✓	Private
Gamma radiometrics					
Potassium (K)	10	On farm	✓	✓	Private
Thorium (Th)	10	On farm	✓	✓	Private
Uranium (U)	10	On farm	✓	✓	Private
Clay content (%)	10	On farm	✓	✓	Private
Sand content (%)	10	On farm	✓	✓	Private
The enhanced vegetation index (EVI)-MODIS	250	NASA	✗	✓	Public
Received Rainfall (January 1st to March 31st)	~ 5000	BOM	✓	✗	Public
Forecasted Rainfall (April 1st to June 30th)	Regional	BOM	✓	✗	Public
Received Rainfall (April 1st to June 30th)	~ 5000	BOM	✗	✓	Public
Forecasted Rainfall (July 1st to August 31st)	~ 5000	BOM	✗	✓	Public

map for July. To calculate the correct amount of required N precisely, we need to know the amount of N which already exists in the soil and this can help to reduce the inputs and prevent fertiliser wastage. Exist N (N^{soil}) can be calculated relying on soil test results. Therefore, soil volume and weight at the top 10 cm were calculated using Eqs. (3) and (4), respectively. Then, N^{soil} was calculated using Eq. (5). Note that the value (10 mg/kg) in Eq. (4) is the nitrate at the top 10 cm which was taken from the soil test and was divided by 1000 to convert it to kg per hectare.

$$\text{Soil volume per hectare}(10\,\text{cm}) = 10000\,\text{m}^2 * 0.1 = 1000\,\text{m}^3/\text{ha} \quad (3)$$

$$\text{Soil weight per hectare} = 1000\,\text{m}^3 * 1.48(\text{g/cm}^3) = 1480\,\text{t/ha} \quad (4)$$

$$N^{soil}(10\,\text{cm}) = 1480\,\text{t/ha} * 10\,\text{mg/kg}/1000 = 14.8\,\text{kg/ha} \quad (5)$$

After determining the amount of existing nitrate in the soil, the calculation of N could be performed for April and July using the following equations:

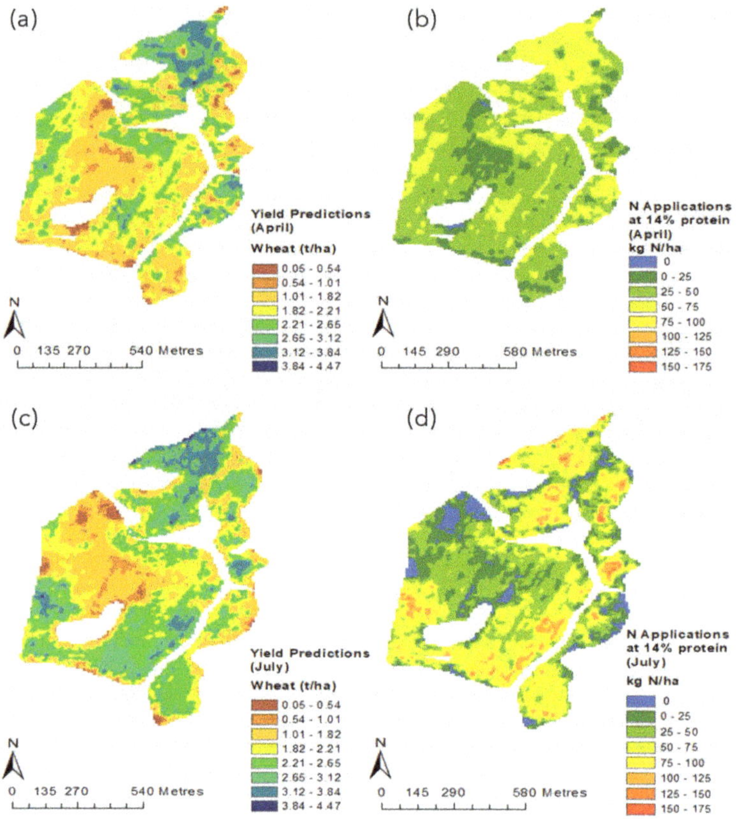

Figure 2 Yield forecast and nitrogen maps. (a) Yield forecast map for April. (b) Nitrogen applications map for April. (c) Yield forecast map for July. (d) Nitrogen applications map for July.

A **Nitrogen requirements based on April forecast:**

Nitrogen requirements were calculated based on the April-forecasted yield map using the following steps:

1 The required N (kg/ha) application to be applied in April was calculated using the following equation:

$$N^{required}(kg/ha) = \text{yield forecast}(kg/ha) * 1.75 * \text{protein}(14\%) * 2 \quad (6)$$

2 Then the N in the soil was subtracted from the results of Eq. (5) to calculate the N application. N^{soil} is the amount of nitrogen which already exists in the soil.

$$\text{Total N application}(kg/ha) = N^{required}(kg/ha) - N^{soil}(kg/ha) \quad (7)$$

3 Finally, the total N application was split into two applications where one has to be applied now and another in July (50/50 split):

$$N\,application\,April(kg/ha) = Total\,N\,application/2 \tag{8}$$

B **Nitrogen requirements can be updated according to the forecast in July as follows:**

1

$$N^{required}(kg/ha) = yield\,forecast(kg/ha) * 1.75 * protein(14\%) * 2 \tag{9}$$

2 Same as Eq. (7), N^{soil} needs to be subtracted from $N^{required}$ for July to calculate the second (topdressing) application.

$$N\,application\,July(kg/ha) = N^{required}(kg/ha) - N\,application\,April(kg/ha) \tag{10}$$

Results showed how N applications can be generated and updated using the resulted yield maps (Fig. 2). In April (Fig. 2a and c), yield was forecasted to be ranging from low at about 0.05 t/ha to high at about 3.9 t/ha. However, these forecasts changed in July to be ranging from 0.1 t/ha to 4.4 t/ha. This is due to more knowledge about in-season rainfall and crop performance from EVI. For example, areas which were forecasted to be moderate to high yield areas in April (> 2 t/ha) were forecasted to produce similar or higher according to the July forecasts (Fig. 2c). Other areas were forecasted to yield less as the conditions changed, with some areas experiencing a huge reduction. Some areas that were forecasted to yield low in April were also forecasted to yield low production in July.

As a result, to get the best outcome regarding fertiliser investment, fertiliser applications should be calculated at pre-sowing stage and then updated later (within-season). For instance, some areas are required to be applied with N in both April and July to get the targeted yield, while other areas (blue patches in Fig. 2b and d) require no N applications due to a very low-crop potential. Some areas (blue-coloured in Fig. 2b) were forecasted to produce between 0.05 t/ha and 0.54 t/ha, meaning that the amount of N in the soil would be sufficient to produce the expected yield. Thus, applied N should be 0 kg N/ha because applying any additional N to the soil would be wasted. However, other areas of the field were forecasted to be low-yielding at the start of the season, but the updated forecast in July showed that these patches were anticipated to increase in yield. Therefore, N fertiliser should be applied in these patches to meet the forecasted yield.

The histograms in Fig. 3 represent the forecasted yield (upper) and the N applications for April and July in Fig. 2. Negative values in both N histograms indicate the above threshold N which are represented as blue areas (Fig. 2b and d) and those were set to be 0 values to indicate that N should not be applied in those areas. Negative values of N in April histogram mean that soil could supply plants with the necessary N and there is no need to apply any extra N. However, negative values in July indicate the amount of N applied in access of what is needed to meet the potential yield. As a result, N applications in July should be adjusted depending on the locations of changes to match the reduction in forecasted yield with the amount of N needed. According to the statistics in this study, over-application can cost about $419.68 in an area of 5.27 ha if fertilisers are applied in July. Depending on how much fertilisers can be wasted per point location, some areas can cost more than others. A point worth mentioning is that some areas which receive high N applications in April and forecasted to produce very low yield in July are considered inevitable N-wastage (over application) because they already received high N applications in April. Over-application problem can be mitigated by increasing the number of forecasting yield models. This allows in adding more splits with less N rate applications.

This model suggests that in-season weather changes can be tracked, and management decisions can be updated regarding fertiliser applications. The advantage of this model is that it was based on an empirical approach to

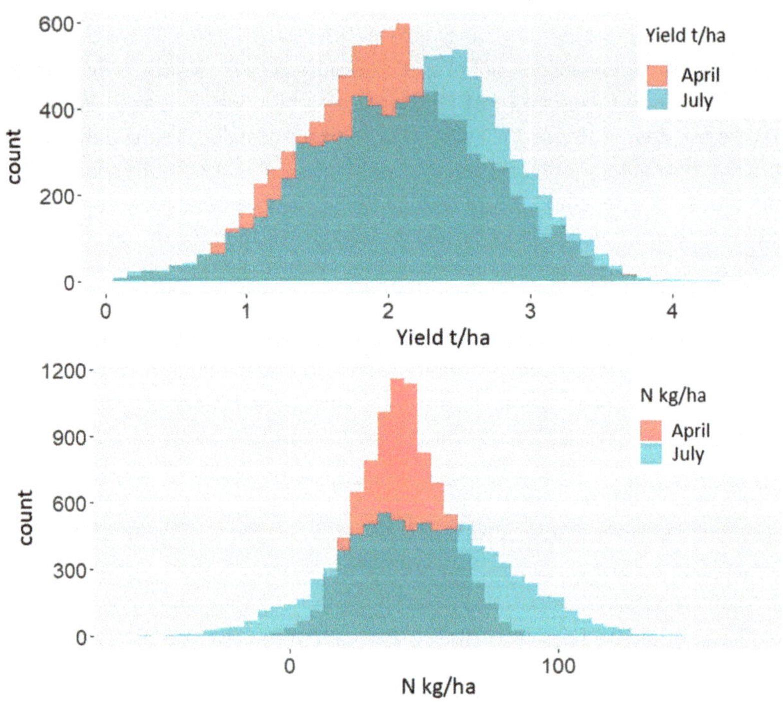

Figure 3 The distribution of forecasted yield in April and July (upper) and the distribution of N in April and July (lower).

forecast crop yield and deriving N requirements based on the resulting maps. Future work should use the forecasted yield to implement field experiments to test the approach over a number of seasons.

8 Case study 2: nitrogen fertiliser decision-making based on soil moisture predictions

Water is the most limiting factor in the rain-fed agricultural areas of WA. Thus, water availability must be taken into account by farmers to calculate yield potential and eventually assess different management decisions such as fertiliser applications. Soil moisture maps were used to calculate yield potential. The resulted yield maps were used to produce N maps in April and July.

8.1 Production of yield potential maps

As mentioned, soil moisture maps were the basis for the calculation of the forecasted yield maps. This was performed by two steps. The first step is to calculate the available soil moisture within the soil profile. We used a soil-water balance model from recent work conducted by Wimalathunge and Bishop (2019) to determine the total available soil moisture to 100 cm in the profile. The soil profile was divided to five layers at 0–5 cm, 5–15 cm, 15–30 cm, 30–60 cm and 60–100 cm. Each layer fills with rainfall, has losses to evapotranspiration (that proportionally decay with depth) and passes down excess flux to the bucket below. Hydraulic properties, including storage capacity and saturated hydraulic conductivity for each layer are derived from pedotransfer functions (PTFs) performed on data gathered from the soil landscapes grid of Australia (SLGA), which provides sand, silt, clay and bulk density estimates on a 100 m grid (Grundy et al., 2015). A total of five soil moisture maps were derived for the five depths mentioned above for the April and July models with a spatial resolution of 100 m. Each group of five soil moisture maps were then used to derive the total soil moisture for the soil profile for April and July.

The produced soil moisture maps in April ranged from 79 mm to 111 mm, which then increased from 111 mm to 146 mm in July due to rainfall (Fig. 4). The increase in soil moisture from April to July in the soil profile generally means that crop yield is expected to increase as well. However, some areas where yield is expected to decrease could be due to the interaction with other variables, such as soil texture or drainage (Passioura, 2008).

The second step requires the use of a simple equation to create the potential forecasted yield maps (Eq. 11). Soil moisture data were included in this equation which replaced the (water supply - evaporation) in the F&S equation. The advantage of this model is that it considers the precipitation,

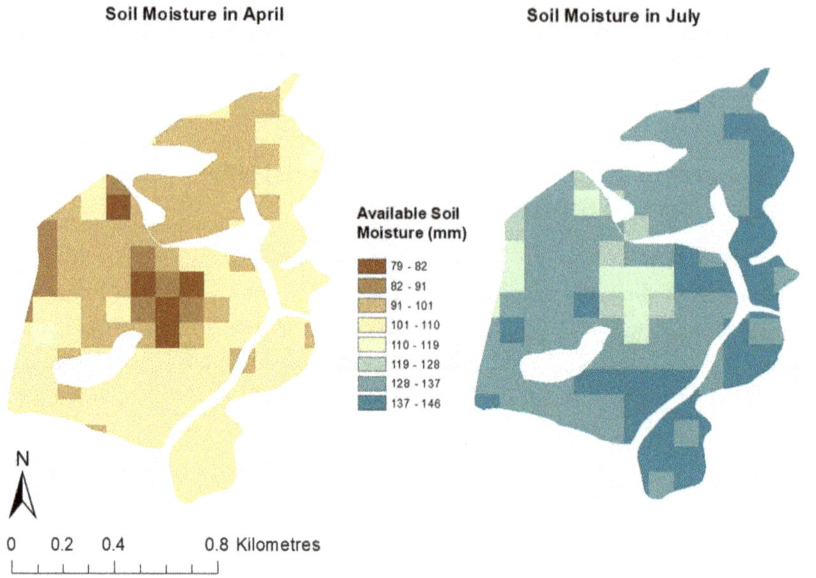

Figure 4 Soil moisture maps for April (left), and July (right).

evapotranspiration, deep drainage and the run off from the soil surface (Wimalathunge and Bishop, 2019).

$$\text{Yield(kg/ha)} = \text{soil moisture(mm)} * 20\text{kg/ha/mm} \tag{11}$$

Where 20 is derived from a water-use efficiency model developed by French and Schultz (1984a) which determined a best-yield benchmark for wheat crops where every 20 kg/ha of wheat is produced by 1 mm of water. It is worth to mention that the soil moisture used in this equation is based on the nowcasts predictions. That means forecasted yield might change depending on the chances of rainfall later in the season and this is the only limitation of this approach.

The forecasted yield maps for April and July are shown in Fig. 5a and c, respectively. The lowest yield forecasted in April is 1.58 t/ha with the highest at 2.24 t/ha. In July, the yield is forecasted to increase (due to increased soil moisture stored in the profile), from 2.2 to 2.9 t/h. As mentioned earlier, these maps give simple guidelines to the Western Australia farmers to calculate the yield potential, assuming there is no limitations such as soil constraints and/or adverse weather conditions (Passioura, 2008).

8.2 Nitrogen applications

Nitrogen requirements were calculated by the same approach described in case study 1 using Eqs. (3)–(10), respectively. According to the N application

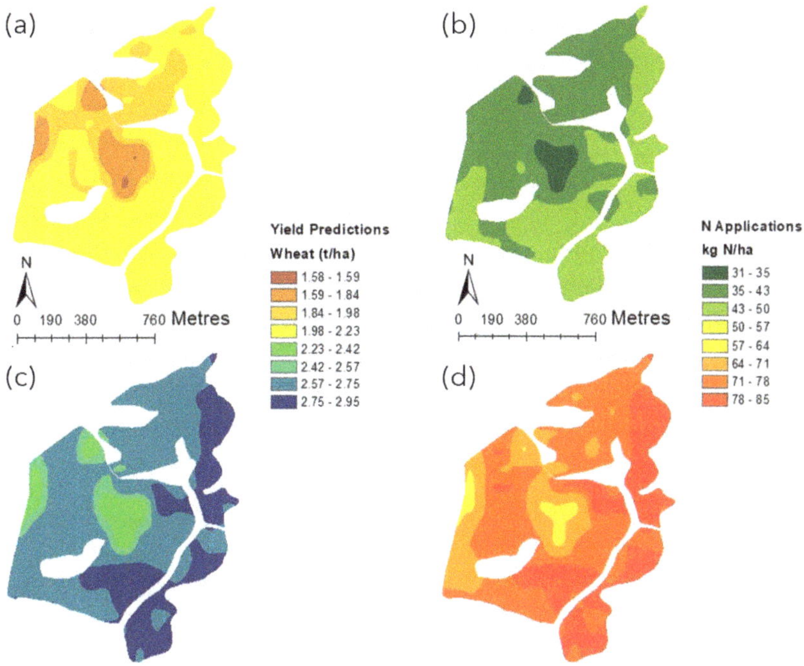

Figure 5 Yield forecast and nitrogen maps. (a) Yield forecast map for April. (b) Nitrogen applications map for April. (c) Yield forecast map for July. (d) Nitrogen applications map for July.

maps, N application in both April and July are profitable and there is no wastage of N throughout the growing season, assuming that there are no environmental and disease constraints. Therefore, the first application was forecasted to be between 31 and 48 kg N/ha to meet the yield goal. In the second application, the demand for N was forecasted to be almost double due to the increase of soil moisture and thus the yield potential. The limitations of this model are that it was based on single variable (soil moisture) to forecast crop yield but in situations without historical yield data this is the best approach.

9 Comparing the two approaches

The two approaches presented were built for the same season and same paddock, but it is necessary to compare between the two approaches to show how the complexity of the model can influence results. Table 2 shows how applications can vary using each model. For example, variation in the forecasted yield was evident between the data-driven and F&S models. Although the mean yield did not change much in the two approaches, the minimum, maximum and standard deviation changed significantly. Much of this is due to the differing spatial resolution of the approaches. Minimum yield using data-driven model could be as low as 0.06 kg/ha in April, that is, much less than the F&S approach which forecasted a minimum yield of 1.58 kg/ha during the same time. The minimum forecasted yield did not change much from April in the data-driven approach, but we can see that F&S forecasts increased considerably to reach a minimum of 2.23 kg/ha in July. This is mainly due to the fact that F&S utilises the most important variable in plant growth which is soil moisture, so the forecasted yield would be dependent on the amount of forecasted soil moisture in the soil. Similarly, the maximum forecasted yield using the data-driven model changed from 3.93 to 4.47 kg/ha for April and July, respectively. However, in the F&S approach there was higher change in the forecasted mean yield compared to the data-driven approach.

The resulted yield forecast maps had a large effect on the N requirements. For example, mean N applications did not vary much in space and time when relying on the data-driven model to forecast yield with an average ranging from 41.60 to 48.51 kg N/ha. However, N applications in July were forecasted to be almost double (74.30 kg N/ha) the N applications of April (41.72 kg N/ha) using the F&S model. Minimum N applications needed in some areas of the paddock were forecasted to be 0 kg N/ha relying on data-driven results. However, the minimum amount of N required was forecasted to be 31.42 kg N/ha using the F&S model. This is again due to yield forecasts which were based on accumulated soil moisture which was forecasted to be higher with progress of the growing season.

Site-specific technologies such as variable rate fertiliser technology and on-the-go yield monitors help farmers and advisors to better manage site-specific land resources (Kachanoski et al., 1996). The main component of site-specific systems is the generation of management maps which shows within field variability, which in turn requires different inputs at different locations of targeted areas. The advancement of agricultural modelling provides new capabilities for collection and data analysis which are becoming more available, and at increasingly higher spatial and temporal resolutions. This can provide better forecasts than the traditional methods, such as soil testing and other rule-of-thumb methods, which rely on basic analysis and experience of individuals.

The two yield-forecasted-based DSSs showed that decisions related to fertiliser applications can be done with diverse data inputs which are largely freely available. For example, F&S-based DSS requires only freely available (public) spatiotemporal data which are mentioned in Table 1. Furthermore, installation of soil moisture probes could help to reduce data inputs by using only soil moisture data for forecasting models. However, if private data such

Table 2 Comparison of performance between two approaches as tools for nitrogen decision-making. Data-driven approach is the empirical space time cube in (Case study 1)

Approach	Period	Mean Y	Min Y	Max Y	SD Y	Mean N	Min N	Max N	SD N
Data-driven	April	1.99	0.06	3.93	0.64	41.60	0	88.94	15.66
	July	2.14	0.11	4.47	0.68	48.51	0	169.02	31.28
F&S	April	2.00	1.58	2.23	0.13	41.72	31.42	47.29	3.20
	July	2.66	2.23	2.94	0.14	74.30	62.60	82.03	3.67

F&S is the approach based on F&S equation (Case study 2). Y is the yield and N is the nitrogen which are both in t/ha and kg/ha, respectively.

as yield maps and radiometric data are available, data-driven-based DSS can provide good results. Previous yield maps can be provided by farmers from previous years which can be the base of decision-making.

Finally, the complexity of any forecasting model can affect the adoption rates by the agricultural community. For example, Carberry et al. (2002) reported that high competency requirements and steep learning curve required by APSIM have prevented users from adopting these types of approaches. It is essential that these DSSs are built with a friendly user interface, which would positively impact their adoption success.

10 Conclusion and future trends

In conclusion, the goal of this chapter is to review some yield-forecasting PA tools that provide farmers with outputs that can assist their fertiliser decision-making. Improved methods for collecting, processing and analysing agricultural data can reduce the uncertainty when making critical decisions related to fertiliser applications. The F&S and data-driven models have been discussed in this chapter in terms of the results and complexity. Although some soil information and yield data were collected from private sources and fed to the data-driven model (which make it complicated to build such model), this approach can be built using only one private data, which is yield data, with other weather and soil data which are freely available. Furthermore, the F&S model can be presented as a simple model to calculate the forecasted yield maps.

Further field experiments are needed to gain evidence of the advantage of these models in different environments and crop types because these models were tested in one environment and on one crop type (wheat). Advisors and growers who make decisions based on scientific approaches can use such models to make N recommendations and update these recommendations as the season progresses.

In agriculture, DSS are usually interactive software-based programs that aid farmers in their decision-making, for example when to sow or when to use fertilisers. Despite the numerous DSSs that are available for agriculture these days, studies show that uptake by farmers is disappointingly low (Rose et al., 2016). Rose et al. (2016) also suggested that DSSs should include usability, cost-effectiveness, performance, relevance to the user, and compatibility with compliance demands.

Further, remote sensing technologies are improving, and it is expected to increase the quality of geospatial data in terms of accuracy and resolution with no or minimum costs which has the major impact on technology users. Therefore, developers should focus more on the delivery of simple and efficient DSSs which can be widely accepted by growers and agricultural advisors.

11 References

Ali, M. H. and Talukder, M. S. U. 2008. Increasing water productivity in crop production–a synthesis. *Agricultural Water Management* 95(11), 1201–13. doi:10.1016/j.agwat.2008.06.008.

Angus, J. F. 2001. Nitrogen supply and demand in Australian agriculture. *Australian Journal of Experimental Agriculture* 41(3), 277–88. doi:10.1071/EA00141.

Angus, J. 2008. Nitrogen fertiliser for wheat-assessing price and weather risks. *Australian Grain* 18, 30.

Angus, J. F. and Grace, P. R. 2017. Nitrogen balance in Australia and nitrogen use efficiency on Australian farms. *Soil Research* 55(6), 435–50. doi:10.1071/SR16325.

Arnott, D. and Pervan, G. 2014. A critical analysis of decision support systems research revisited: the rise of design science. *Journal of Information Technology* 29(4), 269-93. doi:10.1057/jit.2014.16.

Barton, L., Craig, S. and Frances, H. 2016. Where does the nitrogen go soil sources and sinks in Western Australia cropping soils. Available at: https://grdc.com.au/resources-and-publications/grdc-update-papers/tab-content/grdc-update-papers/2016/02/where-does-the-nitrogen-go-soil-sources-and-sinks-in-western-australia-cropping-soils.

Basso, B., Cammarano, D. and Carfagna, E. 2013. Review of crop yield forecasting methods and early warning systems. *Proceedings of the First Meeting of the Scientific Advisory Committee of the Global Strategy to Improve Agricultural and Rural Statistics*. FAO Headquarters, Rome, Italy, pp. 18–19.

Brentrup, F. and Pallière, C. 2010. Nitrogen use efficiency as an agro-environmental indicator. *Proceedings of the OECD Workshop on Agrienvironmental Indicators*, March, pp. 23-6

Broadley, M. R., Escobar-Gutierrez, A. J., Burns, A. and Burns, I. G. 2000. What are the effects of nitrogen deficiency on growth components of lettuce? *The New Phytologist* 147(3), 519–26. doi:10.1046/j.1469-8137.2000.00715.x.

Carberry, P. S., Hochman, Z., McCown, R. L., Dalgliesh, N. P., Foale, M. A., Poulton, P. L., Hargreaves, J. N. G., Hargreaves, D. M. G., Cawthray, S., Hillcoat, N. and Robertson, M. J. 2002. The FARMSCAPE approach to decision support: farmers', advisers', researchers' monitoring, simulation, communication and performance evaluation. *Agricultural Systems* 74(1), 141-77. doi:10.1016/S0308-521X(02)00025-2.

Chlingaryan, A., Sukkarieh, S. and Whelan, B. 2018. Machine learning approaches for crop yield prediction and nitrogen status estimation in precision agriculture: a review. *Computers and Electronics in Agriculture* 151, 61-9. doi:10.1016/j.compag.2018.05.012.

Cook, S. E. and Bramley, R. G. V. 1998. Precision agriculture–opportunities, benefits and pitfalls of site-specific crop management in Australia. *Australian Journal of Experimental Agriculture* 38(7), 753–63. doi:10.1071/EA97156.

Cope, J., Evans, C. and Williams, H. C. 1981. *Soil Test Fertilizer Recommendations for Alabama Crops*.

Creelman, Z., Long, J. and England, D. 2015. Equipping farmers to take a fresh look at rules of thumb. *Rural Extension and Innovation Systems Journal* 11, 151.

Dey, P. 2015. Targeted yield approach of fertiliser recommendation for sustaining crop yield and maintaining soil health. *JNKVV Res.* J 49, 338–46.

Dintwa, E., Tijskens, E., Olieslagers, R., Baerdemaeker, J. D. and Ramon, H. 2004. Calibration of a spinning disc spreader simulation model for accurate site-specific fertiliser application. *Biosystems Engineering* 88(1), 49–62. doi:10.1016/j.biosystemseng.2004.01.001.

Donatelli, M., Magarey, R. D., Bregaglio, S., Willocquet, L., Whish, J. P. M. and Savary, S. 2017. Modelling the impacts of pests and diseases on agricultural systems. *Agricultural Systems* 155, 213–24. doi:10.1016/j.agsy.2017.01.019.

Dunsford, K., Armstrong, R., Tang, C., et al. 2015. Estimating in-crop nitrogen mineralisation in dryland cropping systems of southern Australia. *Proceedings of the 16th Australian Agronomy Conference*, Hobart, Tas.

Ellis, N. R. and Albrecht, G. A. 2017. Climate change threats to family farmers' sense of place and mental wellbeing: a case study from the Western Australian Wheatbelt. *Social Science and Medicine* 175, 161–8. doi:10.1016/j.socscimed.2017.01.009.

Fageria, N. K. 2016. *The Use of Nutrients in Crop Plants*. CRC press.

Filippi, P., Jones, E. J., Wimalathunge, N. S., Somarathna, P. D. S. N., Pozza, L. E., Ugbaje, S. U., Jephcott, T. G., Paterson, S. E., Whelan, B. M. and Bishop, T. F. A. 2019. An approach to forecast grain crop yield using multi-layered, multi-farm data sets and machine learning. *Precision Agriculture* 20(5), 1015–29. doi:10.1007/s11119-018-09628-4.

French, R. and Schultz, J. 1984a. Water use efficiency of wheat in a Mediterranean-type environment. I. The relation between yield, water use and climate. *Australian Journal of Agricultural Research* 35(6), 743–64. doi:10.1071/AR9840743.

French, R. and Schultz, J. 1984b. Water use efficiency of wheat in a Mediterranean-type environment. II. Some limitations to efficiency. *Australian Journal of Agricultural Research* 35(6), 765–75. doi:10.1071/AR9840765.

Gellings, C. W. and Parmenter, K. E. 2016. Energy efficiency in fertilizer production and use. In: Gellings, C. W. (Ed.), *Efficient Use and Conservation of Energy*. Encyclopedia of Life Support Systems, pp. 123–36.

Gourley, C., Melland, A., Waller, R., et al. 2007. Making better fertiliser decisions for grazed pastures in Australia. Department of Primary Industries Victoria, East Melbourne.

Gourley, C. J. P., Dougherty, W. J., Weaver, D. M., Aarons, S. R., Awty, I. M., Gibson, D. M., Hannah, M. C., Smith, A. P. and Peverill, K. I. 2012. Farm-scale nitrogen, phosphorus, potassium and sulfur balances and use efficiencies on Australian dairy farms. *Animal Production Science* 52(10), 929–44. doi:10.1071/AN11337.

Grundy, M. J., Rossel, R. A. V., Searle, R. D., Wilson, P. L., Chen, C. and Gregory, L. J. 2015. Soil and landscape grid of Australia. *Soil Research* 53(8), 835–44. doi:10.1071/SR15191.

Hayman, P. T. and Alston, C. L. 1999. A survey of farmer practices and attitudes to nitrogen management in the northern New South Wales grains belt. *Australian Journal of Experimental Agriculture* 39(1), 51–63. doi:10.1071/EA98120.

Haynes, R. J. and Naidu, R. 1998. Influence of lime, fertilizer and manure applications on soil organic matter content and soil physical conditions: a review. *Nutrient Cycling in Agroecosystems* 51(2), 123–37. doi:10.1023/A:1009738307837.

Hazelton, P. and Murphy, B. 2016. *Interpreting Soil Test Results: What Do All the Numbers Mean?* CSIRO Publishing.

Henchion, M., Hayes, M., Mullen, A. M., Fenelon, M. and Tiwari, B. 2017. Future protein supply and demand: strategies and factors influencing a sustainable equilibrium. *Foods* 6(7), 53. doi:10.3390/foods6070053.

Hoang, V.-N. and Coelli, T. 2011. Measurement of agricultural total factor productivity growth incorporating environmental factors: a nutrients balance approach. *Journal of Environmental Economics and Management* 62(3), 462–74. doi:10.1016/j.jeem.2011.05.009.

Hodges, S. C. 2010. Soil fertility basics. Soil Science Extension, North Carolina State University.

Hunt, J., van Rees, H., Hochman, Z., et al. 2006. Yield prophet®: an online crop simulation service. *Proceedings of the 13th Australian Agronomy Conference*, pp. 10–4.

Kachanoski, R., O'Halloran, I., Aspinall, D., et al. 1996. Delta yield: mapping fertilizer nitrogen requirement for crops. *Better Crops* 80, 20–3.

Kamilaris, A., Kartakoullis, A. and Prenafeta-Boldú, F. X. 2017. A review on the practice of big data analysis in agriculture. *Computers and Electronics in Agriculture* 143, 23–37. doi:10.1016/j.compag.2017.09.037.

Keating, B. A., Carberry, P. S., Hammer, G. L., Probert, M. E., Robertson, M. J., Holzworth, D., Huth, N. I., Hargreaves, J. N. G., Meinke, H., Hochman, Z., McLean, G., Verburg, K., Snow, V., Dimes, J. P., Silburn, M., Wang, E., Brown, S., Bristow, K. L., Asseng, S., Chapman, S., McCown, R. L., Freebairn, D. M. and Smith, C. J. 2003. An overview of APSIM, a model designed for farming systems simulation. *European Journal of Agronomy* 18(3–4), 267–88. doi:10.1016/S1161-0301(02)00108-9.

Khan, S. A., Mulvaney, R. L. and Hoeft, R. G. 2001. A simple soil test for detecting sites that are nonresponsive to nitrogen fertilization. *Soil Science Society of America Journal* 65(6), 1751–60. doi:10.2136/sssaj2001.1751.

Landau, S., Mitchell, R. A. C., Barnett, V., Colls, J. J., Craigon, J., Moore, K. L. and Payne, R. W. 1998. Testing winter wheat simulation models' predictions against observed UK grain yields. *Agricultural and Forest Meteorology* 89(2), 85–99. doi:10.1016/S0168-1923(97)00069-5.

Lawrence, D. N., Cawley, S. T. and Hayman, P. T. 2000. Developing answers and learning in extension for dryland nitrogen management. *Australian Journal of Experimental Agriculture* 40(4), 527–39. doi:10.1071/EA99147.

Li, Z., Liu, H., Zhao, W., Yang, Q., Yang, R. and Liu, J. 2019. Quantification of soil-water balance components. *Hydrology and Earth System Sciences* 23, 4685–706.

Llewellyn, R. S. and Powles, S. B. 2001. High levels of herbicide resistance in rigid ryegrass (*Lolium rigidum*) in the wheat belt of Western Australia. *Weed Technology* 15(2), 242–8. doi:10.1614/0890-037X(2001)015[0242:HLOHRI]2.0.CO;2.

Long, S. P., Marshall-Colon, A. and Zhu, X. G. 2015. Meeting the global food demand of the future by engineering crop photosynthesis and yield potential. *Cell* 161(1), 56–66. doi:10.1016/j.cell.2015.03.019.

Ma, C., Zhang, H. H. and Wang, X. 2014. Machine learning for big data analytics in plants. *Trends in Plant Science* 19(12), 798–808. doi:10.1016/j.tplants.2014.08.004.

Maleki, M. R., Ramon, H., De Baerdemaeker, J. and Mouazen, A. M. 2008. A study on the time response of a soil sensor-based variable rate granular fertiliser applicator. *Biosystems Engineering* 100(2), 160–6. doi:10.1016/j.biosystemseng.2008.03.007.

McDonald, G. 1989. The contribution of nitrogen fertiliser to the nitrogen nutrition of rainfed wheat crops in Australia: a review. *Australian Journal of Experimental Agriculture* 29(3), 455–81. doi:10.1071/EA9890455.

Nendel, C. and Kersebaum, K. C. 2004. A simple model approach to simulate nitrogen dynamics in vineyard soils. *Ecological Modelling* 177(1–2), 1–15. doi:10.1016/j.ecolmodel.2004.01.014.

Ngo, V. M., Le-Khac, N.-A. and Kechadi, M. 2018. An efficient data warehouse for crop yield prediction. arXiv preprint arXiv:1807.00035.

Oliver, M., Bishop, T. and Marchant, B. 2013. *Precision Agriculture for Sustainability and Environmental Protection*. Routledge.

Passioura, J. 2008. Drought resistant crops: wet dream or pipe dream? *Australian Grain* 17, 18.

Probert, M. E., Dimes, J. P., Keating, B. A., Dalal, R. C. and Strong, W. M. 1998. APSIM's water and nitrogen modules and simulation of the dynamics of water and nitrogen in fallow systems. *Agricultural Systems* 56(1), 1-28. doi:10.1016/S0308-521X(97)00028-0.

Rasool, R., Kukal, S. S. and Hira, G. S. 2007. Soil physical fertility and crop performance as affected by long term application of FYM and inorganic fertilizers in rice-wheat system. *Soil and Tillage Research* 96(1-2), 64-72. doi:10.1016/j.still.2007.02.011.

Reid, J. B., Stone, P. J., Pearson, A. J. and Wilson, D. R. 2002. Yield response to nutrient supply across a wide range of conditions: 2. Analysis of maize yields. *Field Crops Research* 77(2-3), 173-89. doi:10.1016/S0378-4290(02)00087-4.

Roberts, T. 2009. The role of fertilizer in growing the world's food. *Better Crops* 93, 12-5.

Robertson, M. J. and Kirkegaard, J. A. 2005. Water-use efficiency of dryland canola in an equi-seasonal rainfall environment. *Australian Journal of Agricultural Research* 56(12), 1373-86. doi:10.1071/AR05030.

Robertson, M., Benson, T. and Shamudzarira, Z. 2000. Simulating nitrogen fertilizer response in low-input farming systems of Malawi. Risk Management Working Paper Series. CIMMYT, Mexico, 8p.

Rochester, I. 2008. Optimizing N fertiliser use and N fertiliser use-efficiency. *Proceedings of the 14th Australian Cotton Conference*, 12-14 August 2008, Broadbeach, Queensland.

Rose, D. C., Sutherland, W. J., Parker, C., Lobley, M., Winter, M., Morris, C., Twining, S., Ffoulkes, C., Amano, T. and Dicks, L. V. 2016. Decision support tools for agriculture: towards effective design and delivery. *Agricultural Systems* 149, 165-74. doi:10.1016/j.agsy.2016.09.009.

Schwenke, G., Beange, L., Cameron, J., Bell, M. and Harden, S. 2019. What soil information do crop advisors use to develop nitrogen fertilizer recommendations for grain growers in New South Wales, Australia? *Soil Use and Management* 35(1), 85-93. doi:10.1111/sum.12469.

Sharma, S. B., Sayyed, R. Z., Trivedi, M. H. and Gobi, T. A. 2013. Phosphate solubilizing microbes: sustainable approach for managing phosphorus deficiency in agricultural soils. *SpringerPlus* 2, 587. doi:10.1186/2193-1801-2-587.

Stone, P. and Hochman, Z. 2004. If interactive decision support systems are the answer, have we been asking the right questions. *New Directions for a Diverse Planet: Proceedings of the 4th International Crop Science Congress*.

Van Duivenbooden, N., Wit, C. T. and Keulen, H. 1995. Nitrogen, phosphorus and potassium relations in five major cereals reviewed in respect to fertilizer recommendations using simulation modelling. *Fertilizer Research* 44(1), 37-49. doi:10.1007/BF00750691.

Whelan, B. 2007. Current status and future directions of PA in Australia. *Proceedings of the 2nd Asian Conference on Precision Agriculture*, Pyeongtaek, Korea, pp. 60-71.

Whelan, B. and Taylor, J. 2013. *Precision Agriculture for Grain Production Systems*. CSIRO publishing.

Whitbread, A. and Hancock, J. 2008. Estimating grain yield with the French and Schultz approaches vs simulating attainable yield with APSIM on the Eyre Peninsula. *Proceedings of the 14th Australian Agronomy Conference*, pp. 21-5.

Wimalathunge, N. S. and Bishop, T. F. A. 2019. A space-time observation system for soil moisture in agricultural landscapes. *Geoderma* 344, 1–13. doi:10.1016/j.geoderma.2019.03.002.

Wong, M., Asseng, S. and Zhang, H. 2005. Precision agriculture improves efficiency of nitrogen use and minimises its leaching at within-field to farm scales. *Precision Agriculture* 05(5), 969.

Živčák, M., Olšovská, K., Slamka, P., Galambošová, J., Rataj, V., Hb, S. and Brestič, M. 2014. Application of chlorophyll fluorescence performance indices to assess the wheat photosynthetic functions influenced by nitrogen deficiency. *Plant, Soil and Environment* 60(5), 210–5. doi:10.17221/73/2014-PSE.

Chapter 8

Developing decision-support systems for crop rotations

Zia Mehrabi, University of British Columbia, Canada

1 Introduction

The simplification of ecosystems for a package of improved seed, synthetic fertilizer and irrigation-driven yield improvement is a design paradigm that has dominated agricultural development (Ramankutty et al., 2018). There are inherent trade-offs in this model. Currently the world comfortably produces enough basic energy to feed all its inhabitants (Food and Agriculture Organization of the United Nations, 2011; Ng et al., 2014). At the same time agricultural lands cover ~38% of the world's land surface area (Foley et al., 2011), contribute to 22% of yearly anthropogenic greenhouse gas emissions (Smith et al., 2014), account for 92% of the human water footprint (Hoekstra and Mekonnen, 2012), result in erosion of ~35 Pg yr^{-1} of soil (Quinton et al., 2010), and lead to widespread nitrate and phosphate pollution (Vitousek et al., 1997), and drive biodiversity loss of 20–30% of local species richness (Newbold et al., 2015). It is now recognized that many of these environmental externalities, in turn risk yield losses to climate change, and pest outbreaks, reducing resilience of the food systems on which humans depend (Bellwood, 2018; Lesk et al., 2016; Ramankutty et al., 2018).

In recent decades, the agricultural sector has fine-tuned the model of optimizing yields through increased efficiency or 'more crop per drop'. These

http://dx.doi.org/10.19103/AS.2020.0069.15

advances have been aided through the improved use of inputs. However these increases in efficiency have not been enough to halt the negative environmental impacts of agriculture on people and planet. This continued degradation has led to calls from food system scientists to move away from simple technological intensification (using synthetic fertilizers, pesticides, improved seeds and water) towards ecological intensification (e.g. the use of diverse crop plantings which help recycle or fix nutrients, manage water and pest outbreaks), as the major pathway towards a more sustainable agricultural future (Coomes et al., 2019). There is some evidence that the transition to sustainable intensification is already taking place in many landscapes (Pretty et al., 2018).

Ecological intensification through increasing the diversity of crops in rotations, cover cropping, intercropping, or increasing non-crop habitat on farmlands, has long been promoted as a major solution to counter the negative impacts of agriculture on environmental and human health (Bommarco et al., 2013; Kremen et al., 2012). This promotion comes from the finding that increasing plant diversity in time and space can increase water-use efficiency, reduce nitrogen leaching, regenerate soils, increase pollination, reduce losses to pests, reduce greenhouse gas emissions, increase nutritional quality of foods and increase the stability of production to climate change (Isbell et al., 2017; Mariotte et al., 2018; Renard and Tilman, 2019). However the composition of elements for the design of ecological systems is not trivial and the benefits of diversity found in ecological studies have largely been studied in separation from the economics of farming. Recent studies which study both show that significant financial trade-offs often exist with managing diverse farming systems (Rosa-Schleich et al., 2019).

The decline in the diversity of crops found on larger farms is tightly linked to the economics of farming and labour. There is a strong inverse relationship between the size of a country's economy and the percentage of people employed in agriculture. Lack of labour forces, an upward shift in farm size, mechanization and a downward shift in the complexity of agricultural ecosystems and the number of crop species grown, are all interlinked. As scale and capital investment increases, the burden of knowledge management, and risk, also increases, further forcing an increase in specialization and a reduction in crop diversity and an increase in continuous cropping cycles (Awokuse and Xie, 2015; Ramankutty et al., 2018; Ricciardi et al., 2018). The focus on optimizing yields in monoculture, with limited focus on ecological intensification, is reflected in modern decision-support systems used by farms for management today. Ecologically based decision-support tools are largely non-existent.

At the core of developing tools for managing more diverse farms, including crop rotations, lies a complex information challenge. Crops interact with each other, their above- and below-ground symbionts (e.g. fungi, bacteria, pollinators and pests), climate, soils, the landscape and farm-management decisions, in extremely context-dependent and site-specific ways. This makes data-driven agronomy and digital extension services difficult and uncertain even in the simplest of agrosystems. Advisory for complex and diverse cropping systems is further exacerbated by a lack of data on local conditions, and lack of data on the outcomes of particular species interactions in response to a wide range of ancillary and seasonally varying factors, a lack of generalizable ecological theory and the gap between economics and ecological inquiry.

In this chapter I will look at the current challenges and opportunities for designing farm-level decision-support tools for diversified agriculture, with a specific focus on developing tools for crop rotations. I will cover the aspects of key information challenges, ecological theory, agronomic models, encoding farmer decisions, finishing the chapter with principles for design pulling together insights from these cases, and future trends and directions.

2 Key information challenges

The landscape of decision-support systems for improved rotations, should be understood within a broader challenge faced by data-driven agronomy itself and the state of play currently in developing predictive tools for aiding farmer's decision-making. The challenges faced in diverse cropping systems add a layer of complexity to these broader challenges through increased heterogeneity of information that needs to be handled. Below are the major information challenges facing data-driven agronomy with a particular focus on optimizing for multiple sustainability objectives.

1 **Information overload.** In the age of big data, the potential for information overload is overwhelming. Farmers, extension agents and researchers contend with agricultural data from the proliferation of sources such as: low-cost sensors; novel satellite data streams; trial databases; weather data; machine data; biodiversity data; market data; supply chain-transparency data; certification data; climate data; farmer-safety data; food-safety data and interest in developing long-term farm-, national- and global-scale datasets across these streams. This leads to cognitive overload, limited time for record keeping and 'analysis paralysis', hindering the use of important information for decision-making.
2 **Data gaps and harmonization.** While there is a growing abundance of data streams, data collection, aggregation and management are costly. This often results in a lack of coverage for some regions, or limited coverage of some variables resulting from aims of isolated data collection initiatives. Data is also collected at varying levels

of granularity, in disparate formats, structures, and through varied protocols, and is of varying degrees of quality. Where data exists, there are often compatibility issues between data sets and no formally agreed or accepted ontologies for translating between data collection efforts.

3. **Tool isolation.** Currently, there is a disconnected ecosystem of sustainable agriculture tools and sustainability assessment frameworks, resulting in a landscape dotted with duplicated efforts, isolated clusters of tools lacking interoperability (e.g. decision tools that are incompatible with on-farm record-keeping software, or carbon calculators which are not integrated with financial accounting software). There are few formal efforts to develop modular services that are plug and play, which ensures the development of decision-support tools involves high up-front costs.
4. **Validation.** Many decision-support systems do not carry estimates of validation of their performance for the indicators they aim to help optimizing for, whether that be improving farmer incomes or reducing pest outbreaks. This lack of validation has led to a wide number of tools with no benchmark as to their reliability. This is exacerbated by the fact that many existing decision-support tools are static and cannot be easily updated based on advancements in farming technology and management innovations, or novel data streams.
5. **Privacy.** The collection and dissemination of data on farmer decision-making processes and on-farm outcomes carries important privacy concerns. Knowledge of how farmers perform on environmental, economic or social outcomes is valuable information that if disclosed to third parties can put farmers at risk of legal action, extortion or financial loss. Taken together with the lack of standards for tool validation, privacy concerns diminish trust in tools.
6. **Inequalities in access.** Information is power. Yet access to information and decision-support tools is mediated by economic or cultural status. The availability and access to digital technology services is unequal globally, and many of the world's poor farmers are currently excluded from the benefits provided by commercial and public decision-support tool providers. At the same time there is an inequality in access to traditional knowledge, and farmer expertise, on outcomes of farm management decisions on sustainability outcomes, and cultural barriers that enforce the integration of traditional knowledge sources with mainstream science.
7. **Supply-driven design.** Cool uses of new data streams and flashy hardware or software interfaces for farm management decision support can be found in abundance, but careful human-centred design of decision-support systems that tackle specific and significant problems faced by farmers are lacking. This is particularly problematic where

disciplinary gaps between natural scientists, engineers and social scientists exist, and ultimately results in the development of tools with poor usability, with low uptake and adoption. There is a need to develop demand-driven design principles and work with farmers directly to design the tools to solve their problems.

3 Ecological theory

Over the last two decades a rapid growth in the ecological study of the costs of benefits of growing different combinations of plants on plant health emerged. We now have strong experimental evidence that plants can culture soils to the benefit or detriment of other plants, either through nutrient depletion, releasing toxic metabolites, or through the recruitment of beneficial or pathogenic symbionts in the root zone (Bever et al., 2012; Inderjit et al., 2011; Van Der Putten et al., 2013). Ecologists call these dynamics between plant interactions, 'plant-soil feedbacks' (PSF), which they study for their roles in ecosystem succession, invasion and species coexistence (Callaway et al., 2004; Klironomos, 2002; Van Der Putten et al., 1993; Schnitzer et al., 2011), but really they are similar in many ways to crop rotations.

One of the theories that has been tested by ecologists is an old idea (recognized as far back as Charles Darwin) that more closely related plant species are likely to share pests and pathogens (Gilbert and Webb, 2007; Webb et al., 2006). This idea of sharing of pests and pathogens led researchers to think that plants that are more closely related will have more negative interactions with each other through the soil (Brandt et al., 2009; Burns and Strauss, 2011; Liu et al., 2012; Sedio and Ostling, 2013). However, the most comprehensive assessment to date found that this pattern is not general - and fails to hold across a large range of flowering plants, life histories, life cycles and even within groups of recently diverged species (Mehrabi and Tuck, 2015), suggesting that using relatedness to predict which plants will do better in rotations may be of limited use (Mehrabi and Tuck, 2015; Ingerslew and Kaplan, 2018).

Newer advancements in ecology have hypothesized that fast-growing resource-exploitative species ('fast species') with highly decomposable tissues replenish nutrients quicker and have higher fertilizing effects on soil than slow-growing resource conservative species. Positive effects of fast species on the growth of subsequent plants could be explained by their fertilizing effects and soil chemistry legacies of high plant-available N, or due to proliferation of microbes involved in nutrient mobilization in fast soils (Baxendale et al., 2014; Grigulis et al., 2013; Ke et al., 2015). However there are a number of other processes associated with fast species such as the proliferation of pathogens (Veresoglou et al., 2013), losses of beneficial fungi (Grigulis et al., 2013; Hoeksema et al., 2010; Orwin et al.,

2010), phytotoxic effects of highly decomposable tissue inputs (Bonanomi et al., 2011, 2006) and disruptions or lags to recycling of plant materials (Hobbie, 2015), which would drive plant-soil feedbacks in the opposite direction, with plants performing worse on soil cultured by faster species. Separating out the effects of each of these factors will require targeted experimentation.

While ecologists still struggle to pin down reasons why species perform better or worse when grown in sequence, case-by-case insights have been developed which can help specific agricultural systems. For example, in limiting negative effects of below-ground pests, such as nematodes, fungi or bacteria (Mariotte et al., 2018; Silva et al., 2018), and through maximizing positive effects of plants which build symbionts with microbes in the soil and enhance nutrient-use efficiency (Bender and van der Heijden, 2015), or increase above-ground resistance to pests (Pineda et al., 2017). Some of these case studies are particularly important, such as with take-all in wheat, where new genetic varieties have been identified which create pathogen-suppressive soils and can be rotated with more susceptible varieties, reducing yield losses by 3 t/ha^{-1} (Mehrabi et al., 2016). It is this depth of understanding of the ecology of soil biology that makes the research on plant-soil-feedback mechanisms so appealing with the idea that one day we may be able to engineer soils to help agricultural systems perform better by manipulating soil biology.

4 Agronomic models

Alongside ecological research on plant-soil-feedbacks is agronomic research on 'soil sickness', a phenomenon which results from continuous cropping of the same species on soil. Crop rotation is the central method used by humans to overcome the observed yield declines seen with successive monocultures (Bennett et al., 2012; Dick, 1992; Lawes, 1895; Raaijmakers et al., 2009). Much empirical work has been conducted on which sequences of crops lead to optimal outcomes.

One well-known agronomist rule of thumb is the use of N_2-fixing plants in rotations. Recent synthesis shows benefits to rotations for cereals rotated with grain legumes, with yield increases of approx. 29% relative to continuous cropping of cereals. However these benefits are only observed in systems where nitrogen application rates are low, becoming negligible in systems with N applications > 150 kg/ha (Cernay et al., 2018). The benefits of legume rotations in low-input systems, and their benefits for farmer livelihoods have been documented elsewhere in specific country case-studies (Snapp et al., 2010), but scaling these benefits to high input synthetically fertilized systems, in, for example, the EU or North America to reduce dependency on N inputs is less clear.

A range of other agronomic meta-analyses have been conducted on the benefits and costs of rotations for pest control, product quality, input use reduction, production stability, improvement in soil organic carbon and soil quality, biodiversity, greenhouse gas emissions and economic profitability (Beillouin et al., 2019). In some meta-analyses the benefits of rotation for specific outcomes are reported to be clear, although often crop differences are masked by pooled analysis across systems and crop types. For example, organic systems have been reported to have higher yield and profits relative to conventional rotations when grown in longer and more diverse rotations (Crowder and Reganold, 2015; Ponisio et al., 2015). Longer rotations have also been reported to produce strong positive effects on microbial richness (Venter et al., 2016) and enhanced microbial N and C and microbial activity (Lori et al., 2017).

The benefits of rotation can vary immensely by crop and system type. For example greenhouse warming potential and emissions intensity of soybean and maize rotations is greater than in continuous maize crops, but lower in smaller grain crop combinations of barley and pea compared to continuous barley (Sainju, 2016). While in some crops, the effects seem clearer, for example, in canola in North America diverse rotations for every 3–4 years are required to maintain yields (Assefa et al., 2018), in many other cases the responses are highly contextual, because the benefits of rotation depend on interactions between crop species combinations and environmental conditions. For example, in wheat systems wheat yields after break crops are on average 0.5–1 t/ha more productive after oats, and 1.2 t/ha more productive after grain legumes, than continuous crops, but the benefits only hold for the first 2 years of break crop, and are less beneficial after 3 years except in drought (Angus et al., 2015).

In addition to problems of generalizability of empirical data, meta-analyses of agronomic models of 'soil sickness' also do not report the predictive skill of particular crop associations for specific outcomes (e.g. yield, biodiversity, climate mitigation, biodiversity impacts, etc.) , and there are significant biases in geographic and system-level coverage. A recent systematic review of 99 meta-analyses assessing diversification, including studies of rotation, found that the majority of reported benefits of crop diversification covered only 10 key crops: pea, cowpea, bean, soybean, oat, rice, sorghum, barley, wheat and maize, with most studies based in North America and Europe (Beillouin et al., 2019).

Despite these shortcomings, the accumulation of experimental agronomic data is pushing us to a space where, if the coverage of systems and crop types increases, we may be able to make empirically based predictions of the likely outcomes of different rotation sequences in the near future. Data-driven models of optimal crop rotation pairs could therefore be within reach, even if general ecological theory does not yet exist to ground it. Incorporation of these empirical data, with insights from operations research-based optimization studies, alongside formalization of farmer decision-making, and mechanistic modelling for rotations, remains a new frontier of this work (Box 1).

Box 1 A plethora of approaches

A number of statistical, rule-based, mathematical, and mechanistic models have been developed for aiding recommendations of crop sequences for crop rotations (Beillouin et al., 2019; Dury et al., 2012). These models have in general not been integrated or widely adopted by farmers or extension agents. These models include:

- Recommendations for rotations based on rules set by experts, which cover all theoretical permutations of crops, then filter sequences based on thresholds for outcomes of interest (e.g. nitrogen leaching, soil erosion, weed and pest management), which may be calculated from simple spreadsheet or crop-simulation models parameterized by field trials, and ranked based on likely economic benefits (Bachinger and Zander, 2007). These models help to formalize the inclusion of multiple objectives in models where little data exists, although, as currently formulated, they are inflexible in dealing with non-linear or complex crop or sequence plans (Castellazzi et al., 2008).
- Recommendations for sequences based on empirical observations for particular crop pairs or groups. As covered in Section 4, a number of trials have been compiled into meta-databases which cover the known relationship between particular crop interactions and particular outcomes from an agronomic perspective (Beillouin et al., 2019). These models can be used as a basis to make recommendations for specific crop selections. However currently the scope of recommendation is limited in crop type and geographic coverage.
- Recommendations based on the maximization of a set of single or multiple objectives (e.g. gross margins, profits, labour, land, market demand, water, waste, food supply, pesticide use), given particular a priori constraints. These constraints include yields dependent on previous crops (El-Nazer and McCarl, 1986), fixed known allocations of land per crop (Detlefsen and Jensen, 2007), forbidden crop sequences (Haneveld and Stegeman, 2005), pre-desired planting principles (Forrester et al., 2018), known demand and stocking lengths (Costa et al., 2014) or incorporation of spatial constraints such as blocks holding homogenous management practices (Akplogan et al., 2013). Coefficients in these models can be modelled probabilistically to account for the influence of stochastic factors such a climate feedbacks (Itoh et al., 2003). While these models have tackled the optimization problems of rotation recommendations, they

are time-consuming to set up and are bounded by study-specific constraints and objectives.

- Recommendations based on mechanistic models. Mechanistic or process-based crop models can incorporate dynamic responses of outcomes of interest to external factors, such as climate and markets. Platforms for simulating different realizations of process-based models have been developed (Bergez et al., 2013) that allow for running simulations of the impacts of choice of rotations on outcomes such as water use (Chacoff and Aizen, 2006), yields and nitrogen dynamics (Kollas et al., 2015). The major problem with process-based models for rotation modelling is the lack of representation of biotic interactions between crop species, data on specific non-major crops, soil chemistry dynamics, particularly with organic matter, and interactions with specific management practices (Kollas et al., 2015).

There is a clear need to combine insights across these different modelling approaches into the next generation of decision-support systems for rotations. We need a better observation of farm management data, more biologically realistic mechanistic models, adaptive recommendations based on personalized time-varying constraints and real-time data on climate, markets and pest outbreaks. To ensure adoption, these models must be integrated into culturally and technologically accessible interfaces designed in iterative participatory processes with farmers.

5 Encoding farmer decisions

Ecologists, agronomists and economists' representations of how farmers make decisions on rotations represent different ways of knowing about the world, which may not represent each other or model the way farmer makes rotation decisions. Academics have documented models of farmer decision-making in disparate fields, but how close has this brought us to effective decision support?

Economic models designed to maximize utility have been built around survey-driven decision trees developed with farmers to understand how farmers cope with risk (Adesina and Sanders, 1991), and in visualization of cropping choice outcomes for illiterate farmers (Collins et al., 2013). Notably however, there is a lot of contention about the usefulness of purely rational economic models in representing real-world decision-making. Operational decision-making theory is one school of thought that may help to overcome this by attempting to understand the mental models of farmers when they are undertaking rotations, documenting the process of information selection by farmers and exploring methods of dealing with time variant objectives (Martin-Clouaire, 2017)

A key branch of studying farmer behaviour to date has been founded on the idea that decision-making can be encoded into a set of conditional IF-THEN statements, which together represent a 'model for action' derived from both objectives of the farmer and plans of schedules to realize those objectives. This method has been particularly useful in mapping out the temporal and spatial dynamics of farmers crop-planning decisions, and have been used for encoding vastly different farming systems, for example, in locations such as France and Cameroon governed by very different social and biophysical contexts (Aubry et al., 1998; Dounias et al., 2002). Spatial representations of these models for action have also been undertaken (Dury et al., 2013), and there have recently been efforts to join economic models with these methods, along with biophysical crop models as a way to model short-, medium- and long-term components of decision-making (Robert et al., 2018).

A complementary approach, recognized in the early 1980s by anthropologists (Chibnik, 1980), is the use of statistical learning to try and understand patterns in farmer decision-making, without depending on explicit encoding of farmers intentions or plans. This purely relies on inferring likely constraints on decision-making from realizations of plan execution. There is currently untapped opportunity to use of modern statistical methods to understand the heterogeneous nature of farmer decisions, particularly for rotations. This will be aided if time series of management decisions can be obtained, and likely offers a powerful basis for developing personalized decision support.

While these methods are useful, perhaps the most overlooked step in understanding farmers' models for optimal crop sequences selection is to put more time into understanding farmers' ways of knowing. This is typically achieved by employing methods of qualitative data collection and observation, such as interviews, surveys and ethnography developed in anthropology and sociology. Farmers' own ways of knowing about rotations help illustrate existing concepts and tools used for planning, and can help build an understanding of what farmers think grow well together - and why (Mohler and Johnson, 2009). Expectedly farmers also hold a breadth and depth of information not incorporated or reconciled with formal scientific literature (Bentley and Thiele, 1999). The integration of both farmers' traditional and emergent knowledge of what crops work best together, as well as the knowledge of the well-being generated from decision processes made outside those optimizing financial gain, remains perhaps the largest gap in our understanding of developing effective decision-support systems in farming.

6 Design principles

The fundamental hurdle facing data-based decision-support systems for rotations is not so much a technical feat that you can solve at your computer, as it is access and adoption of tools you build. Solving this challenge requires a clear understanding of what farmers' key pain points are with respect to managing their farm and their rotations, and creating useful solutions to these problems. Some tips are given below:

Safeguard interests. Data-driven agronomy is a fast-moving field, and due to many commercial interests in this area, farmers are bombarded with surveys and market research aimed at building tools for decision support, which can be extremely extractive in nature and lead to survey fatigue. You should work with farmers with whom you have strongly aligned interests and a long-term commitment also. Decision-support tools should be designed and to empower users through data and knowledge and to safeguard against the risks posed by overreliance and loss of resilience, traditional knowledge or adaptive capacity. You will need to think about the ethics of your project, have clear protocols to minimize risk and have your data management plans approved by a recognized ethics board. You should have strong privacy and data-sharing agreements in place.

Know your user. Understand the population of farmers you want to work with through surveys, in depth semi-structured interviews, focus groups, design workshops and ethnographic methods such as non-participatory observation. Focus on the broader normative stance, preferred knowledge sources and formats, demographics, institutional context, political context, what makes individuals that you are working with tick, why they farm, what their greatest issues are and what the broader constraints in realizing their objectives as a farmer are. We are not focussed on rotations here but a more general understanding of the user group.

Define a model. Get a deep view on how your farming group currently models rotations, which can substantially differ from ecological, economic and agronomic representations of how academics or researchers like to model how farmers make decisions about sequential cropping patterns. How do they currently map out their rotations? Do they use spreadsheets, wheel diagrams, pen and paper, mental maps based on weather, or celestial-based cues? Intuition? And when do they make their plans, which actors play roles and constrain their option space, and what kinds of decisions happen in and out of season? Have they tried other solutions to deal with this problem in the past, and how did they arrive at their current solution? Do they see rotation selection as a problem that needs solving, or are there more important things on their agenda they would rather have help with?

Focus in. Many complementary methods exist for digesting and prioritizing data generated through working with farmers, for example, through affinity diagrams, card sorting, content analysis, persona development, journey mapping, network analysis, decision trees, fuzzy cognitive mapping and so on. Using such analytic methods to simplify the goal of the tool development to focus in on one main problem to solve for your farmer group, while maintaining the nuances of the subsets of problems required for solving that problem, is a major milestone in the design process.

State and test your assumptions. All along the development cycle you will find yourself making assumptions about the ways farmers operate. It is important to explicitly state these assumptions. They should be built with your farmers directly, and as new assumptions arise they should be tested through future engagements. The development of usable decision-support tools will be marked by rapid iterations across ideas and assumptions that are tested and validated continuously.

Be forward looking. Sometimes it is not possible to develop decision-support systems or features due to lack of empirical data - this is particularly the case for crop sequence decision-support systems for vegetable crops. This may require a re-focus on features that are not top of your agenda as a crop rotation specialist or developer - and may even seem tangential. It is your job to understand which features are of key utility to the farming objectives and to build tools which service their needs - which can mean offering short-term utility in financial management, or lifting paperwork burden for certifications, but at the same time generating the data required for a rotation feature to be built and delivered to farmers at a later date.

Check the market. Once you have focussed in, it makes sense to do comparative analysis of existing solutions offered to solve rotation choice problems, or subsets of the problem such as crop planning tools, both in terms of their features and their information architecture. If you have the resources conduct user testing on these alternative solutions to identify what your user group does and does not like about them, this will help you define a clearer idea of which design is likely to work for you.

Interoperate. It makes little sense to develop a standalone decision-support tool, for example, for crop rotations - because decision-making in agricultural systems results from the integration of multiple sources of information, and processes. If farmers have existing tools and models for financial management, or labour management, fertilizer management, or other farm activities, then your rotation tool should be able to speak to those models. A common ontology and the development of translators between alternative semantic representations of farm objects and models are essential, and you should be using existing work as building blocks for your operation where they exist (e.g. standardized crop lists and trait information with taxonomically accepted names, weather data, soil data, plant and pest-trait databases).

Optimize. Bad rotation choice decisions can impact negatively on the environment and human health efforts should be made to provide an understanding of the trade-offs that a given choice of sequence will likely have, so that informed, rather than prescribed, decisions can be made. It is unlikely that all of the criteria a farmer is using to make choices will have available data and, so a focus on key proxies for main variables of importance (e.g. nutrient balance, water balance, yield, pesticide use etc.) here is a good starting point.

Personalize. Farming systems and individual farmer's models of how their farm works are incredibly diverse. The only solution to deal with this problem of multiple possible combinations of social, biophysical and value-based factors is to develop personalized rotation sequence decision-support services. These services by their nature are data hungry, and this means design has to account for data gathering as a key feature. The utility in personalized services is that they can readily account for adaptive and flexible shifts in farmers' operations, decisions and learning processes.

Make accessible. One of the key challenges with developing decision-support tools for the majority of farmers worldwide is the large inequality in the distribution of technology availability, access and utilization to different farmers - whether in service coverage, mobile phone ownership, internet access, the cost of data, and education, age culture, or language barriers which limit the usage of these services. Developing a solution that overcomes this inequality in decision support deserves a central place in the development of new decision-support tools.

7 Outlook

Crop rotations are a formative component of agricultural practices and are a key path towards improving multiple farming outcomes, for people and the planet. Improved decision-support systems for crop rotations therefore hold great potential for improving food system sustainability. New opportunities exist to gather and collect on-farm data at unprecedented temporal and spatial resolution and frequency. These developments herald an age of personalized decision-support tools able to provide adaptive recommendations for multiple sustainability indicators, based on real time data on climate, markets, and pests. Realizing this opportunity requires multiple information challenges be solved. Researchers will need to link insights from the development of alternative models for crop rotation across diverse disciplines and subject fields, to design tools that are interoperable with the wider farm data ecosystem, and that are built to both gather data and internally validate predictions over time. All of these efforts must be combined with improved human-centred design, participatory methods that seek to understand farmer's way of knowing, and a long-term view to development that builds lasting partnerships with farmers, improves access and trust, empowers users through ownership of their data and maintains farmers agency in the design and governance of the decision-support ecosytems that serve them.

8 Where to look for further information

This chapter covered a lot of material and you might be wondering where to go from here? Much is covered in the references, with the exception of the design thinking. However there are some good books and resources to get you started, such as: IDEO's Human Centred Design Toolkit (https://www.ideo.com/post/design-kit); as well as books such as The Design Thinking Playbook (https://www.design-thinking-playbook.com/playbook-en) and This is Service Design Doing (http://www.thisisservicedesignthinking.com/).

9 References

Adesina, A. A. and Sanders, J. H. 1991. Peasant farmer behavior and cereal technologies: stochastic programming analysis in Niger. *Agric. Econ.* 5, 21-38. doi:10.1016/0169-5150(91)90034-I.

Akplogan, M., De Givry, S., Métivier, J. P., Quesnel, G., Joannon, A. and Garcia, F. 2013. Solving the crop allocation problem using hard and soft constraints. *RAIRO-Oper. Res.* 47(2), 151-72. doi:10.1051/ro/2013032.

Angus, J. F., Kirkegaard, J. A., Hunt, J. R., Ryan, M. H., Ohlander, L. and Peoples, M. B. 2015. Break crops and rotations for wheat. *Crop Pasture Sci.* 66(6), 523-52. doi:10.1071/CP14252.

Assefa, Y., Vara Prasad, P. V. V., Foster, C., Wright, Y., Young, S., Bradley, P., Stamm, M. and Ciampitti, I. A. 2018. Major management factors determining spring and winter canola yield in North America. *Crop Sci.* 58(1), 1-16. doi:10.2135/cropsci2017.02.0079.

Aubry, C., Papy, F. and Capillon, A. 1998. Modelling decision-making processes for annual crop management. *Agric. Syst.* 56(1), 45-65. doi:10.1016/S0308-521X(97)00034-6.

Awokuse, T. O. and Xie, R. 2015. Does agriculture really matter for economic growth in developing countries ? *Can. J. Agric. Econ.* 63(1), 77-99. doi:10.1111/cjag.12038.

Bachinger, J. and Zander, P. 2007. ROTOR, a tool for generating and evaluating crop rotations for organic farming systems. *Eur. J. Agron.* 26(2), 130-43. doi:10.1016/j.eja.2006.09.002.

Baxendale, C., Orwin, K. H., Poly, F., Pommier, T. and Bardgett, R. D. 2014. Are plant-soil feedback responses explained by plant traits? *New Phytol.* 204(2), 408-23 doi:10.1111/nph.12915.

Beillouin, D., Ben-Ari, T. and Makowski, D. 2019. A dataset of meta-analyses on crop diversification at the global scale. *Data Br.* 24, 103898. doi:10.1016/j.dib.2019.103898.

Bellwood, P. 2018. The search for ancient DNA heads east. *Science* 361(6397), 31-2. doi:10.1126/science.aat8662.

Bender, S. F. and van der Heijden, M. G. A. 2015. Soil biota enhance agricultural sustainability by improving crop yield, nutrient uptake and reducing nitrogen leaching losses. *J. Appl. Ecol.* 52(1), 228-39. doi:10.1111/1365-2664.12351.

Bennett, A. J., Bending, G. D., Chandler, D., Hilton, S. and Mills, P. 2012. Meeting the demand for crop production: the challenge of yield decline in crops grown in short rotations. *Biol. Rev.* 87(1), 52-71. doi:10.1111/j.1469-185X.2011.00184.x.

Bentley, J. W. and Thiele, G. 1999. Bibliography: farmer knowledge and management of crop disease. *Agric. Hum. Values* 16(1), 75-81. doi:10.1023/A:1007558919244.

Bergez, J. -E., Chabrier, P., Gary, C., Jeuffroy, M. H., Makowski, D., Quesnel, G., Ramat, E., Raynal, H., Rousse, N., Wallach, D., Debaeke, P., Durand, P., Duru, M., Dury, J., Faverdin, P., Gascuel-Odoux, C. and Garcia, F. 2013. An open platform to build, evaluate and simulate integrated models of farming and agro-ecosystems. *Environ. Modell. Softw.* 39, 39–49. doi:10.1016/j.envsoft.2012.03.011.

Bever, J. D., Platt, T. G. and Morton, E. R. 2012. Microbial population and community dynamics on plant roots and their feedbacks on plant communities. *Annu. Rev. Microbiol.* 66, 265–83. doi:10.1146/annurev-micro-092611-150107.

Bommarco, R., Kleijn, D. and Potts, S. G. 2013. Ecological intensification: harnessing ecosystem services for food security. *Trends Ecol. Evol. (Amst.)* 28(4), 230–8. doi:10.1016/j.tree.2012.10.012.

Bonanomi, G., Sicurezza, M. G., Caporaso, S., Esposito, A. and Mazzoleni, S. 2006. Phytotoxicity dynamics of decaying plant materials. *New Phytol.* 169(3), 571–8. doi:10.1111/j.1469-8137.2005.01611.x.

Bonanomi, G., Incerti, G., Barile, E., Capodilupo, M., Antignani, V., Mingo, A., Lanzotti, V., Scala, F. and Mazzoleni, S. 2011. Phytotoxicity, not nitrogen immobilization, explains plant litter inhibitory effects: evidence from solid-state 13C NMR spectroscopy. *New Phytol.* 191(4), 1018–30. doi:10.1111/j.1469-8137.2011.03765.x.

Brandt, A. J., Seabloom, E. W. and Hosseini, P. R. 2009. Phylogeny and provenance affect plant-soil feedbacks in invaded California grasslands. *Ecology* 90(4), 1063-72. doi:10.1890/08-0054.1.

Burns, J. H. and Strauss, S. Y. 2011. More closely related species are more ecologically similar in an experimental test. *Proc. Natl. Acad. Sci. U.S.A.* 108(13), 5302-7. doi:10.1073/pnas.1013003108.

Callaway, R. M., Thelen, G. C., Rodriguez, A. and Holben, W. E. 2004. Soil biota and exotic plant invasion. *Nature* 427(6976), 731–3. doi:10.1038/nature02322.

Castellazzi, M. S., Wood, G. A., Burgess, P. J., Morris, J., Conrad, K. F. and Perry, J. N. 2008. A systematic representation of crop rotations. *Agric. Syst.* 97(1–2), 26–33. doi:10.1016/j.agsy.2007.10.006.

Cernay, C., Makowski, D. and Pelzer, E. 2018. Preceding cultivation of grain legumes increases cereal yields under low nitrogen input conditions. *Environ. Chem. Lett.* 16(2), 631–6. doi:10.1007/s10311-017-0698-z.

Chacoff, N. P. and Aizen, M. A. 2006. Edge effects on flower-visiting insects in grapefruit plantations bordering premontane subtropical forest. *J. Appl. Ecol.* 43(1), 18–27. doi:10.1111/j.1365-2664.2005.01116.x.

Chibnik, M. 1980. *The Statistical Behavior Approach: the Choice between Wage Labor and Cash Cropping in Rural Belize, Agricultural Decision Making*. Academic Press, Inc. doi:10.1016/b978-0-12-078882 8.50009-x.

Collins, A. J., Vegesana, K. B., Seiler, M. J., O'Shea, P., Hettiarachchi, P. and McKenzie, F. 2013. Simulation and mathematical programming decision-making support for smallholder farming. *Environ. Syst. Decis.* 33(3), 427-39. doi:10.1007/s10669-013-9460-7.

Coomes, O. T., Barham, B. L., MacDonald, G. K., Ramankutty, N. and Chavas, J.-P. 2019. Leveraging total factor productivity growth for sustainable and resilient farming. *Natl. Sustain.* 2(1), 22–8. doi:10.1038/s41893-018-0200-3.

Costa, A. M., dos Santos, L. M. R., Alem, D. J. and Santos, R. H. S. 2014. Sustainable vegetable crop supply problem with perishable stocks. *Ann. Oper. Res.* 219, 265–83. doi:10.1007/s10479-010-0830-y.

Crowder, D. W. and Reganold, J. P. 2015. Financial competitiveness of organic agriculture on a global scale. *Proc. Natl. Acad. Sci. U.S.A.* 112(24), 7611–6. doi:10.1073/pnas.1423674112.

Detlefsen, N. K. and Jensen, A. L. 2007. Modelling optimal crop sequences using network flows. *Agric. Syst.* 94(2), 566–72. doi:10.1016/j.agsy.2007.02.002.

Dick, R. P. 1992. A review: long-term effects of agricultural systems on soil biochemical and microbial parameters. *Agric. Ecosyst. Environ.* 40(1–4), 25–36. doi:10.1016/0167-8809(92)90081-L.

Dounias, I., Aubry, C. and Capillon, A. 2002. Decision-making processes for crop management on African farms. Modelling from a case study of cotton crops in northern Cameroon. *Agric. Syst.* 73(3), 233–60. doi:10.1016/S0308-521X(01)00077-4.

Dury, J., Schaller, N., Garcia, F., Reynaud, A. and Bergez, J. E. 2012. Models to support cropping plan and crop rotation decisions. A review. *Agron. Sustain. Dev.* 32(2), 567–80. doi:10.1007/s13593-011-0037-x.

Dury, J. Ô., Garcia, F., Reynaud, A. and Bergez, J. E. 2013. Cropping-plan decision-making on irrigated crop farms: a spatio-temporal analysis. *Eur. J. Agron.* 50, 1–10. doi:10.1016/j.eja.2013.04.008.

El-Nazer, T. and McCarl, B. A. 1986. The choice of crop rotation: a modeling approach and case study. *Am. J. Agric. Econ.* 68(1), 127–36. doi:10.2307/1241657.

Foley, J. A., Ramankutty, N., Brauman, K. A., Cassidy, E. S., Gerber, J. S., Johnston, M., Mueller, N. D., O'Connell, C., Ray, D. K., West, P. C., Balzer, C., Bennett, E. M., Carpenter, S. R., Hill, J., Monfreda, C., Polasky, S., Rockström, J., Sheehan, J., Siebert, S., Tilman, D. and Zaks, D. P. M. 2011. Solutions for a cultivated planet. *Nature* 478(7369), 337–42. doi:10.1038/nature10452.

Food and Agriculture Organization of the United Nations. 2011. The state of food insecurity in the world, organization. Available at: https://doi.org/ISBN978-92-5-106927-1.

Forrester, R. J., Rodriguez, M., Forrester, R. and Rodriguez, M. 2018. An integer programming approach to crop rotation planning at an organic farm. *UMAP J.* 38, 5–25.

Gilbert, G. S. and Webb, C. O. 2007. Phylogenetic signal in plant pathogen-host range. *Proc. Natl. Acad. Sci. U.S.A.* 104(12), 4979–83. doi:10.1073/pnas.0607968104.

Grigulis, K., Lavorel, S., Krainer, U., Legay, N., Baxendale, C., Dumont, M., Kastl, E., Arnoldi, C., Bardgett, R. D., Poly, F., Pommier, T., Schloter, M., Tappeiner, U., Bahn, M. and Clément, J. C. 2013. Relative contributions of plant traits and soil microbial properties to mountain grassland ecosystem services. *J. Ecol.* 101(1), 47–57. doi:10.1111/1365-2745.12014.

Haneveld, W. K. K. and Stegeman, A. W. 2005. Crop succession requirements in agricultural production planning. *Eur. J. Oper. Res.* 166(2), 406–29. doi:10.1016/j.ejor.2004.03.009.

Hobbie, S. E. 2015. Plant species effects on nutrient cycling: revisiting litter feedbacks. *Trends Ecol. Evol. (Amst.)* 30(6), 357–63. doi:10.1016/j.tree.2015.03.015.

Hoeksema, J. D., Chaudhary, V. B., Gehring, C. A., Johnson, N. C., Karst, J., Koide, R. T., Pringle, A., Zabinski, C., Bever, J. D., Moore, J. C., Wilson, G. W. T., Klironomos, J. N. and Umbanhowar, J. 2010. A meta-analysis of context-dependency in plant response to inoculation with mycorrhizal fungi. *Ecol. Lett.* 13(3), 394–407. doi:10.1111/j.1461-0248.2009.01430.x.

Hoekstra, A. Y. and Mekonnen, M. M. 2012. The water footprint of humanity. *Proc. Natl. Acad. Sci. U.S.A.* 109(9), 3232-7. doi:10.1073/pnas.1109936109.

Inderjit, Wardle, D. A., Karban, R. and Callaway, R. M. 2011. The ecosystem and evolutionary contexts of allelopathy. *Trends Ecol. Evol. (Amst.)* 26(12), 655-62. doi:10.1016/j.tree.2011.08.003.

Ingerslew, K. S. and Kaplan, I. 2018. Distantly related crops are not better rotation partners for tomato. *J. Appl. Ecol.* 55(5), 2506-16. doi:10.1111/1365-2664.13156.

Isbell, F., Adler, P. R., Eisenhauer, N., Fornara, D., Kimmel, K., Kremen, C., Letourneau, D. K., Liebman, M. and Polley, H. W. 2017. Benefits of increasing plant diversity in sustainable agroecosystems. *J. Ecol.* 105(4), 871-9. doi:10.1111/1365-2745.12789.

Itoh, T., Hiroaki, I. and Naneseji, T. 2003. A model of crop planning under uncertainty in agricultural management. *Int. J. Prod. Econ.* 81, 555-8.

Ke, P. J., Miki, T. and Ding, T. S. 2015. The soil microbial community predicts the importance of plant traits in plant-soil feedback. *New Phytol.* 206(1), 329-41. doi:10.1111/nph.13215.

Klironomos, J. N. 2002. Feedback with soil biota contributes to plant rarity and invasiveness in communities. *Nature* 417(6884), 67-70. doi:10.1038/417067a.

Kollas, C., Kersebaum, K. C., Nendel, C., Manevski, K., Müller, C., Palosuo, T., Armas-Herrera, C. M., Beaudoin, N., Bindi, M., Charfeddine, M., Conradt, T., Constantin, J., Eitzinger, J., Ewert, F., Ferrise, R., Gaiser, T., Cortazar-Atauri, I. Gd de, Giglio, L., Hlavinka, P., Hoffmann, H., Hoffmann, M. P., Launay, M., Manderscheid, R., Mary, B., Mirschel, W., Moriondo, M., Olesen, J. E., Öztürk, I., Pacholski, A., Ripoche-Wachter, D., Roggero, P. P., Roncossek, S., Rötter, R. P., Ruget, F., Sharif, B., Trnka, M., Ventrella, D., Waha, K., Wegehenkel, M., Weigel, H. J. and Wu, L. 2015. Crop rotation modelling-a European model intercomparison. *Eur. J. Agron.* 70, 98-111. doi:10.1016/j.eja.2015.06.007.

Kremen, C., Iles, A. and Bacon, C. 2012. Diversified farming systems : an agroecological, systems-based alternative to modern industrial agriculture. *Ecol. Soc.* 17(4), 44.

Lawes, J. B. 1895. The Rothamsted experiments. *Trans. Highl. Agric. Soc. Scotland, 5th Ser.* 7, 11-354.

Lesk, C., Rowhani, P. and Ramankutty, N. 2016. Influence of extreme weather disasters on global crop production. *Nature* 529(7584), 84-7. doi:10.1038/nature16467.

Liu, X., Liang, M., Etienne, R. S., Wang, Y., Staehelin, C. and Yu, S. 2012. Experimental evidence for a phylogenetic Janzen-Connell effect in a subtropical forest. *Ecol. Lett.* 15(2), 111-8. doi:10.1111/j.1461-0248.2011.01715.x.

Lori, M., Symnaczik, S., Mäder, P., De Deyn, G. and Gattinger, A. 2017. Organic farming enhances soil microbial abundance and activity–A meta-analysis and meta-Regression. *PLoS ONE* 12(7), e0180442. doi:10.1371/journal.pone.0180442.

Mariotte, P., Mehrabi, Z., Bezemer, T. M., De Deyn, G. B., Kulmatiski, A., Drigo, B., Veen, G. F. C., van der Heijden, M. G. A. and Kardol, P. 2018. Plant-soil feedback: bridging natural and agricultural sciences. *Trends Ecol. Evol. (Amst.)* 33(2), 129-42 doi:10.1016/j.tree.2017.11.005.

Martin-Clouaire, R. 2017. Modelling operational decision-making in agriculture. *Agric. Sci.* 8(7), 527-44. doi:10.4236/as.2017.87040.

Mehrabi, Z. and Tuck, S. L. 2015. Relatedness is a poor predictor of negative plant - soil feedbacks. *New Phytol.* 205(3), 1071-5.

Mehrabi, Z., McMillan, V. E., Clark, I. M., Canning, G., Hammond-Kosack, K. E., Preston, G., Hirsch, P. R. and Mauchline, T. H. 2016. *Pseudomonas* spp. diversity is negatively associated with suppression of the wheat take-all pathogen. *Sci. Rep.* 6, 29905. doi:10.1038/srep29905.

Mohler, C. L. and Johnson, S. E. 2009. *Crop Rotation on Organic Farms A Planning Manual.*

Newbold, T., Bennett, D. J., Choimes, A., Collen, B., Day, J., Palma, A. De, Díaz, S., Edgar, M. J., Feldman, A., Garon, M., Harrison, M. L. K., Alhusseini, T., Echeverria-london, S., Ingram, D. J., Itescu, Y., Kattge, J., Kemp, V., Kirkpatrick, L., Kleyer, M., Laginha, D., Correia, P., Martin, C. D., Meiri, S., Novosolov, M., Pan, Y., Phillips, H. R. P., Purves, D. W., Robinson, A., Simpson, J., Tuck, S. L., Weiher, E., White, H. J., Ewers, R. M. and Mace, G. M. 2015. Global effects of land use on local terrestrial biodiversity. *Nature* 520, 45–50. doi:10.1038/nature14324.

Ng, M., Fleming, T., Robinson, M., Thomson, B., Graetz, N., Margono, C., Mullany, E. C., Biryukov, S., Abbafati, C., Abera, S. F., Abraham, J. P., Abu-Rmeileh, N. M. E., Achoki, T., AlBuhairan, F. S., Alemu, Z. a., Alfonso, R., Ali, M. K., Ali, R., Guzman, N. A., Ammar, W., Anwari, P., Banerjee, A., Barquera, S., Basu, S., Bennett, D. a., Bhutta, Z., Blore, J., Cabral, N., Nonato, I. C., Chang, J.-C., Chowdhury, R., Courville, K. J., Criqui, M. H., Cundiff, D. K., Dabhadkar, K. C., Dandona, L., Davis, A., Dayama, A., Dharmaratne, S. D., Ding, E. L., Durrani, A. M., Esteghamati, A., Farzadfar, F., Fay, D. F. J., Feigin, V. L., Flaxman, A., Forouzanfar, M. H., Goto, A., Green, M. a., Gupta, R., Hafezi-Nejad, N., Hankey, G. J., Harewood, H. C., Havmoeller, R., Hay, S., Hernandez, L., Husseini, A., Idrisov, B. T., Ikeda, N., Islami, F., Jahangir, E., Jassal, S. K., Jee, S. H., Jeffreys, M., Jonas, J. B., Kabagambe, E. K., Khalifa, S. E. A. H., Kengne, A. P., Khader, Y. S., Khang, Y.-H., Kim, D., Kimokoti, R. W., Kinge, J. M., Kokubo, Y., Kosen, S., Kwan, G., Lai, T., Leinsalu, M., Li, Y., Liang, X., Liu, S., Logroscino, G., Lotufo, P. a., Lu, Y., Ma, J., Mainoo, N. K., Mensah, G. a., Merriman, T. R., Mokdad, A. H., Moschandreas, J., Naghavi, M., Naheed, A., Nand, D., Narayan, K. M. V., Nelson, E. L., Neuhouser, M. L., Nisar, M. I., Ohkubo, T., Oti, S. O., Pedroza, A., Prabhakaran, D., Roy, N., Sampson, U., Seo, H., Sepanlou, S. G., Shibuya, K., Shiri, R., Shiue, I., Singh, G. M., Singh, J. a., Skirbekk, V., Stapelberg, N. J. C., Sturua, L., Sykes, B. L., Tobias, M., Tran, B. X., Trasande, L., Toyoshima, H., van de Vijver, S., Vasankari, T. J., Veerman, J. L., Velasquez-Melendez, G., Vlassov, V. V., Vollset, S. E., Vos, T., Wang, C., Wang, S. X., Weiderpass, E., Werdecker, A., Wright, J. L., Yang, Y. C., Yatsuya, H., Yoon, J., Yoon, S.-J., Zhao, Y., Zhou, M., Zhu, S., Lopez, A. D., Murray, C. J. L. and Gakidou, E. 2014. Global, regional, and national prevalence of overweight and obesity in children and adults during 1980–2013: a systematic analysis for the Global Burden of Disease Study 2013. *Lancet* 384(9945), 766–91. doi:10.1016/S0140-6736(14)60460-8.

Orwin, K. H., Buckland, S. M., Johnson, D., Turner, B. L., Smart, S., Oakley, S. and Bardgett, R. D. 2010. Linkages of plant traits to soil properties and the functioning of temperate grassland. *J. Ecol.* 98(5), 1074–83. doi:10.1111/j.1365-2745.2010.01679.x.

Pineda, A., Kaplan, I. and Bezemer, T. M. 2017. Steering soil microbiomes to suppress aboveground insect pests. *Trends Plant Sci.* 22(9), 770–8. doi:10.1016/j.tplants.2017.07.002.

Ponisio, L. C., Gonigle, L. K. M., Mace, K. C., Palomino, J., Valpine, P. De and Kremen, C. 2015. Diversification practices reduce organic to conventional yield gap. *Proc. Biol. Sci.* 282(1799), 20141396. doi:10.1098/rspb.2014.1396.

Pretty, J., Benton, T. G., Bharucha, Z. P., Dicks, L. V., Flora, C. B., Godfray, H. C. J., Goulson, D., Hartley, S., Lampkin, N., Morris, C., Pierzynski, G., Prasad, P. V. V., Reganold, J., Rockström, J., Smith, P., Thorne, P. and Wratten, S. 2018. Global assessment of agricultural system redesign for sustainable intensification. *Natl. Sustain.* 1(8), 441–6. doi:10.1038/s41893-018-0114-0.

Quinton, J. N., Govers, G., Van Oost, K. and Bardgett, R. D. 2010. The impact of agricultural soil erosion on biogeochemical cycling. *Nat. Geosci.* 3(5), 311–4. doi:10.1038/ngeo838.

Raaijmakers, J. M., Paulitz, T. C., Steinberg, C., Alabouvette, C. and Moënne-Loccoz, Y. 2009. The rhizosphere: a playground and battlefield for soilborne pathogens and beneficial microorganisms. *Plant Soil* 321(1-2), 341–61. doi:10.1007/s11104-008-9568-6.

Ramankutty, N., Mehrabi, Z., Waha, K., Jarvis, L., Kremen, C., Herrero, M. and Rieseberg, L. H. 2018. Trends in global agricultural land use: implications for environmental health and food security. *Annu. Rev. Plant Biol.* 69, 789-815. doi:10.1146/annurev-arplant-042817-040256.

Renard, D. and Tilman, D. 2019. National food production stabilized by crop diversity. *Nature* 571(7764), 257–60. doi:10.1038/s41586-019-1316-y.

Ricciardi, V., Ramankutty, N., Mehrabi, Z., Jarvis, L. and Chookolingo, B. 2018. How much of the world's food do smallholders produce? *Glob. Food Sec.* 17, 64–72. doi:10.1016/j.gfs.2018.05.002.

Robert, M., Thomas, A., Sekhar, M., Raynal, H., Casellas, É., Casel, P., Chabrier, P., Joannon, A. and Bergez, J. É. 2018. A dynamic model for water management at the farm level integrating strategic, tactical and operational decisions. *Environ. Modell. Softw.* 100, 123–35. doi:10.1016/j.envsoft.2017.11.013.

Rosa-Schleich, J., Loos, J., Mußhoff, O. and Tscharntke, T. 2019. Ecological-economic trade-offs of Diversified Farming Systems – a review. *Ecol. Econ.* 160, 251–63. doi:10.1016/j.ecolecon.2019.03.002.

Sainju, U. M. 2016. A global meta-analysis on the impact of management practices on net global warming potential and greenhouse gas intensity from cropland soils. *PLoS ONE* 11(2), e0148527. doi:10.1371/journal.pone.0148527.

Schnitzer, S. A., Klironomos, J. N., HilleRisLambers, J., Kinkel, L. L., Reich, P. B., Xiao, K., Rillig, M. C., Sikes, B. A., Callaway, R. M., Mangan, S. A., van Nes, E. H. and Scheffer, M. 2011. Soil microbes drive the classic plant diversity-productivity pattern. *Ecology* 92, 1385–92.

Sedio, B. E. and Ostling, A. M. 2013. How specialised must natural enemies be to facilitate coexistence among plants? *Ecol. Lett.* 16(8), 995–1003. doi:10.1111/ele.12130.

Silva, J. C. Pd, Medeiros, F. H. Vd and Campos, V. P. 2018. Building soil suppressiveness against plant-parasitic nematodes. *Biocontrol Sci. Technol.* 28(5), 423–45. doi:10.1080/09583157.2018.1460316.

Smith, P., Bustamante, M., Ahammad, H., Clark, H., Dong, H., Elsiddig, E. A., Haberl, H., Harper, R., House, J., Jafari, M., Masera, O., Mbow, C. and Ravindranath, N. H. 2014. Agriculture, forestry and other land use (AFOLU). In: Edenhofer, O., Pichs-Madruga, R., Sokona, Y., Farahani, E., Kadner, S., Seyboth, K. and Adler, A. (Eds), *Climate Change 2014: Mitigation of Climate Change. Contribution of Working Group III to the Fifth Assessment Report of the Intergovernmental Panel on Climate Change*. Cambridge University Press, Cambridge, UK.

Snapp, S. S., Blackie, M. J., Gilbert, R. A., Bezner-Kerr, R. and Kanyama-Phiri, G. Y. 2010. Biodiversity can support a greener revolution in Africa. *Proc. Natl. Acad. Sci. U.S.A.* 107(48), 20840–5. doi:10.1073/pnas.1007199107.

Van Der Putten, W. H., Van Dijk, C. and Peters, B. A. M. 1993. Plant-specific soil-borne diseases contribute to succession in foredune vegetation. *Nature* 362(6415), 53–6. doi:10.1038/362053a0.

Van Der Putten, W. H., Bardgett, R. D., Bever, J. D., Bezemer, T. M., Casper, B. B., Fukami, T., Kardol, P., Klironomos, J. N., Kulmatiski, A., Schweitzer, J. A., Suding, K. N., Van de Voorde, T. F. J. and Wardle, D. A. 2013. Plant-soil feedbacks: the past, the present and future challenges. *J. Ecol.* 101(2), 265–76. doi:10.1111/1365-2745.12054.

Venter, Z. S., Jacobs, K. and Hawkins, H. J. 2016. The impact of crop rotation on soil microbial diversity: a meta-analysis. *Pedobiologia* 59(4), 215–23. doi:10.1016/j.pedobi.2016.04.001.

Veresoglou, S. D., Barto, E. K., Menexes, G. and Rillig, M. C. 2013. Fertilization affects severity of disease caused by fungal plant pathogens. *Plant Pathol.* 62(5), 961–9. doi:10.1111/ppa.12014.

Vitousek, P. M., Mooney, H. A., Lubchenco, J. and Melillo, J. M. 1997. Human domination of Earth's ecosystems. *Science* 277(5325), 494–9. doi:10.1126/science.277.5325.494.

Webb, C. O., Gilbert, G. S. and Donoghue, M. J. 2006. Phylodiversity-dependent seedling mortality, size structure, and disease in a Bornean rain forest. *Ecology* 87(Suppl. 7), S123–31. doi:10.1890/0012-9658(2006)87[123:PSMSSA]2.0.CO;2.

Chapter 9

Decision-support systems for pest monitoring and management

B. Sailaja, Ch. Padmavathi, D. Krishnaveni, G. Katti, D. Subrahmanyam, M. S. Prasad, S. Gayatri and S. R. Voleti, ICAR-Indian Institute of Rice Research, India

1 Introduction

Pests including insects, diseases and weeds have major effects on crop production in many regions of the world. Crops are damaged by more than 10 000 species of insects, 30 000 species of weeds, 1000 species of nematodes and around 100 000 diseases. However, only 10% of all the identified pest species are considered as major pests responsible for destroying around 20% of total global crop production annually (Dhaliwal et al., 2015). Farmers are often confused about identifying pests, and so are not able to apply the right control measures at the right time during crop cultivation. Effective pest management requires a wide range of information from different sources. Pest-management decisions are based on knowledge of the host crop, pest identification, characteristics and life cycles, as well as natural enemies of pests and their interactions in an ecosystem. Other information includes host plant resistance, agronomic practices such as crop rotations and other cultural and agronomic operations which may affect pest life cycles and proliferation. Pest management also needs to take into account the range of cultural, biological, physical and chemical pest-control techniques available within an integrated pest management (IPM) programme. IPM therefore requires a complex

http://dx.doi.org/10.19103/AS.2020.0069.18

decision-making process. The availability of the relevant information organized in one place significantly increases the chance of pest management being effective. The development of decision-support systems (DSS) is thus a key tool in reducing the losses from pests.

Advances in information technology (IT) have opened up many opportunities to address the challenges in knowledge-based agricultural pest management. Tools such as databases, inference mechanisms, graphic user interfaces, object-oriented and open source programs, artificial intelligence (AI), machine/deep learning, Internet of Things (IoT), data mining techniques, and so on are widely available for use in the development of support systems for growers. The major challenges in developing these systems and examples of DSS are discussed in this chapter.

2 Pest identification

Farmers often struggle either to identify a particular pest or the symptoms of damage they cause to crops. Many symptoms are similar and may have a range of potential causes. In addition to the ability to match a pest or symptom to a matching image in a database, for example, additional information on the location, soil, weather parameters, crop variety and genotype, age and stage of crop and so on may be needed to correctly diagnose the problem and suggest a solution. The information required and the challenges involved in a correct diagnosis highlight the need for expert systems and image-based diagnostic techniques to support farmers.

An expert system is a computer program that emulates the decision-making ability of a human expert (Jackson, 1998). It uses the knowledge base of human expertise for problem solving. The inference engine is an automated reasoning system that evaluates the current state of the knowledge base, applies relevant rules and then adds new knowledge into the knowledge base (Hayes-Roth et al., 1983). There are many expert systems available for most major crops. The most popular expert system, PLANT was developed by the University of Illinois for diagnosis of soybean diseases and insect damage. A unique feature of the system is that it uses two types of decision rules (Michalski et al., 1983):

- rules based on expert diagnostic knowledge and
- rules obtained through inductive learning from several hundred cases of diseases

Examples of other expert systems for particular crops are:

- Corn – the corn/maize disease remote diagnostic system (Xinxing et al., 2012)

- CPEST for coffee (Mansingha et al., 2007)
- DCAS for sugarcane (Nwagu et al., 2018)
- Rice Doctor: developed by the International Rice Research Institute (IRRI) (http://www.knowledgebank.irri.org)
- VEGES: a multilingual expert system for the diagnosis of pests, diseases and nutritional disorders of six greenhouse vegetables: pepper, lettuce, cucumber, bean, tomato and aubergine (Yialourisa et al., 1997)

These systems are complemented by local initiatives/technologies.

An example of a locally developed DSS is the standalone rice expert system developed to identify pest and disease problems in rice crops by the Indian Institute of Rice Research (IIRR) as part of the Indian Council of Agriculture Research (ICAR) (Sailaja et al., 2004) and upgraded to the web-based Ricexpert system (www.Ricexpert.in). This system uses a user questionnaire and an inference mechanism to conduct formal reasoning based on user answers and corresponding rules in the knowledge base. There are 90 rules for identifying insect pests and 105 rules for identifying diseases. There is a separate module with recommended biological, cultural or chemical methods of control as well as recommendations for resistant varieties. Another system is the rice crop doctor expert system. This is an initiative by the National Institute of Agricultural and Extension Management (MANAGE) in the Indian Ministry of Agriculture to diagnose pests and diseases for rice crops and suggest preventive as well as curative measures (Yelapure and Kulkarni, 2012). Other systems developed in India include AMRAPALIKA for mango (Rajkishore, 2006), AGREX (Ganesan, 2007) and the Cassava Expert Protect System (http://ctcritools.in/).

Since mobile phones are now widely used by the farmers, and internet connectivity is available in many areas at affordable prices, technologies have moved from web to mobile with advanced AI capabilities. Image-based diagnostics is now AI-based. Image classification and processing techniques take into account features such as shape, colour, texture and morphology. Artificial neural networks (ANNs) are widely used for image diagnostics. An example is 'eSagu', an IT-based personalized agro-advisory system aiming to deliver high-quality personalized (farm-specific) agro-expert advice directly to farmers. In this system, farmers send digital photographs of crops together with other information to a central location where experts use agricultural information systems to tailor agronomic (including pest control) advice which can then be sent back to the farmers' mobiles (http://www.esagu.in).

The German agricultural tech startup PEAT has developed the Plantix app which uses images to detect plant diseases using AI and machine-learning techniques (https://plantix.net/en/; http://www.fao.org/e-agriculture). A smart phone collects the image which is matched with a server image before the system offers plant health diagnosis advice. An ANN has also been used for diagnosing diseases of grapes and apples (Jhuria et al., 2013). Avishkar, a mobile-based AI module, has been developed for rice pest detection using basic details of the location and crop combined with machine-learning algorithms (http: //www. icar-iirr.org).

Disease diagnosis can also be undertaken using spectral reflectance patterns. The normalized difference vegetation index (NDVI) is a commonly used indicator in remote sensing that identifies vegetation and measures a plant's overall health. The development of drone technology now allows for a more precise localized measurement that complements area wide assessments of plant health from images taken by satellites or aircraft. The NDVI-based vegetation indices derived from hyperspectral data have been used, for example, to detect stress in wheat caused by pests (Genc et al., 2008). Hyperspectral remote sensing has been used to detect damage to rice plants caused by the brown planthopper (BPH) in both field and glasshouse conditions. The study revealed that the variation in plant reflectance due to BPH damage was smaller at shorter wavelengths (350–730 nm) and larger at longer wavelengths, viz., NIR (740–925 nm) followed by mid infrared (MIR) (926–1800 nm) (Prasannakumar et al., 2013).

3 Pest monitoring

While accurately identifying pests is fundamental to their effective management, it is important to have effective monitoring of insect populations, ideally with the ability to predict when populations might reach damaging levels. It is clearly better to prevent populations reaching these levels rather than respond to damage once it has started to occur on a significant scale.

Insect pests are usually monitored through a variety of pheromone traps, sticky traps, pitfall traps, suction traps and light traps. Data from trap catches can assist in decision-making. As an example, the first appearance of a pest can initiate more comprehensive pest scouting and the implementation of an IPM programme to limit the pest. However, manual daily inspection of these traps is labour-intensive and requires expert staff to correctly identify insect species found in these traps. To overcome these problems, intelligent smart traps have been developed that provide automated monitoring using computer vision, object recognition and artificial intelligence (www.trapview.com). The pests caught in these traps are recognized using integrated cameras with cloud-based processing software, categorized and automatically counted. Similar automated traps are being used in apple orchards where growers receive information by wireless transmission from traps that record insect activity using cameras and other devices (https://www.goodfruit.com). Internet-based communication technology is used to send the data to the producer's smartphone. The farmer receives daily reports of insect counts and species types together with alert messages if action needs to be taken. These traps help in both efficient monitoring and in prompting appropriate pest-management techniques to farmers and growers.

Recent advances in technology have also reduced the reliance on traps through the development of pest detection sensors that help in the early detection of insect pests and diseases by providing real-time data from the field. These include image sensors which capture images of pests in the field, acoustic sensors which monitor pest noise levels, thermographic sensors that capture infrared radiation emitted from the plant surface (to detect changes due to disease infection) and fluorescence sensors tracking changes in chlorophyll and photosynthetic activity related to pest activity (https://www.farmmanagement.pro/).

Advances in the field of remote sensing offer ample scope to use this technology for pest monitoring, damage assessment and provision of data to pest management decision-support systems (Gogoi et al., 2018; Sudha Rani et al., 2018; Shanmugapriya et al., 2019). Area-wide pest monitoring is traditionally done through regular field surveys which are time-consuming, labour-intensive and error prone. Remote sensing is a rapid, non-invasive and efficient technique, which obtains data on spectral properties from satellites and ground-based platforms using hyperspectral and multispectral radiometers to detect pest incidence and potential damage (Huang et al., 2012). Pest damage has been linked with spectral indices based on leaf pigments (Riedell and Blackmer, 1999; Yang and Cheng, 2001; Prabhakar et al., 2012). Hyperspectral reflectance patterns have been used to assess damage caused by brown planthoppers and leaf folders in rice. Based on the correlation and multinomial logistic regression (MLR) analysis, sensitive bands specific to leaf folder and brown planthopper damage have been identified as hyperspectral indices. These indices can be used to assess area wide damage caused by these two insects using satellite images (Padmavathi et al., 2014). LANDSAT data in near infrared bands (MSS6 and MSS7) has been used to discriminate rust affected wheat from healthy wheat (Nagarajan et al., 1984). Remote-sensing techniques like optical and video imaging in near infrared and microwave regions have been used to quantify the nocturnal flight behaviour of pod borer, *Helicoverpa armigera*, in pigeonpea (Riley et al., 1992). Fitzgerald (2000) demonstrated that multispectral remote sensing (MRS) would allow farmers to detect early infestation of mites in large-scale cotton fields due to colour shifts and changes in canopy appearance over time. These areas were later located with GPS by field scouts to verify the population and advise farmers for pesticide application.

Pest distribution is the most authoritative way to visualize the presence and extent of plant pests.

Geographical information systems (GIS) can be effectively used for preparing pest/disease distribution maps. District-level maps have been prepared to show changes in the intensity and distribution of different rice diseases over three decades using production survey data under the All India Coordinated Rice Improvement Project (AICRIP) (Laha et al., 2016; http://www.iirr-geoportal.in). The potential risk of pest and disease damage to rice crop was studied at the province level using a GIS to monitor the relationship between environmental and climatic variables. By overlaying data such as temperature, humidity and stage of crop, it is possible to identify patterns which will help in predicting the occurrence of the diseases. Crop-pest interactions will change significantly with climate change leading to an impact on pest distribution and crop losses (https://www.geospatialworld.net). Similarly, a geo-media package has been used in identifying key pests and delineating endemic districts in India using historical data from production-oriented survey reports (POS, 1990–2000) (Sailaja et al., 2002).

4 Pest forecasting

Accurate forecasting of pest attacks is clearly preferable so that sufficient time is available for farmers to plan strategies with maximum efficiency. In nature, pests are regulated by biotic factors like parasitoids, predators and pathogens along with abiotic factors like temperature, relative humidity, rainfall and wind speed (Hance et al., 2007; Thomson et al., 2010). Pest populations fluctuate depending on the host plant, crop stage, season, location and management interventions. The complex interactions among these various factors and ecosystem components determine pest population dynamics. Understanding pest population dynamics provides the foundation for forecasting the increases in pest populations to economically damaging levels and taking action either to prevent those increases or deal with infestations in a timely, planned and effective way.

Pest forecasting in temperate regions often involves tracking hatching of overwintering eggs or the first adult emergence from overwintering pupae (Collier et al., 1991; Trnka et al., 2007). The emergence typically takes place over a relatively short period of time and is not too difficult to monitor. In tropical parts of the world, where weather conditions permit continuous breeding of pests most of the time, forecasting relates to the first appearance of the pest in the crop (Krishnaiah et al., 1997) or in identifying activity in an adjoining area with a recorded history of pest infestation at levels causing serious economic damage (Otuka et al., 2005). Successful forecasting techniques are based in part on monitoring and in part on insect phenology and population models using knowledge of the biology and ecology of the pests concerned. Other models are built on understanding conditions causing diseases and on understanding of disease epidemiology.

Decision-support systems (DSSs) have been developed in many countries since the 1990s to assist with the management of plant diseases such as potato late blight, apple scab, cereal leaf diseases, strawberry diseases and grape downy mildew (Pavan et al., 2011; Shtienberg, 2013). Forecasters such as BLITECAST (Krause et al., 1975), FAST (Madden et al., 1978), and the apple scab predictive system (Jones et al., 1980) are examples of early tools that were designed to assist farmers with decisions relating to disease management. In the Netherlands around one-third of potato growers use commercially available DSSs to assist with management of late blight (Cooke et al., 2011).

Insects are poikilothermic organisms which cannot regulate their internal temperature and hence their development is dependent on the surrounding environment. Temperature is a critical abiotic factor affecting the development, survival and reproduction of insect species and is a key factor in predicting potential outbreaks (Gilbert and Raworth, 1996; Legg et al., 2000; Tobin et al., 2003). The rate of insect development is affected by the temperature to which insects are exposed (Campbell et al., 1974). Insects require a certain amount of heat (degree days) to develop from one life stage to the other (Gordan, 1999). Quantification of the relationship between insect development and temperature is helpful in predicting the seasonal prevalence and population dynamics of insects. The degree-day model, also known as the thermal summation model, is based on a linear relationship between temperature and the rate of development of insect pests (Campbell et al., 1974). As some temperatures are lethal to insect development, it is obvious that development must be a non-linear temperature function at the temperature extremes. Non-linear development rate functions based on enzyme kinetics have been developed to describe high-temperature (Johnson and Lewin, 1946) and low-temperature (Hultin et al., 1955) inhibition, as well as both extremes (Sharpe and DeMichele, 1977; Schoolfield et al., 1981; Ikemoto, 2005; Shi et al., 2011). Using both linear and non-linear models, lower and upper development threshold temperatures (T_{min} and T_{max}) along with the optimum threshold temperature (T_{opt}) can be calculated to estimate total thermal units (degree days) required for the development of a pest.

Phenology models predict the timing of events in the development of an insect and are critical tools needed for predicting, evaluating and understanding the dynamics of pest populations in agroecosystems under a variety of environmental conditions. Accurate predictions, however, require accurate recording of the temperatures experienced by the organisms as well as the duration of development (Morgan, 1991; Danks, 2000). The degree-day models have long been used as part of the DSS to help growers predict spray timing or when to begin pest scouting (Welch et al., 1978; Higley et al., 1986). Phenology models are also used as part of pest risk analysis to predict invasive alien species for taking up quarantine measures (Baker, 1991; Jarvis and Baker, 2001).

Phenology-based degree-day models have been developed in a number of countries. The University of California Statewide IPM programme has an inventory of insect development data on their website for use in degree-day models (http://www.ipm.ucdavis.edu/MODELS). A well-known example of an interactive model is the North Carolina State University Animal and Plant Health Inspection Service Plant Pest Forecasting System (NAPPFAST; Magarey et al., 2007, 2015). This has been developed with an internet-based graphical user interface to link interactive templates of pest incidence with weather databases. Insect Life Cycle Modeling (ILCYM) software is a generic open-source computer-aided tool that facilitates the development of process-based, temperature-driven and age-structured insect phenology models for prediction of pest activity in specific agroecological zones (Sporleder et al., 2014; Mujica et al., 2017). These models can predict insect species distribution and provide risk mapping when integrated with GIS.

A web-enabled decision-support system called 'Crop Pest DSS' has been developed in India under the National Agricultural Innovation Project (NAIP) for pest prediction in rice and cotton-based cropping systems (www.crida.in:8080/naip). This website enables the user to calculate degree days and thus forecast pest activity (Fig. 1). Phenology-based degree-day models have been developed and validated for the major insect pests of cotton and rice to forecast pest population dynamics as a basis for appropriate IPM strategies

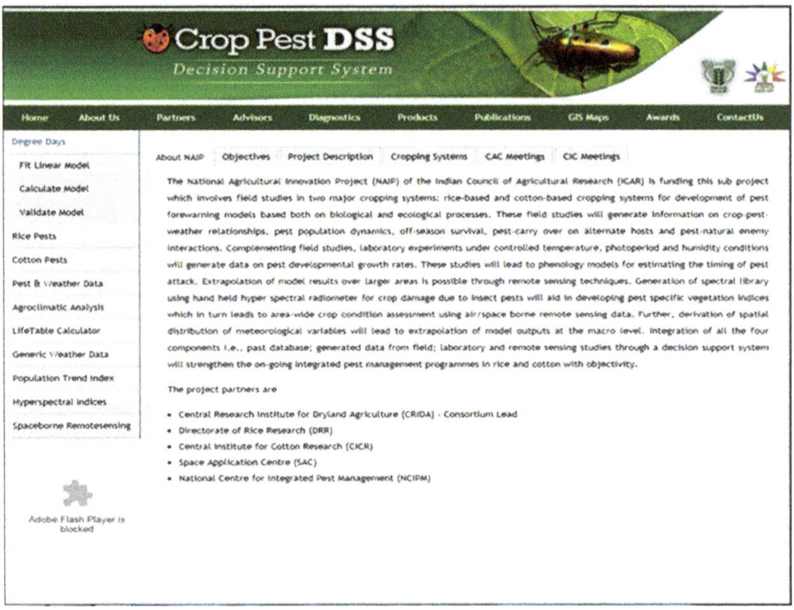

Figure 1 Home page of crop pest DSS.

(Prasad and Prabhakar, 2011; Padmavathi et al., 2013; Sreedevi et al., 2013). These use historical data for weather patterns and pest incidence in major locations across India. This data can be used for forecasting of pest activity in these locations to inform extension specialists and farmers.

Ecological life tables are useful tools in the study of population dynamics of insects having discrete generations. Data from long-term population studies can be used to construct population models that relate adequately to biological reality. Apart from generating population estimates, life table analysis helps in recognizing the independent factors such as parasitoids, predators, pathogens and weather factors responsible for increase and decrease in pest numbers from generation to generation (Morris, 1963; Varley and Gradwell, 1970; Padmavathi et al., 2008). A well-known example is the DYMEX modelling package that was developed to facilitate construction of mechanistic, process-based multi-cohort population models based on an understanding of pest life cycles (Sutherst et al., 2000; Maywald et al., 2007). Pest population models help in determining the optimal IPM control strategy for a given situation (Plant and Mange, 1987). An example of a comprehensive population dynamics model is HElicoverpa Armigera and Punctigera Simulation (HEAPS) which was developed in Australia. This incorporates the spatial structure of the habitat and pest population and explicitly simulates adult movement within a regional cropping system. HEAPS includes modules for the spatial representation of the region, moth movement, oviposition, pest and crop development and pest mortality (Fitt et al., 1995).

Plant pathologists have developed a large number of disease models based on understanding the mobilization of primary inoculum, the production, spread and efficiency of secondary inoculum, or both (Rossi et al., 2009; Bregaglio and Donatelli, 2015). These are typically based on four sub-domains in the modelling of plant disease epidemics (Donatelli et al., 2017):

1 the production of primary inoculums and the occurrence of primary infections
2 the development of secondary infection cycles during the cropping season
3 the interactions between epidemic development and crop physiological processes
4 the impact of agricultural management practices on disease development

These models can be used to simulate a polycyclic fungal plant epidemic and to quantify its impact on crop growth.

As part of the National Initiative on Climate Resilient Agriculture (NICRA) Project, the National Centre for Integrated Pest Management (ICAR-NCIPM) in India has developed a mobile app called 'Pest Prediction Empirical Model Based

System (PestPredict)' to generate predictions of insects and disease outbreaks by relating weather input data to model equations predicting pest population growth. A rule-based system (RBS) helps in the prediction of insect pests in rice, *Spodoptera litura* of groundnut and early blight of tomato (PestPredict-RBS) (Vennila et al., 2016).

Pest phenology models and population models that help in predicting pest populations are not holistic in nature. In order to know the impact of these biotic constraints on crop growth and yield, it is essential to quantify the complex interactions that take place between crop-pest and pathogen systems, ecosystem functions and processes, biophysical flows, soil and weather parameters and other multi-trophic interactions. This integration of crop models with pest and disease models is made possible with generic simulators that help to identify key processes. There have been many efforts to use crop growth models to simulate the effect of pest damage on crop growth and yield by linking the damage effect of pest population levels to the physiological rates and state variables of these models.

A generic approach to simulate the damage effects of single or multiple pests has been attempted using crop growth models such as CERESRice (which is a part of the DSSAT crop model) in the Philippines (Pinnschmidt et al., 1995) and InfoCrop in India (Chander et al., 2007; Reji et al., 2008; Yadav and Chander, 2010). GENEPEST, a generic crop growth model including the damage mechanisms of pests, has been made available online on the APSnet Plant Health Instructor platform (Savary and Willocquet, 2014). BioMA includes a module for modelling damage to plants and a module to simulate the impact of disease control via agricultural management (Donatelli, 2014; Bregaglio and Donatelli, 2015). The DYMEX-APSIM link has been successfully used to model wheat rust (*Puccinia striiformis*) growth (Whish et al., 2015).

5 Integrated pest management (IPM)

IPM programmes depend heavily on accurate and timely information about pest dynamics. An IPM DSS requires information such as pest identification, pest life histories (cycles), sampling and decision-making criteria, pest developmental models linked to weather networks, biorational pest control methods and currently available pesticides, together with safety issues and environmental impacts. Management guidelines for a particular insect pest include a population density, usually denoted as the 'action threshold' or 'economic threshold'. As long as the pest density remains below this threshold, no action is needed. However, if the insect population density exceeds this level, control action to deal with the pest is recommended. The goal of the economic threshold is to prevent a pest population from reaching the point where its damage causes monetary losses that are equal to the cost of control

(Seiter, 2018). Information and Communication Technology (ICT) provides a wide range of DSS tools which are being used extensively in IPM programmes all over the world (Waheed et al., 2003; Sailaja and Padmakumari, 2009).

An early IPM DSS was SOYBUG, an expert system for Florida farmers to help control four important insect pests of soybeans: the velvet bean caterpillar, stink bug, corn earworm and soybean looper (Beck et al., 1989). SOYBUG integrates a variety of threshold rules based on crop phenology and yield data and gives specific recommendations for pesticides and application rates. A more recent decision-support tool is PESTMAN to help landowners manage weeds in Texas and New Mexico. This model allows the user to identify the appropriate technology to use and calculates the economic impact of the treatment (https://blackland.tamu.edu).

The EntomoLOGIC decision tool is derived from the SIRATAC decision-support system used in the Australian cotton industry from 1976 to 1993 to reduce the risk associated with pest management using chemical pesticides. This was developed by CSIRO in collaboration with the University of Western Sydney (Hearn and Bange, 2002). Advances in hand-held computing have resulted in expanding the development of the system for use with Palm OS hand-held devices, allowing it to be used widely by cotton growers in Australia (Bange et al., 2004). RICEPEST, a model simulating yield loss due to several rice pests under a range of specific production situations in tropical Asia has been developed by IRRI in the Philippines (Willocquet et al., 2002). The BlightPro DSS for potato and tomato late blight management was developed in the United States to integrate pathogen information (mefenoxam sensitivity and host preference), the effects of weather, host resistance and fungicide application on disease development in order to improve in-season disease management (http://blight.eas.cornell.edu/blight/). A recent development is dynamic websites that include interactive models, GIS-based decision systems, real-time weather and market information which provide real-time information and decision-making tools for farmers, for example, developed by the European Federation for Information Technology in Agriculture, Food and the Environment (EFITA) (www.efita.net).

6 Case studies

6.1 Rice expert system

A web-based rice expert system has been developed for diagnosing insect pest and disease problems of the rice crop (www.ricexpert.in; Sailaja et al., 2016). This expert system for rice crop was developed using Microsoft SQL as the back end and ASP.Net as the front end. The main components of this expert system are:

- A knowledge base
- An inference engine
- A user interface

The knowledge base contains the knowledge necessary to understand, formulate and solve problems. The knowledge base was created by entering facts and rules in tabular form (21 tables in all). The knowledge base has 90 rules for identifying insect pests and 105 rules for identifying diseases. The inference engine is the brain of the expert system. An inference mechanism was developed using a Microsoft .net program (https://dotnet.microsoft.com/) to conduct formalized reasoning to address user questions and develop corresponding rules in the knowledge base (mainly the 'If…then' type statements). Two inference methods, forward and backward chaining, were used in the programme. In total, there were 16 stored procedures and 23 active server pages (ASPs) designed for developing this application. The user interface (Fig. 2) consists of a series of questions and answers to diagnose a problem, the ability to browse major pests/diseases/varieties, as well as access information on crop protection measures, commonly used pesticides for rice and how best to use them.

The questionnaire menu of the expert system (Fig. 3) begins by collecting information about the location using drop-down menus. At the second level it collects information on weather followed by crop details such as variety, crop stage and so on. At the third level, questions gather information on the various symptoms in the field encountered by the user. This sequence is designed to access the right answer from the knowledge base with regard to

Figure 2 Home page of rice expert system.

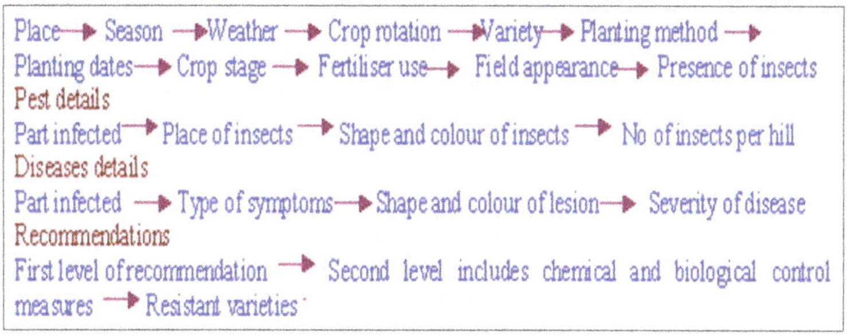

Figure 3 Flow chart of different processes to diagnose insect pest and diseases.

disease or insect pest problems. After the input from the user to questions on field symptoms, details and images of potential diseases or insect pests will start appearing on screen to help the user narrow down identification. At the final level, the engine generates various control measure options as recommendations for controlling the disease or insect pest. The sequence of questions for identifying the stem borer is shown in Fig. 4. In addition to pest/disease identification, the system also provides advice on crop nutrient deficiency. The system is not only used to diagnose pest or disease problems but also maintains a database of newly emerging rice pests. This data can be further analysed to identify location-specific pest/disease problems, movement of pest/diseases and so on.

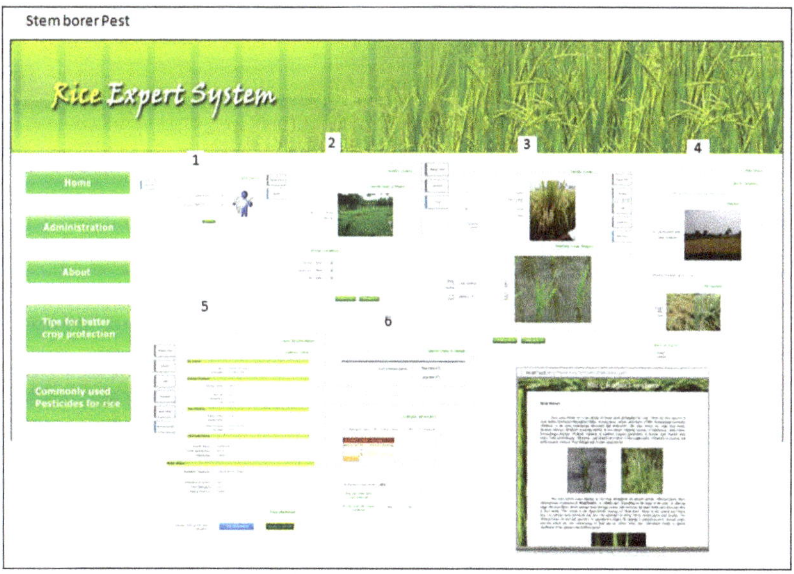

Figure 4 Flow questions to diagnose insect pest stem borer.

6.2 Crop pest DSS

Crop pest DSS was developed utilizing phenology-based degree-day models to predict the timing of insect activity and stage for decision-making in the management of key pests of rice (http://www.crida.in:8080/naip/index.jsp). The system provides key information on rice pests, identification, degree-day models, a pest-related weather database, agroclimatic analysis, a life table calculator and population trend index needed for critical interventions in IPM (Fig. 5).

The rice leaf folder, *Cnaphalocrocis medinalis*, Guenée (Lepidoptera: Pyralidae), is the most widely distributed and commonly found foliage feeder of rice in Southeast Asia. Initial studies focussed upon life cycle analysis to

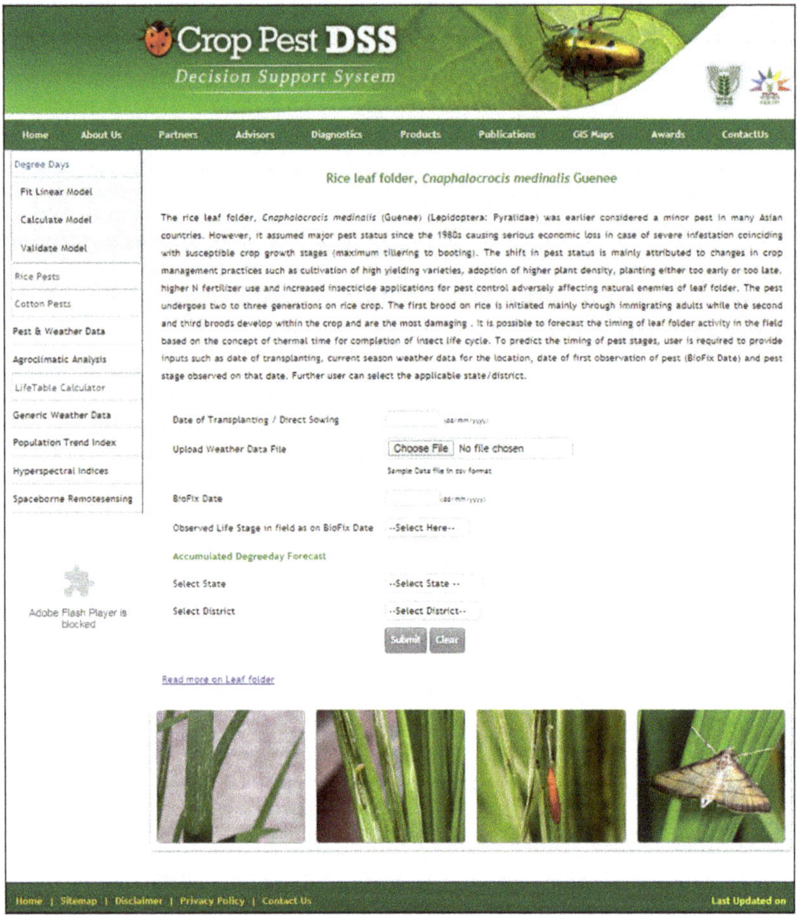

Figure 5 Rice leaf folder page - crop pest DSS.

identify key mortality factors at different stages of the pest life cycle. In the case of the leaf folder, larval and pupal parasitoids were found to cause 32–34% mortality. Temperature was found to be the most important abiotic factor directly influencing development, survival and reproduction, with an indirect impact on generation time and population growth rate.

Response to temperature was assessed by exposing rice leaf folder eggs to seven constant temperatures (18°C, 20°C, 25°C, 30°C, 32°C, 34°C and 35°C) and allowing them to develop into adults. At each of the mentioned temperature survival and duration of development was noted. Linear regressions were used to determine the relationship between developmental rate (1/duration) and temperature and to estimate intercept (a) and slope (b). After determining the lower temperature threshold for each stage, the thermal constant (the number of degree days required for complete development) was estimated from the reciprocals of the fitted regression line (b^{-1}). This allowed the total degree days required for the development of each stage and from egg to adult to be estimated. A thermodynamic Sharpe-Schoolfield-Ikemoto model (SSI model) estimated intrinsic optimum temperature for the development of the leaf folder as 24.2°C with a thermal constant of 445 degree days. Lower (TL) and upper (TU) threshold temperatures were estimated at 11.2°C and 36.4° C, respectively (Fig. 6). Based on these parameters an insect phenology model was developed to predict leaf folder population dynamics in the field (Padmavathi et al., 2013).

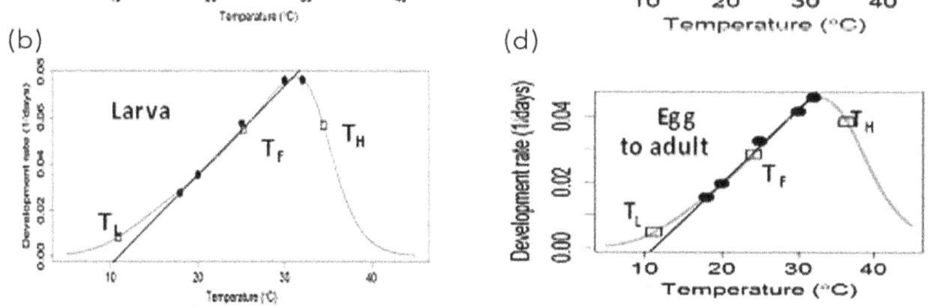

Figure 6 Linear and non-linear thermodynamic Optim SSI model fitted to the temperature-dependent development of *Cnaphalocrocis medinalis*. Circles indicate data points. Open circles indicate data points outside the range of the linear model. The curved line indicates the values of developmental rate predicted by the Optim SSI model whereas the straight line denotes the values of linear fitting. The three open squares denote the predicted mean developmental rates at T_L, T_F and T_H.; (Source: Padmavathi et al., 2013.

Temperature thresholds and thermal constants are potential indicators of the distribution and abundance of leaf folders. The accumulated degree days (ADD) form the base for developing a phenology model to predict populations of leaf folder in the field. These were validated by field tests at three locations using light traps to confirm the activity of rice leaf folder in relation to daily maximum and minimum temperatures, including timing of first appearance (Table 1). Predicted and observed peak catches of rice leaf folder moths in light traps were compared across seasons (Fig. 7) (Padmavathi et al., 2013). The degree-day model was also validated using historic weather data.

Table 1 Leaf folder forecast using crop pest DSS

Leaf folder forecast - summary data		
Generation	Insect stage	Expected starting date of stage
I	Egg	August 28
	Larva	September 01
	Pupa	September 18
	Adult	September 24
II	Egg	September 29
	Larva	October 03
	Pupa	October 20
	Adult	October 27
III	Egg	October 31
	Larva	October 23
	Pupa	November 03
	Adult	November 16

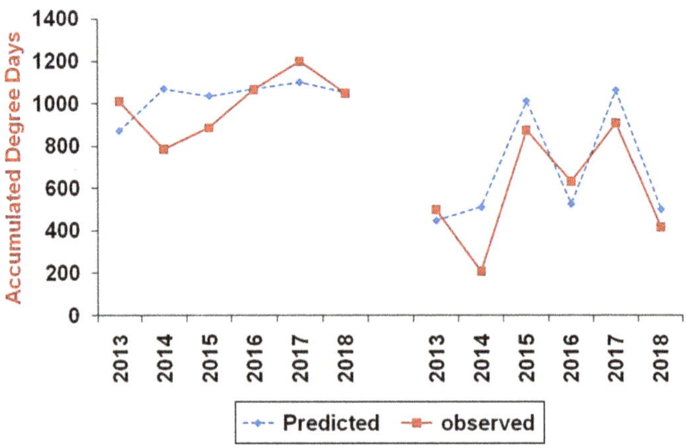

Figure 7 Validation of phenology-based DD model - predicted and observed peak catches of leaf folder moths in light trap.

Rice plants with varying levels of damage symptoms from leaf hoppers (LH) and brown plant hoppers (BPH) were sampled from farmers' fields located at several locations in India. Spectral reflectance was recorded using a hand-held hyperspectral radiometer (350–2500 nm). Correlation and multinomial logistic regression (MLR) analysis were performed to identify sensitive bands specific to LF and BPH. Principle component analysis (PCA) was performed to identify optimum band combinations which were used to build MLR models, which were validated using independent data sets. Spectral reflectance at bands 540 nm, 670 nm, 760 nm, 1454 nm and 1784 nm were found sensitive to BPH, while they were at 390 nm, 675 nm, 780 nm, 1130 nm and 1560 nm for LF. Using the reflectance values at these sensitive bands in MLR, a new set of hyperspectral indices were identified for assessing BPH and LF damage (Fig. 8). Classification accuracy of the models showed promising results. The percent correct classification was in the range between 38–76 for LF and 52–72 for BPH. The new set of hyperspectral indices developed in this study would be useful in remotely assessing damage of LF and BPH using aircraft or satellite platforms. Results of this study are useful for future applications in the field of area-wide pest damage assessment using remote sensing (Padmavathi et al., 2014).

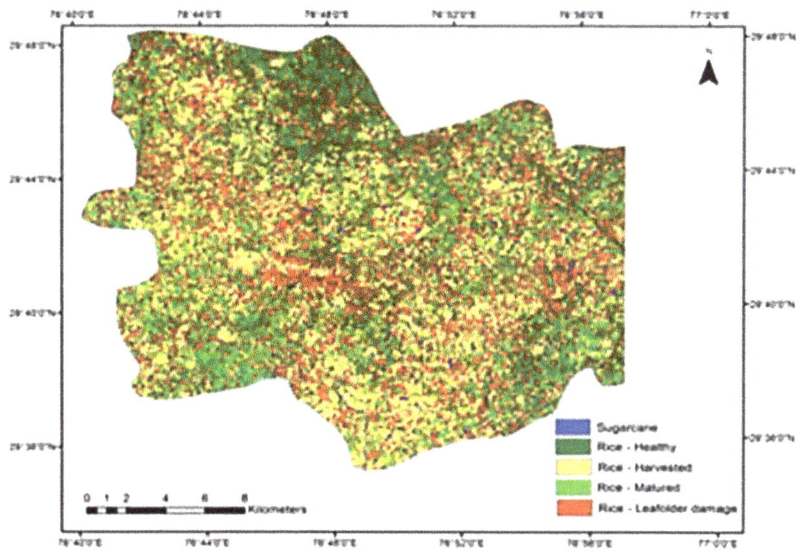

Figure 8 Classified satellite image of rice leaf folder damage in Haryana – analysis showed that about 27% of the area was infested with leaf folder damage, and the area under healthy and harvested fields was about 39% and 33%, respectively.

6.3 IIRR Geo-Portal

IIRR Geo-Portal ((http://iirr-geoportal.in/) has been created to visualize intensity and distribution of rice pests over three decades. Using rice pest survey data from 1981 to 2010, a district-based database was created by categorizing diseases as low (L:<5%), low to medium (L-M:6-15%), medium (M: 16-25%), medium to severe (M-S: 26-50%) and severe (S:>50%). The data were grouped into three decades: 1981-1990, 1991-2000 and 2001-2014. A district average incidence

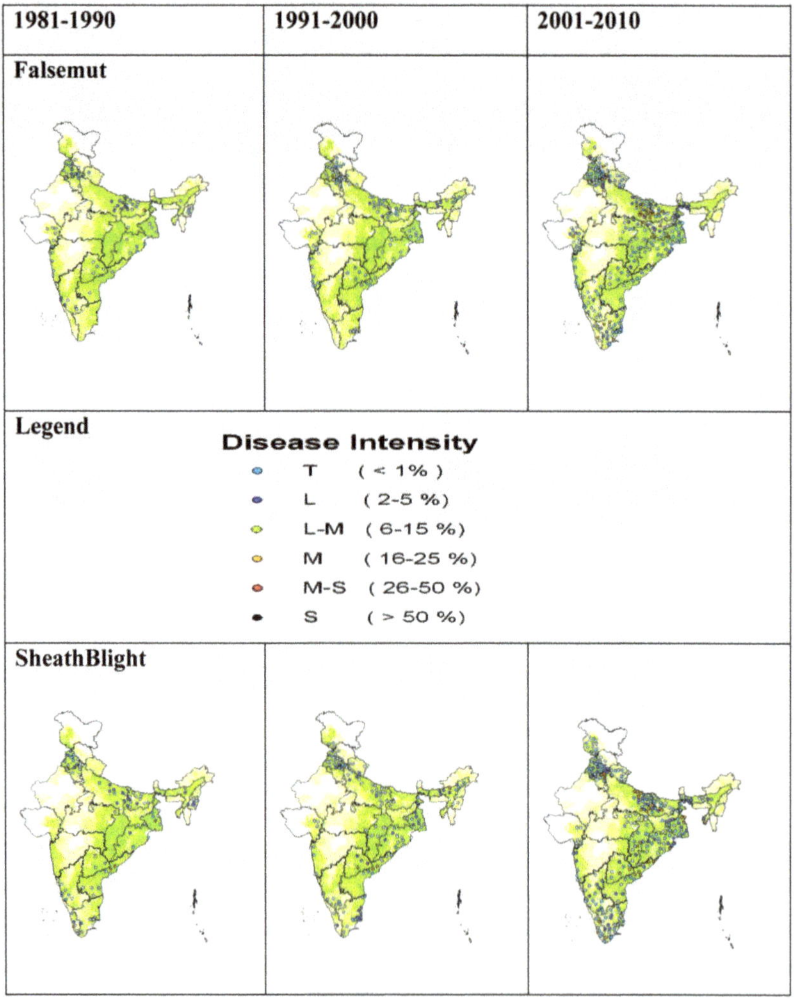

Figure 9 Intensity and distribution of false smut and sheath blight diseases of rice crop during three decades (1981-1990, 1991-2000 and 2001-2010); Source: www.iirr-geoportal.in.

was then calculated in each block. The data were then converted into numerical format by allotting scores on the scale of 1-5. Block-wise disease distribution maps were generated using ArcMap software (Fig. 9). The analysis of historical data for different rice diseases showed significant increases in different rice diseases both in terms of their geographic distribution and intensity.

6.4 Mobile rice IPM app

A mobile app called Rice IPM has been developed for 11 major insect pests, 10 diseases and 18 weed species as well as nutrient deficiencies in crops (http://www.icar-iirr.org). The app includes the means for identification of pests and symptoms of damage (both text and image) at different stages of their life cycle. It also includes suggested recommended management practices like resistant varieties, planting methods, cultural and chemical control methods. If a farmer cannot identify symptoms of damage or disease, he/she can access an image gallery of visible symptoms, choose the image closest to the visible symptom and access appropriate pest management practices easily using the IPM app. If the farmer wants to know about pests at a particular stage of rice cultivation, he/she can select the option 'Stages of Rice Crop' and get the information on different pests and related management practices (Fig. 10). This app is downloadable from Google play store or the IIRR website (https://play.google.com/store/apps/details?id=org.iirr.varipirusasyarakshana&hl=en). The app has been successfully tested and installed in around 100 farmers' mobile phones for further testing in the field.

Figure 10 Home page, image gallery of visible symptoms, and identification and management pages of mobile rice IPM.

6.5 Mobile-based AI module for identification of rice pests

An AI-based mobile app (called Aavishkar) has been developed using machine learning to identify pests and automatically generate advice on their management (http://www.icar-iirr.org). This app prompts for basic information on crop and location details and uploads of images of pests or their damage symptoms. The app analyses the image and information and indicates the name of the pest followed by the advice on IPM strategies to deal with the pest using both text and voice (Fig. 11). Images of pests are processed using advanced image processing techniques and machine/deep learning to detect the pest and assign an accuracy of identification (indicated as a percentage). Tensor flow has been used for training of image detection and python script was used for developing the app. The app was developed using around 600 images. It is currently being tested and will be uploaded soon in Google play store.

7 Summary and future trends

This chapter highlights the role of DSS for pest monitoring and management through information technology like remote sensing, GIS technologies, spectral indices, image-based diagnostics, DSS, expert systems and phenology-based degree-day models. Applications range from rule-based models to

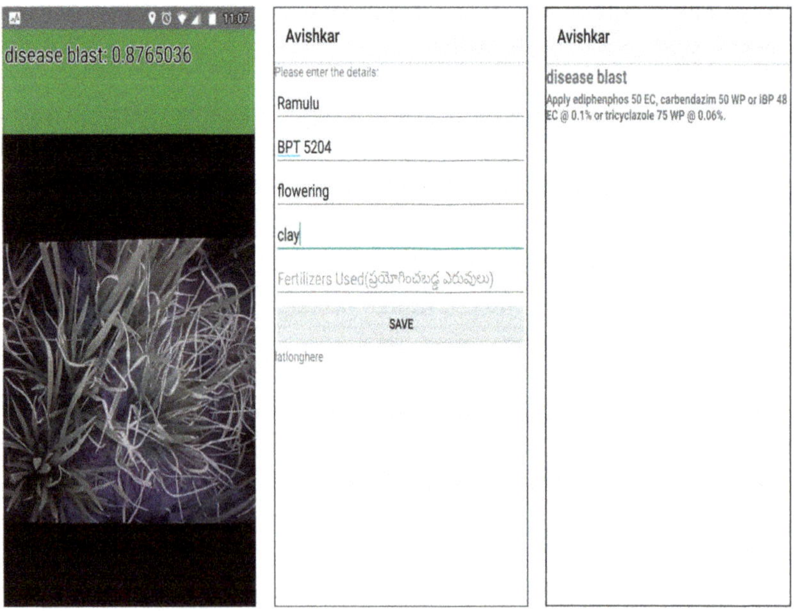

Figure 11 Home page, camera click and advisory pages of 'Aavishkar' app.

phenology models to help in pest forecasting. The chapter has included case studies of rice expert systems, crop pest DSS using phenology-based degree-day models, and mobile-based artificial intelligence (AI) modules for the identification of rice pests. These DSSs can be developed effectively using advanced software tools available from open source platforms or using commercial software packages. However, it is important that the majority of existing DSS are not holistic in nature, mostly concentrating on a single pest/ disease rather than the whole crop production system. Their ability to predict potential pest outbreaks also remains limited. Further developments in the range of monitoring techniques and energy-efficient, low-cost wireless sensor networks capable of transmitting large volumes of data in real time, combined with further advances in big data analytics capable of finding patterns in this mass of data, will improve the accuracy of DSS in identification, prediction and tailored management advice specific to a location and per the farmer's needs.

8 Where to look for further information

8.1 Research papers

1 Damos, P. 2015. Modular structure of web-based decision support systems for integrated pest management. A review. *Agronomy for Sustainable Development* 35(4), 1347–72. doi:10.1007/s13593-015-0319-9.
2 Donatelli, M., Magarey, R. D., Bregaglio, S., Willocquet, L., Whish, J. P. M. and Savary, S. 2017. Modelling the impacts of pests and diseases on agricultural systems. *Agricultural Systems* 155, 213–24. doi:10.1016/j.agsy.2017.01.019.
3 Small, I. M., Joseph, L. and Fry, W. E. 2015. Development and implementation of the BlightPro decision support system for potato and tomato late blight management. *Computer and Electronics in Agriculture* 115, 57–65.
4 Taechatanasat, P. and Armstrong, L. 2014. Decision support system data for farmer. ECU Publications Post.

8.2 Organisations

1 ICAR - Central Research Institute for Dryland Agriculture, Hyderabad, India, http://www.crida.in.
2 ICAR - Indian Institute of Rice Research, Hyderabad, India, http://www.icar-iirr.org.
3 ICAR - National Centre for Integrated Pest Management, New Delhi, India, http://www.ncipm.res.in.

3 ICAR - Central Research Institute for Dryland Agriculture, Hyderabad, India, http://www.crida.in.
4 Indian Institute of Information Technology(IIIT), Hyderabad, India, http://www.IIIt.ac.in.
5 International Rice Research Institute (IRRI), Manila, The Philippines, http://www.irri.org.

9 References

Aavishkar - a mobile-based AI module for identification of rice pests. Developed during Smart Rice Hackathon 2019 (24 hours continuous) organised by ICAR-IIRR, 09-10 February 2019. Available at: http://icar-iirr.org/IIRR_News Letter_1_2019.pdf.

An initial method for predicting the spatial distribution of pest and disease using GIS. Available at: https://www.geospatialworld.net (accessed on 29 July 2019).

A web based Rice Expert System. Available at: http://www.Ricexpert.in (accessed on 06 July 2019).

A web enabled decision support system: "Crop Pest DSS" for pest prediction in rice and cotton based cropping systems. Available at: http://www.crida.in:8080/naip.

Baker, C. R. B. 1991. The validation and use of a life-cycle simulation model for risk assessment of insect pests. *EPPO Bulletin* 21(3), 615–22. doi:10.1111/j.1365-2338.1991.tb01295.x.

Bange, M. P., Deutshcer, S. A., Larsen, D., Linsley, D. and Whiteside, S. 2004. A handheld decision support system to facilitate improved insect pest management in Australian cotton systems. *Computers and Electronics in Agriculture* 43(2), 131–47. doi:10.1016/j.compag.2003.12.003.

Beck, H. W., Jones, P. and Jones, J. W. 1989. SOYBUG: an expertsystem for soybean insect pest management. *Agricultural Systems* 30(3), 269–86. doi:10.1016/0308-521X(89)90091-7.

Blackland Research & Extension Center - Texas A&M Agrilife. Pestman pest tool aids in management decisions. Available at: https://blackland.tamu.edu/news-archive/pestman-pest-tool-aids-in-management-decisions/ (accessed on 08 September 2019).

Bregaglio, S. and Donatelli, M. 2015. A set of software components for the simulation of plant airborne diseases. *Environmental Modelling and Software* 72, 426–44. doi:10.1016/j.envsoft.2015.05.011.

Campbell, A., Frazer, B. D., Gilbert, N., Gutierrez, A. P. and Mackauer, M. 1974. Temperature requirements of some aphids and their parasites. *Journal of Applied Ecology* 11(2), 431–8. doi:10.2307/2402197.

Chander, S., Kalra, N. and Aggarwal, P. K. 2007. Development and application of crop growth simulation modelling in pest management. *Outlook on Agriculture* 36(1), 63–70. doi:10.5367/000000007780223704.

Collier, R. H., Finch, S. and Phelps, K. 1991. A simulation model for forecasting the timing of attack of *Delia radicum* on cruciferous crops. *EPPO Bulletin* 21(3), 419–24. doi:10.1111/j.1365-2338.1991.tb01271.x.

Cooke, L. R., Schepers, H. T. A. M., Hermansen, A., Bain, R. A., Bradshaw, N. J., Ritchie, F., Shaw, D. S., Evenhuis, A., Kessel, G. J. T., Wander, J. G. N., Andersson, B., Hansen, J. G., Hannukkala, A., Nærstad, R. and Nielsen, B. J. 2011. Epidemiology and integrated control of potato late blight in Europe. *Potato Research* 54(2), 183–222. doi:10.1007/s11540-011-9187-0.

Danks, H. V. 2000. Measuring and reporting life cycle duration in insects and arachnids. *European Journal of Entomology* 97(3), 285–303. doi:10.14411/eje.2000.046.

Dhaliwal, G. S., Jindal, V. and Mohindru, B. 2015. Crop losses due to insect pests: global and Indian scenario. *Indian Journal of Entomology* 77(2), 165–8. doi:10.5958/0974-8172.2015.00033.4.

Donatelli, M. 2014. BioMA – biophysical model application framework. Available at: https://en. wikipedia.org/wiki/BioMA (accessed on 28 August 2019).

Donatelli, M., Magarey, R. D., Bregaglio, S., Willocquet, L., Whish, J. P. M. and Savary, S. 2017. Modelling the impacts of pests and diseases on agricultural systems. *Agricultural Systems* 155, 213–24. doi:10.1016/j.agsy.2017.01.019.

E-agriculture, Plantix app. Available at: http://www.fao.org/e-agriculture (accessed on 27 August 2019).

European Federation for Information Technology in Agriculture, Food and the Environment (EFITA), 26 September 2019. Available at: http://www.effita.net.

Farm management, farming revolution: the use of sensors in crop pest detection. Available at: https://www.farmmanagement.pro/farming-revolution-the-use-of-sensors-in-crop-pest-detection (accessed on 06 August 2019).

Fitt, G. P., Dillon, M. L. and Hamilton, J. G. 1995. Spatial dynamics of *Helicoverpa* populations in Australia: simulation modelling and empirical studies of adult movement. *Computers and Electronics in Agriculture* 13(2), 177–92. doi:10.1016/0168-1699(95)00024-X.

Fitzgerald, G. 2000. Bug checking for mites – from the sky. *Australian Cottongrower* 21, 29–31.

Ganesan, V. 2007. Decision support system 'Crop-9-DSS' for identified crops. *World Academy of Science, Engineering and Technology: International Journal of Biological, Biomolecular, Agricultural, Food and Biotechnological Engineering* 1(12), 186–8.

Genc, H., Genc, L., Turhan, H., Smith, S. E. and Nation, J. L. 2008. Vegetation indices as indicator of damage by the sunn pest (Hemiptera: Scutelleridae) to field grown wheat. *Academic Journals* 7(2), 173–80.

Gilbert, N. and Raworth, D. A. 1996. Insects and temperature: a general theory. *Canadian Entomologist* 128(1), 1–13. doi:10.4039/Ent1281-1.

Gogoi, N. K., Deka, B. and Bora, L. C. 2018. Remote sensing and its use in detection and monitoring plant diseases: *a review. Agricultural Reviews* 39(4), 307–13. doi:10.18805/ag.R-1835.

Good Fruit Grower. Remote pest management with automated traps with an electronic trap and wireless network, growers can spend less time scouting in the orchard. Available at: https://www.goodfruit.com/remote-pest-managment-with-automated-traps/ (accessed on 04 August 2019).

Gordan, H. T. 1999. Growth and development ofinsects. In: Huffaker, C. B. and Gutierrez, A. P. (Eds), *Ecological Entomology* (2nd edn.). Wiley, New York, NY, p. 82.

Hance, T., van Baaren, J., Vernon, P. and Boivin, G. 2007. Impact of extreme temperatures on parasitoids in a climate change perspective. *Annual Review of Entomology* 52(1), 107–26. doi:10.1146/annurev.ento.52.110405.091333.

Hayes-Roth, F., Waterman, D. and Lenat, D. 1983. *Building Expert Systems*. Addison Wesley. ISBN:0-201-10686-8.

Hearn, A. B. and Bange, M. P. 2002. SIRATAC and CottonLOGIC: persevering with DSSs in the Australian cotton industry. *Agricultural Systems* 74(1), 27–56. doi:10.1016/S0308-521X(02)00019-7.

Higley, L. G., Pedigo, L. P. and Ostlie, K. R. 1986. DEGDAY: a program for calculating degree-days, and assumptions behind the degree-day approach. *Environmental Entomology* 15(5), 999–1016. doi:10.1093/ee/15.5.999.

Huang, W., Luo, J., Zhang, J., Zhao, J., Zhao, C. and Wang, J. 2012. Crop disease and pest monitoring by remote sensing. In: Escalante Ramirez, B. (Ed.), *Remote Sensing: Applications*. Intech Publishers, Rijeka, Croatia, pp. 31–75. Available at: https://www.intechopen.com/books/remote-sensing-applications/crop-disease-and-pest-monitoring-by-remote-sensing.

Hultin, E., Uusitalo, E., Larsen, I. and Prydz, H. 1955. Influence of temperature on the rate of enzymatic processes. *Acta Chemica Scandinavica* 9, 1700–10. doi:10.3891/acta.chem.scand.09-1700.

ICTs for Agricultural and Rural Development (ICTARD). International Institute of Information Technology. Available at: http://www.esagu.in (accessed on 05 August 2019).

IIRR Geo-Portal. Available at: http://iirr-geoportal.in/ (accessed on 12 August 2019).

Ikemoto, T. 2005. Intrinsic optimum temperature for development of insects and mites. *Environmental Entomology* 34(6), 1377–87. doi:10.1603/0046-225X-34.6.1377.

Jackson, P. 1998. *Introduction to Expert Systems* (3rd edn.). Addison Wesley, p. 2.

Jarvis, C. H. and Baker, R. H. A. 2001. Risk assessment for nonindigenous pests. I. Mapping the outputs of phenology models to assess the likelihood of establishment. *Diversity and Distributions* 7(5), 223–35. doi:10.1046/j.1366-9516.2001.00113.x.

Jhuria, M., Prashant Kumar, A. and Borse, R. 2013. Image processing for smart farming: detection of disease and fruit grading. *IEEE Second International Conference on Image Information Processing(ICIIP)*, Shimla, India. doi:10.1109/iciip.6707647.

Johnson, F. H. and Lewin, I. 1946. The growth rate of *E. coli* in relation to temperature, quinine and coenzyme. *Journal of Cellular and Comparative Physiology* 28(1), 47–75. doi:10.1002/jcp.1030280104.

Jones, A. L., Lillevik, S. L., Fisher, P. D. and Stebbins, T. C. 1980. A microcomputer- based instrument to predict primary apple scab infection periods. *Plant Disease* 64(1), 69–72. doi:10.1094/PD-64-69.

Krause, R. A., Massie, L. B. and Hyre, R. A. 1975. Blitecast – computerized forecast of potato late blight. *Plant Disease Reporter* 59, 95–8.

Krishnaiah, N. V., Pasalu, I. C., Padmavathi, L., Krishnaiah, K. and Ram Prasad, A. S. R. 1997. Day degree requirement of rice yellow stem borer, *Scirpophaga incertulas* (Walker). *Oryza* 34, 185–6.

Laha, G. S., Sailaja, B., Srinivas Prasad, M., Ladhalakshmi, D., Krishnaveni, D., Ram, S., Prakasam, V., Yugander, A., Kannan, C., Valarmathi, P. and Ravindra Babu, V. 2016. Changes in rice disease scenario in India: an analysis from production oriented survey. Technical Bulletin No 91. ICAR-IIRR, Hyderabad, India, p. 95.

Legg, D. E., Van Vleet, S. M. and Lloyd, J. E. 2000. Simulated predictions of insect phonological events made by using mean and median functional lower developmental thresholds. *Journal of Economic Entomology* 93(3), 658–61. doi:10.1603/0022-0493-93.3.658.

Madden, L., Pennypacker, S. P. and MacNab, A. A. 1978. FAST, a forecast system for *Alternaria solani* on tomato. *Phytopathology* 68(9), 1354–58. doi:10.1094/Phyto-68-1354.

Magarey, R. D., Fowler, G. A., Borchert, D. M., Sutton, T. B., Colunga-Garcia, M. and Simpson, J. A. 2007. NAPPFAST: an internet system for the weather based mapping of plant pathogens. *Plant Disease* 91(4), 336–45. doi:10.1094/PDIS-91-4-0336.

Magarey, R. D., Borchert, D. M., Fowler, G. A. and Hong, S. C. 2015. The NCSU/APHIS plant pest forecasting system (NAPPFAST). In: Venette, R. C. (Ed.), *Pest Risk Modelling and Mapping for Invasive Alien Species*, pp. 82–96. doi:10.1079/9781780643946.0082.

Mansingha, G., Reichgelt, H. and OseiBrysonc, K.-M. 2007. CPEST: an expert system for the management of pests and diseases in the Jamaican coffee industry. *Expert System with Applications* 32(3), 184–92.

Maywald, G. F., Kriticos, D. J., Sutherst, R. W. and Bottomley, W. 2007. *Dymex Model Builder Version 3: User's Guide*. Hearne Scientific Software Pty. doi:10.1371/journal.pone.0040313.g006.

Michalski, R. S., Devis, J. H., Bhist, V. S. and Sinclair, J. B. 1983. A computer-based advisory system for diagnosing soyabean diseases in Illinois. *Plant Disease* 67(4). doi:10.1094/PD-67-459.

Mobile rice IPM. Available at: https://play.google.com/store/apps/details?id=org.iirr.varipirusasyarakshana& hl=en.

Morgan, D. 1991. The sensitivity of simulation models to temperature changes. *EPPO Bulletin* 21(3), 393–7. doi:10.1111/j.1365-2338.1991.tb01267.x.

Morris, R. F. 1963. The dynamics of epidemic spruce budworm populations. *Memoirs of the Entomological Society of Canada* 95(31), 1–332. doi:10.4039/entm9531fv.

Mujica, N., Sporleder, M., Carhuapoma, P. and Kroschel, J. 2017. A temperature dependent phenology model for *Liriomyza huidobrensis* (Diptera: Agromyzidae). *Journal of Economic Entomology* 110(3), 1333–44. doi:10.1093/jee/tox067.

Nagarajan, S., Bahadur, P. and Nayak, S. K. 1984. Contemplating the management of brown rust resistance genes to mitigate the spread of *Puccinea recondita* f. sp. *Tritici*. *Indian Phytopathology* 37, 490–7.

Nwagu, C. K., Omankwu, O. C. and Inyiama, H. 2018. A review of expert systems in agriculture. *International Journal of Computer Science and Information Security (IJCSIS)* 16(4), 126–9.

Otuka, A., Watanabe, T., Suzuki, Y., Matsumura, M., Furuno, A. and Chino, M. 2005. Real-time prediction system for migration of rice planthoppers *Sogatella furcifera* (Horváth) and *Nilaparvata lugens* (Stål) (Homoptera: Delphacidae). *Applied Entomology and Zoology* 40(2), 221–9. doi:10.1303/aez.2005.221.

Padmavathi, Ch., Katti, G., Padmakumari, A. P. and Pasalu, I. C. 2008. Mortality and fertility life tables of leaf folder *Cnaphalocrocis medinalis* Guenee (Lepidoptera: Pyralidae) in rice. *Entomon* 33(1), 1–13.

Padmavathi, C., Katti, G., Sailaja, V., Padmakumari, A. P. K., Jhansilakhsmi, V., Prabhakar, M. and Prasad, Y. G. 2013. Temperature thresholds and thermal requirements for the development of rice leaf folder, *Cnaphalocrocis medinalis*. *Journal of Insect Science* 13, 96. doi:10.1673/031.013.9601.

Padmavathi, Ch., Katti, G., Padmakumari, A. P. K., Jhansilakshmi, V., Prabhakar, M. and Prasad, Y. G. 2014. Development of decision support system for the management of rice leaf folder, *Cnaphalocrocis medinalis* Guenee. Technical Bulletin No. 76. Directorate of Rice Research, Hyderabad, India, 40p. ISBN:978-81-928249-0-1.

Pavan, W., Fraisse, C. W. and Peres, N. A. 2011. Development of a web-based disease forecasting system for strawberries. *Computers and Electronics in Agriculture* 75(1), 169–75. doi:10.1016/j.compag.2010.10.013.

Pinnschmidt, H. O., Batchelor, W. D. and Teng, P. S. 1995. Simulation of multiple species pest damage in rice using CERES-Rice. *Agricultural Systems* 48(2), 193–222. doi:10.1016/0308-521X(94)00012-G.

Plant, R. E. and Mange, M. 1987. Modeling and simulation in agricultural pest management. *SIAM Review* 29(2), 235–61. doi:10.1137/1029043.

Prabhakar, M., Prasad, Y. G., Thirupathi, M., Sreedevi, G., Dharajothi, B. and Venkateswarlu, B. 2011. Use of groundbase hyperspectral remote sensing for detection of stress in cotton caused by leafhopper (Hemiptera: Cicadellidae). *Computers and Electronics in Agriculture* 79(2), 189–98. doi:10.1016/j.compag.2011.09.012.

Prabhakar, M., Prasad, Y. G. and Rao, M. N. 2012. Remote sensing of biotic stress in crop plants and its application for pest management. In: Venkateswarlu, B., et al. (Eds), *Crop Stress and Its Management*. Springer Science, Dordrecht, the Netherlands, pp. 517–45.

Prasad, Y. G. and Prabhakar, M. 2011. Pest monitoring and forecasting. In: Abrol, D. B. and Shankar, U. (Eds), *Integrated Pest Management Principles and Practice*. CABI Publishing, UK, pp. 41–57.

Prasad, R., Ranjan, K. R. and Sinha, A. K. 2006. AMRAPALIKA: an expert system for the diagnosis of pests, diseases, disorders in Indian mango. *Knowledge-Based Systems* 19(1), 9–21. doi:10.1016/j.knosys.2005.08.001.

Prasannakumar, N. R., Chandar, S., Sahoo, R. N. and Gupta, V. K. 2013. Assessment of Brown Plant Hopper, (*Nilaparvata lugens*) [Stål], damage in rice using hyperspectral remote sensing. *International Journal of Pest Management* 59(3), 180–8. doi:10.1080/09670874.2013.808780.

Rajkishore P., Kumar R. R. and Sinha, A. K. (2006). 'AMRAPALIKA: An expert system for the diagnosis of pests, diseases, disorders in Indian mango'. Knowl.-Based Syst., 19(1), 9–21.

Reji, G., Chander, S. and Aggawal, P. K. 2008. Simulating rice stem borer, *Scirpophaga incertulas* Wlk., damage for developing decision support tools. *Crop Protection* 27(8), 1194–9. doi:10.1016/j.cropro.2008.02.005.

Rice Doctor. Available at: http://www.knowledgebank.irri.org/decision-tools/rice-doctor/diagnostic-tool (accessed on 25 September 2019).

Riedell, W. E. and Blackmer, T. M. 1999. Leaf reflectance spectra of cereal aphid damaged wheat. *Crop Science* 39(6), 1835–40. doi:10.2135/cropsci1999.3961835x.

Riley, J. R., Armes, N. J., Reynolds, D. R. and Smith, A. D. 1992. Nocturnal observation on flight behavior of *Helicoverpa armigera* (Lepidoptera: Noctuidae) in the post-rainy season in central India. *Bulletin of Entomological Research* 82(2), 243-56. doi:10.1017/S0007485300051798.

Rossi, V., Giosuè, S. and Caffi, T. 2009. Modelling the dynamics of infections caused by sexual and asexual spores during Plasmopara viticola epidemics. *Journal of Plant Pathology* 91, 615–27.

Sailaja, B. and Padmakumari, A. P. K. 2009. 'Application of Information Technology in Rice IPM', eco-friendly management of pests and diseases in rice and rice based cropping systems. DRR Winter School Lectures Compendium No. 45/2009. Directorate of Rice Research, Hyderabad, India, 348p.

Sailaja, B., Katti, G. and Laha, G. S. 2002. GIS as a tool for the analysis of pest incidence and productivity of rice in India – a new approach. In: *Resources Management in Plant Protection during Twenty First Century*, pp. 191–6.

Sailaja, B., Krishnaveni, D., Padma Kumari, A. P. and Mishra, B. 2004. Developing computer expert for management of pathogens and insect pests in rice. *Indian Journal of Mycology and Plant Pathology* 34(2), 495–9.

Sailaja, B., Padmakumari, A. P. K., Krishnaveni, D., shaik, N. M., Nagarjuna Kumar, R., Gayatri, S. and Sankar, G. R. M. 2016. Web based expert system for identifying pests and disease problems of rice crop. In: Roy, A. K. (Ed.), *Emerging Technologies of the 21st Century*. New India Publishing Agency, pp. 355–64. ISBN:978-93-83305-33-9.

Savary, S. and Willocquet, L. 2014. Simulation modeling in botanical epidemiology and crop loss analysis. *APSnet Education Center. The Plant Health Instructor*. doi:10.1094.PHI-A-2014-0314-01.

Schoolfield, R. M., Sharpe, P. J. H. and Magnuson, C. E. 1981. Non-linear regression of biological temperature-dependent rate models based on absolute reaction-rate theory. *Journal of Theoretical Biology* 88(4), 719–31. doi:10.1016/0022-5193(81)90246-0.

Seiter, N. 2018. Integrated pest management: what are economic thresholds, and how are they developed? *Farmdoc Daily* (8), 197. Department of Agricultural and Consumer Economics, University of Illinois at Urbana-Champaign.

Shanmugapriya, P., Rathika, S., Ramesh, T. and Janaki, P. 2019. Applications of remote sensing in agriculture: a review. *International Journal of Current Microbiology and Applied Sciences* 8(1), 2270–83. doi:10.20546/ijcmas.2019.801.238.

Sharpe, P. J. H. and DeMichele, D. W. 1977. Reaction kinetics of poikilotherm development. *Journal of Theoretical Biology* 64(4), 649–70. doi:10.1016/0022-5193(77)90265-x.

Shi, P. J., Ikemoto, T., Egami, C., Sun, Y. and Ge, F. 2011. A modified program for estimating the parameters of the SSI model. *Environmental Entomology* 40(2), 462–9. doi:10.1603/EN10265.

Shtienberg, D. 2013. Will decision-support systems be widely used for the management of plant diseases? *Annual Review of Phytopathology* 51, 1–16. doi:10.1146/annurev-phyto-082712-102244.

Sporleder, M., Tonnang, H., Juarez, H., Carhuapoma, P., Gonzales, J. C., Mendoza, D., Simon, R. and Kroschel, J. 2014. ILCYM – insect life cycle modeling. A software package for developing temperature based insect phenology models with applications for local, regional and global analysis of insect population and mapping. Manual for ILCYM, version 3.0. International Potato Center, Lima, Peru.

Sree Visakham cassava expert system. Available at: http://ctcritools.in/cassavaexpert/diagnostics.php (accessed on 06 August 2019).

Sreedevi, G., Prasad, Y. G., Prabhakar, M., Rao, G. R., Vennila, S. and Venkateswarlu, B. 2013. Bioclimatic thresholds, thermal constants and survival of mealybug, *Phenacoccus solenopsis* (Hemiptera: Pseudococcidae) in response to constant temperatures on *Hibiscus*. *PLoS ONE* 8(9), e75636. doi:10.1371/journal.pone.0075636.

Sudha Rani, D., Venkatesh, M. N., Naga Satya Sri, Ch. and Anand Kumar, K. 2018. Remote sensing as pest forecasting model in agriculture. *International Journal of Current Microbiology and Applied Sciences* 7(3), 2680–9. doi:10.20546/ijcmas.2018.703.310.

Sutherst, R. W., Maywald, G. F. and Russell, B. L. 2000. Estimating vulnerability under global change: modular modelling of pests. *Agriculture, Ecosystems and Environment* 82(1–3), 303–19. doi:10.1016/S0167-8809(00)00234-6.

The BlightPro DSS for potato and tomato late blight management. Available at: http://blight.eas.cornell.edu/blight/ (accessed on 25 September 2019).

The cassava expert protect system. Available at: http://ctcritools.in/cassavaexpert/diagnostics.php (accessed on 24 October 2019).

Thomson, L. J., Macfadyen, S. and Hoffman, A. A. 2010. Predicting the effect of climate change on natural enemies of agricultural pests. *Biological Control* 52(3), 296-306. doi:10.1016/j.biocontrol.2009.01.022.

Tobin, P. C., Nagarkatti, S. and Saunders, M. C. 2003. Phenology of grape berry moth (Lepidoptera: Tortricidae) in cultivated grape at selected geographic locations. *Environmental Entomology* 32(2), 340-6. doi:10.1603/0046-225X-32.2.340.

Trapview. Automated pest monitoring system. Available at: http://www.trapview.com (accessed on 26 September 2019).

Trnka, M., Muska, F., Semeradvoa, D., Dubrovsky, M., Kocmankova, E. and Zalud, Z. 2007. European corn borer life stage model: regional estimates of pest development and spatial distribution under present and future climate. *Ecological Modelling* 207(2-4), 61-84. doi:10.1016/j.ecolmodel.2007.04.014.

UC IPM. Models: insects, mites, diseases, plants, and beneficials. Research and IPM. Available at: http://www.ipm.ucdavis.edu/MODELS (accessed on 28 September 2019).

Varley, G. C. and Gradwell, G. R. 1970. Recent advances in insect population dynamics. *Annual Review of Entomology* 15(1), 1-24. doi:10.1146/annurev.en.15.010170.000245.

Vennila, S., Singh, J., Wahi, P., Bagri, M., Das, D. K. and Srinivasa Rao, M. 2016. Web enabled weather based prediction for insect pests of rice. ICAR-National Research Centre for Integrated Pest Management, New Delhi, India, 50p.

Waheed, I. B., Coop, L. and Kogan, M. 2003. Integrated pest management and internet-based delivery systems. *Neotropical Entomology* 32, 373-83.

Welch, S. M., Croft, B. A. and Michels, M. F. 1978. PETE: an extension phenology modeling for management of muti-species pest complex. *Environmental Entomology* 7, 487-94.

Whish, J. P. M., Herrmann, N. I., White, N. A., Moore, A. D. and Kriticos, D. J. 2015. Integrating pest population models with biophysical crop models to better represent the farming system. *Environmental Modelling and Software* 72, 1-10.

Willocquet, L., Savary, S., Fermandez, L., Elazegui, F. A., Castilla, N., Zhu, D., Tang, Q., Huang, S., Lin, X., Singh, H. M. and Srivastava, R. K. 2002. Structure and validation of RICEPEST, a production situation-driven, crop growth model simulating rice yield response to multiple pest injuries for tropical Asia. *Ecological Modelling* 153(3), 247-68. doi:10.1016/S0304-3800(02)00014-5.

Xinxing, L., Zhang, L., Fu, Z., et al. 2012. The corn disease remote diagnostic system in China. *Journal of Food, Agriculture and Environment* 10, 617-20.

Yadav, D. S. and Chander, S. 2010. Simulation of rice plant hopper damage for developing pest management decision support tools. *Crop Protection* 29(3), 267-76. doi:10.1016/j.cropro.2009.10.005.

Yang, C. M. and Cheng, C. H. 2001. Spectral characteristics of rice plants infested by brown plant hoppers. *Proceedings of the National Science Council, Republic of China. Part B: Life Sciences* 25, 180-6.

Yelapure, S. J. and Kulkarni, R. V. 2012. Literature review on expert system in agriculture. *International Journal of Computer Science and Information Technologies* 3(5), 5086–9.

Yialourisa, C. P., Passamb, H. C., Sideridisa, A. B. and Metin, C. 1997. VEGES – a multilingual expert system for the diagnosis of pests, diseases and nutritional disorders of six greenhouse vegetables. *Computers and Electronics in Agriculture* 19(1), 55–67. doi:10.1016/S0168-1699(97)00032-X.

Chapter 10

Developing decision support systems for improving data management in agricultural supply chains

Gerhard Schiefer, University of Bonn, Germany

1 Introduction

Food production involves one of the most complex organizational models of enterprise cooperation. The core is constituted by the actors involved in the flow of food products reaching from production agriculture to various levels of food processing, trade and retail, which, in addition, reach from rural agricultural regions to urban environments.

The actors involved in the production flow are summarized as members of an agricultural or food supply value chain. A food supply value chain (or shorter: a food value chain or food chain, a term used in the remainder of the chapter) is a theoretical construct for demonstrating the flow of products from agriculture to retail and consumers. However, a food value chain with a fixed group of enterprises with a continuous trade relationship is more the exception than the rule. The usual food chain view is comprised of a network of enterprises with dynamically changing trade relationships reaching from agriculture to consumers, resulting in a dynamically changing organization of food supply chains and their participating enterprises (Fig. 1).

However, whatever the flow of products, coordination is needed for assuring that production and distribution processes along the chain deliver

http://dx.doi.org/10.19103/AS.2020.0069.20

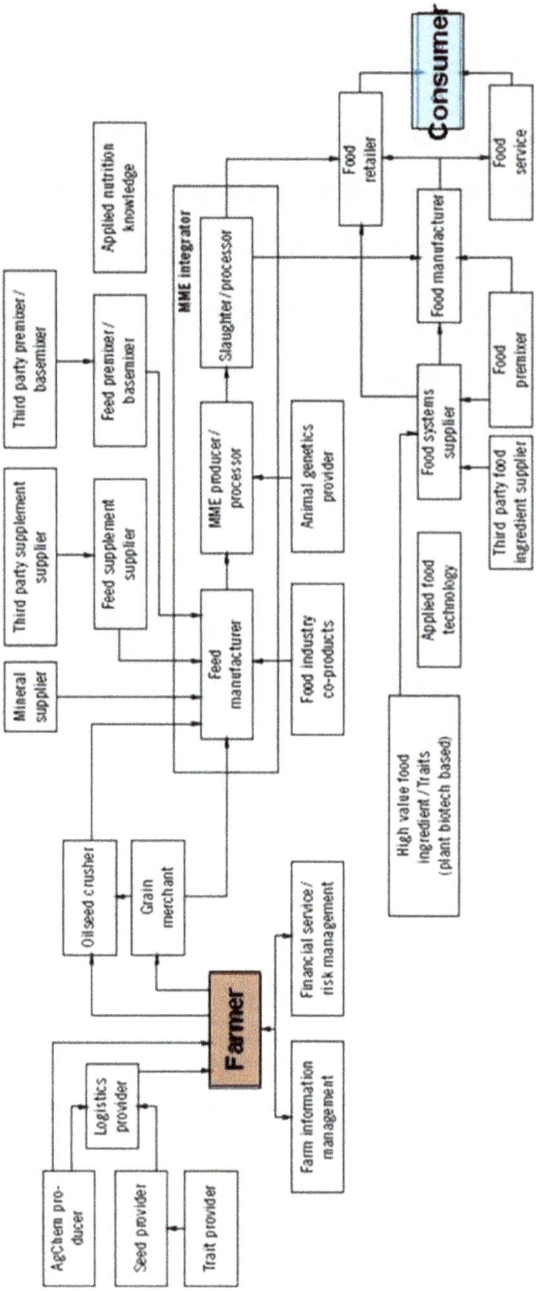

Figure 1 Model of a food value chain. Source: author.

certain product qualities at consumers' end or interact with the environment for environmental protection according to society's interests. These 'classical' coordination needs have been discussed among stakeholders or interested groups in business, policy and research since a long time ago.

With the increasing mutual dependency of enterprises within the food chain especially in quality production and the emerging discussions on sustainability, the flow of data along the chain and their coordination are receiving increasing attention not the least because of increasing requirements for transparency along the food chain and toward consumers. Data coordination has proved to be quite complex. The point of data use might differ from the point of data collection. A critical point concerns agriculture as many of the data needed at later stages of the chain depend on issues where agriculture could provide information. This is the basis for coordination needs, which do not just have to agree on data collection and data use but also on agreements regarding the allocation of costs and benefits.

The flow of data is closely connected with the sequence of decisions that are being made along the food chain. Decisions require data, and the functioning of the sequence of decisions depends on the availability of data where needed. Decisions at one stage of the food chain might provide data for subsequent stages and influence activities and decisions at other stages. Coordination needs have to assure that the data for decisions are available where and when needed and that decisions do not contradict each other but contribute to the realization of joint objectives.

This constellation characterizes the difficulties in the management of data across the food value chain or in arriving at a coordination of decisions that serve the needs of all enterprises concerned. As the food sector builds on food value chains, the difficulties in data management and decision making in food value chains translate into difficulties of the whole food sector in moving forward in new initiatives. It is with this background that the food sector is one of the slowest sectors in adapting new sector encompassing developments including the utilization of modern information and communication technology or the realization of joint initiatives toward improvements in the sustainability of production and trade.

This chapter discusses the issues and provides a framework for supporting data management and decision support in food value chains. Starting from a discussion of the decision situation in food value chains, the chapter outlines a selection of tools for decision support and concludes with a presentation of some decision problems in data management and food chain organization.

2 Decisions in supporting data management

2.1 Decision situation

In dealing with the complexity of data management and decision situations in food chains, enterprises may face two alternatives in making decisions on activities and developments. They may ask for a decision in issues that:

a) are not dependent on decisions of other enterprises or will not affect other enterprises in the chain (complete decision autonomy); examples concern internal efficiency or enterprise-based quality management; or
b) affect other enterprises, are dependent on other enterprises or are only of relevance, if other enterprises cooperate (limited decision authority); examples include data communication or quality and environmental management across the chain.

The first situation is the traditional enterprise decision situation and will not be dealt with in this chapter. Our focus is on the second situation where any solution has to consider the *interdependencies* within the food chain.

For dealing with such a situation, food chains could agree on a so-called 'chain captain', usually a strong partner in the chain who organizes and coordinates joint decisions. This approach is traditionally favored in situations where a) one partner has a very dominant role such as a regionally dominant dairy company or slaughterhouse and b) partners have close and maybe even exclusive trade relationships. Cases in point were the initial 'IKB chain initiatives' (van Trijp et al., 1997) in the Netherlands where partners in 'almost closed' chains were agreeing on advanced production standards. The drawback of this model is its limited flexibility of aligning agricultural production variations with the need for serving market agreements.

The alternative to following a 'chain captain' is characterized by a network situation where trade relationships along the value chain are dynamically evolving within a food (chain) network situation depending on production and market developments, that is where enterprises change trading partners in response to production and market developments. In this scenario, enterprises are faced with a number of coordination problems including

a) the coordination of agreements on joint action as, for example, on the delivery of products that serve certain quality or environmental needs, and
b) the coordination of agreements on the distribution of costs and gains.

Most of today's discussions on food chain developments focus on the first line of coordination and omit a discussion of the second line. However, the second line of coordination is of crucial relevance for realizing developments that need to find acceptance with enterprises in the long run. It is more the rule than the exception that the costs for realizing certain product characteristics or data collection and processing efforts occur at stages that are different from the stages where the gains are realized. A typical example for engagements in product characteristics involves 'animal welfare' where a higher level of animal welfare is provided through agriculture while higher prices with consumers are realized at retail. The 'animal welfare' example is also representative for an unbalanced situation in data collection and use as most data have to be collected at agriculture, which have to carry the collection costs while the added value of transparency with consumers is realized at retail stage.

In a network situation, the coordination of joint action as well as reaching agreements on the allocation of costs and gains is not easy to reach. The sector usually needs an external facilitator for being able to move forward. The government might take over the facilitator role through regulatory activities. Alternatives are new organizational chain models where partners have agreed on following specific management rules in dealing with coordination needs.

They are usually linked to product or communication standards that provide the base for a common understanding. Product standards may be managed by certain groups of the chains (as demonstrated by the example of GlobalG.A.P., a standard management activity evolving from retail groups; www.globalgap.org), jointly by all groups of the value chain (as demonstrated by the example of Q&S, a standard management group initially evolving from agriculture, www.q-s.de/en/), or by independent standardization organizations (as the example of the ISO standards involving the quality management standards ISO9001, www.iso.org). Communication standards need to build on a) a common technology such as the blockchain technology (Galvez et al., 2018) and b) standards in the infrastructure of communication as provided, for example, by the EPCIS standards of the GS/1 standardization organization (www.gs1.org).

Apart from regulatory initiatives by government, the engagement in standards in the food sector will lead to different levels of adherence, with some enterprises confirming to standards and others not. If companies adhering to different levels of standards engage in trade relationships, the emerging product characteristics will usually be determined by the lower level of standards. A related situation concerns data communication where companies might engage on similar levels of data exchange agreements. If not, the company with lower data exchange engagement determines the data level arriving at consumers' end (Fig. 2).

2.2 Decision model approaches

It is obvious that in a chain coordination situation, classical decision models building on mathematical relationships will not work. Decision support is

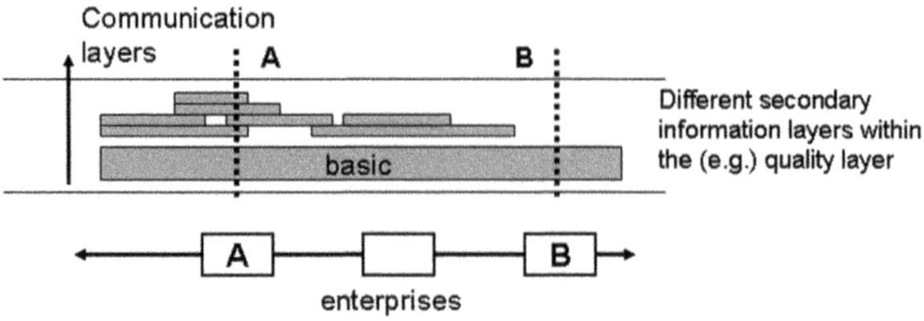

Figure 2 Quality layer alternatives. Source: author.

needed for a multi-player situation where partners *need to agree* on the various coordination levels, that is, on focus, allocation of activities and distribution of costs and gains. A suitable approach builds on the identification, evaluation and selection of decision alternatives. The following approaches might look simple on first sight, but they are not, as they require substantial knowledge of the complexity of a decision situation and the underlying objectives that reach beyond pre-determined mathematical relationships.

The identification of alternatives might be difficult to achieve but is not dependent on coordination needs. The difficulty is primarily due to the need for specifying boundaries of the decision space. It is obvious that in a long-term view (strategic view) the boundaries are further apart than in a short-term view. The long-term view may consider fundamental changes, while the short-term view is looking more at adaptations of existing solutions to changes in the economic, social and natural environment.

Once the boundaries have been set, there are no 'best' approaches for specifying decision alternatives. It usually builds on more or less structured approaches linking creativity with formal analysis. Long-term views require more creativity, while short-term views might be able to look at eliminating identified deficiencies. For the short-term view various tools have received high visibility, especially in connection with process improvements. They include the 'Failure Mode and Effect Analysis (FMEA)' (Hu-ChenLiu et al., 2013) and the application of the 'HACCP principle' (Hazard and Critical Control Point, www.fao.org/docrep/005/y1390e/y1390e09.htm), which is very much used in, but not restricted to, food safety assurance activities. As an example, the FMEA can be efficiently used in situations where one wants to identify deficiencies in food chain data communication and use and decide about improvements and improvement priorities.

The evaluation of alternatives is crucial. Partners have to agree on common characteristics (usually on multiple characteristics) that could be related to their objectives and would have to be evaluated by each partner regarding his/her individual objectives. As chain companies might arrive at different preferences for alternatives, evaluation needs to be followed by a negotiation process for arriving at a common decision. If the evaluation is based on monetary terms, a joint decision could be easier to reach than if evaluation is based on different characteristics with different scales. Assuming that the monetary advantage of a certain decision alternative is 'best', considering a complete value chain reaching from agriculture to consumers, it would be considered a 'fair' decision if gains are allocated among partners according to engagement in costs.

For supporting the decision process, a number of decision tools have been developed. The availability of different decision tools signals that there is no best tool but that each of the tools considers different aspects. It is therefore common that users will build their decision not just on the results of a single tool but apply different ones and arrive at a decision by evaluating the different 'solutions' selecting one of them, or find a suitable combination. We will present two basic tools, decision tables and the SWOT analysis, which provide insights into principal approaches.

3 Decision tools

3.1 Decision tables

Decision tables relate decision alternatives to a) future scenarios and b) objectives, considering different possible realizations of objectives due to external influences (scenarios) (Fig. 3). As an example, an objective could represent money return from the realization of an alternative in a certain scenario considering gains and costs. A decision table provides transparency and allows users to identify the alternative that best fits own interests, captured in a 'decision rule'. Literature has developed various decision rules for finding the 'best' alternative (McFadden, 2001). In the example, the rule assumes that the expected objective value (EV) of an alternative and its maximal value across alternatives are used for selecting the best alternative. However, such a rule can provide only suggestions. The final decision is the responsibility of the enterprise that will follow its own 'rule'.

The decision table can be adapted to situations where a multiple of objectives needs to be considered. It is an established procedure to 'translate' each of the objective values related to an alternative and a single scenario into

Decision alternative	Scenario 1	Scenario 2	Scenario 3	Expected objective value
Alternative a	Obj.value a1	Obj.value a2	Obj.value a3	EV (alternative a): EVa
Alternative b	Obj.value b1	Obj.value b2	Obj.value b3	EV (alternative b): EVb
Alternative c	Obj.value c1	Obj.value c2	Obj.value c3	EV (alternative c): EVc
Probability p of scenario occurrence	p1	p2	p3	Rule: max {EVa, EVb, EVc} for selecting alternative

Figure 3 Decision table.

a common scale, for example, a scale reaching from 0 to 10. This transfer of an objective value into a standardized scale reaching from very low (0) to very high (10) involves an evaluation of the objective value regarding its positioning within a sensible range.

The scale values of the different objective values are added, considering weights that express their relevance for the company (Fig. 4). This results in a single scale value, related to the alternative in the selected scenario. The further procedure is then similar to a single objective situation building on decision rules.

3.2 SWOT analysis

The view of the SWOT analysis differs from the view of the decision table. It involves a dynamic view, starting in:

a) a first step from a present situation and moving in; and
b) a second step to a situation one would encounter after alternatives have been realized.

It is less concrete than the decision table but links its evaluation to the situation of the enterprise (or chain). Each step is captured in a cross table, which lists the strength and weaknesses of the enterprise or chain and the opportunities and threats to the enterprise or chain related to available decision alternatives.

Strengths represent an enterprise's (chain's) success criteria, weaknesses its deficiency criteria. Opportunities and threats are related to decision alternatives that, if realized, could improve the strengths or increase/decrease weaknesses of an enterprise (chain). The comparison of strengths and weaknesses in the two cross tables (present situation and future situation) provides the base for finding a decision. The decision support might be called a 'soft' support as it does not include mathematical relationships or the formulation of results in numeric terms. However, by analyzing the tables, decision makers get an impression of the comparative advantages and disadvantages of different alternatives based on his/her implicit evaluation of relevance (Fig. 5).

It is common to capture the two cross tables in a two-dimensional graph that displays a single value balance of strengths and weaknesses and a single value

Calculation of a cell scale value (Alternative X, Scenario Y) in an example for two objectives:
Objective 1 (Alternative X, Scenario Y): objective 1 value --> scale value 1 (value 0-10)
Objective 2 (Alternative X. Scenario Y): objective 2 value --> scale value 2 (value 0-10)
Summary cell scale value (Alternative X, Scenario Y): w*scale value 1 + (1-w)*scale value 2

Figure 4 Calculation scheme for a cell in the decision table for two objectives.

Opportunities for enterprise (chain)	Strengths of enterprise (chain)
> List of opportunities ………	> List of strengths ………
Threats to enterprise (chain)	**Weaknesses of enterprise (chain)**
> List of risks (threats) ………	> List of weaknesses ………

Figure 5 Cross table of SWOT analysis.

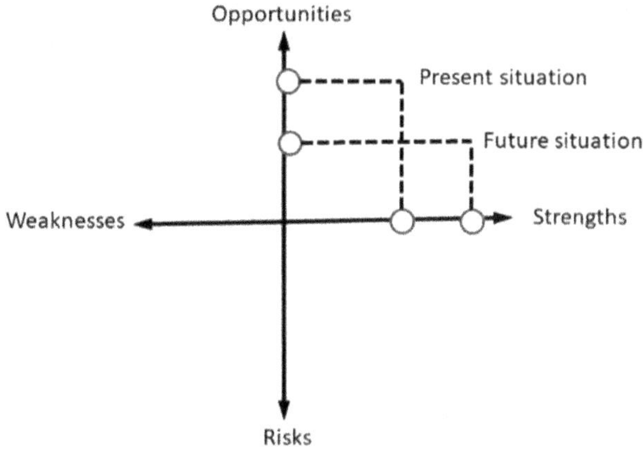

Figure 6 SWOT analysis graph. Source: author.

balance of opportunities and threats for the present and the future situation, both translated into a basic scale. Such balances are based on decision makers' (rough) estimates of relationships and weights (Fig. 6).

In the graph, the future situation (after the implementation of the decision alternative) is characterized by a lower level of opportunities (as one opportunity has been realized already) and a higher level of strengths due to the realization of the alternative. If different alternatives are up for decision, the consequences of the alternatives can be captured in different SWOT tables and integrated into the graph. The graph could be a base for further discussions. As an example, if two different decision alternatives would lead to two different future situations, the future situation with the highest contribution to an enterprise's (or chain's) strength might look most attractive. However, if it is linked to a lower level of remaining opportunities, an enterprise might lose flexibility for future activities. This asks for a detailed analysis of future needs and expectations.

Suitable decision alternatives identified with a SWOT analysis can be integrated into a decision table for reaching a final decision building on a more concrete specification of objective values, for example, in monetary terms.

3.3 Multi-criteria decision tools

With many different characteristics to consider in deciding about decision alternatives, a variety of additional supporting approaches summarized as 'multi-criteria decision analysis' have been developed. Most of them are beyond the discussion in this chapter. They all involve the need for certain mathematical calculations of varying complexity. However, for integrating different decision makers into a joint decision process with a number of decision alternatives, the analytical hierarchy process (AHP) has gained some prominence (Saaty and Vargas, 2001). It is basically focusing on supporting the negotiation process through a structured process in which the different decision makers engage in pairwise comparisons of different alternatives. The difficulty is in the decomposition of the decision situation into a number of sub-decisions that can be dealt with independently. The AHP involves a complex mathematical model that summarizes the results of the sub-decisions into an overall 'group decision'. This outline demonstrates the complexity of the approach, which usually is supported by computer-based model calculations.

4 Principal case studies

4.1 Data management for transparency

A specific decision situation concerns data exchange between enterprises along the chain and toward consumers. It is a core feature for realizing transparency and trust, which is gaining increasing attention by consumers and which depends on the availability of suitable data and their consideration in suitable decision activities. We will not elaborate on problems that may arise because of data manipulations in sourcing and/or along the chain. There is a lot of discussion going on. The relevance of both issues can be reduced through technological developments. Data sourcing can be improved through emerging sensor technologies (Grossi et al., 2019) and data communication through blockchain concepts (Galvez et al., 2018).

The remaining problems concern agreements on data collection and data exchange, which require coordination among participants. The coordination problems resemble the general coordination problems in production issues. Enterprises have to agree on what data to collect, where to collect, when to collect, what to communicate, how to communicate regarding format and timing and where to use, by whom and what for. The relevance of the last issue, the clarification of what data are being used for, has frequently been underestimated. There are examples especially from quality management where data collection in agriculture failed because farmers did not trust retailers' communicated interest and where afraid, data could be used at their disadvantage. In a network situation, there might be different levels of agreements, which at the end will provide different levels of data for retail and its ability to inform consumers.

One of the major difficulties in improving transparency and trust is due to the fact that most information that might be relevant for consumers is to be collected in agriculture while retail is the main benefiter. Examples presently in discussions concern the use of pesticides in plant products, especially vegetables and fruits and the consideration of animal welfare issues. These differences are one of the major barriers in improving data communication along chains as agriculture is not sufficiently benefiting from its data collection and the related costs. Furthermore, in present organizational schemes for data exchange, agriculture is no longer considered owner of data as soon as they have left its premises. As a consequence, agriculture has no longer influence on what its data are being used for, a major obstacle for improvements in data communication.

That is a situation where new data collection and communication alternatives have to be identified and evaluated by all concerned. Sensor-based technology might help cutting costs for data collection and transfer. An appropriate allocation of costs, benefits and ownership asks for more radical new opportunities. One of the opportunities discussed in literature that could provide support in solving the cost and benefit problem is the establishment of 'data markets'. Agriculture would offer data (e.g. through an online platform), and trading partners could make offers for receiving data. This alternative could easily deal with differences in data intensity. Such approaches could lead to differences in product offers in retail shops, products with less and products with more information on issues of interest to consumers and in consequence, with differences in prices. Groups with already-intensive data collection initiatives in agriculture such as GlobalG.A.P. have for long discussed the need for differing products in retail subject to the availability of product information.

A simplified version of a data market has been established in the 'book-and-claim' concept that has been used in soybean production and sales (https://greenpalm.org/about-greenpalm/why-greenpalm-makes-a-difference/book-and-claim-supply-chain-model). In the production of high-quality soybeans, the quality information captured in a quality label (representing a data cluster) is being separated from the actual products. Products and labels are then marketed separately. The products are channeled into the regular soybean market without any link to the initial label information (Fig. 7).

At the end of the chain, buyers of soybean could purchase a fitting number of labels and attach them to the products, selling them as 'high-quality' products. This approach of separating data from products works well, if data refer to quality issues that are not linked to measurable product characteristics but refer to credence attributes such as fair trade or environmental-conscious production. It assures that the sales quantity of 'high-quality products' matches

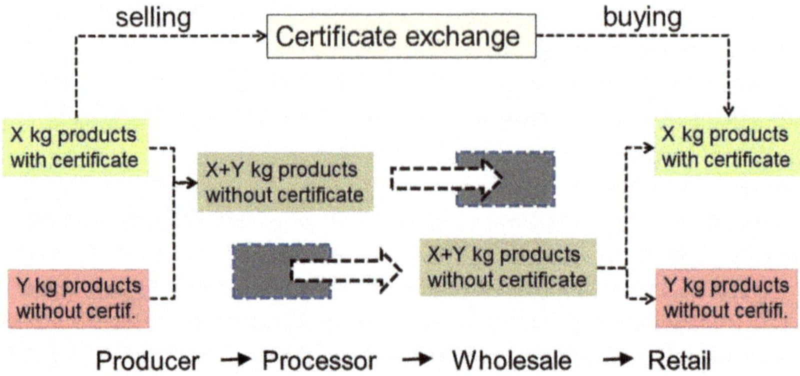

Figure 7 Book-and-claim approach. Source: author.

the production quantity. In cases where data refer to measurable product characteristics, data markets have to assure that traded data remain linked to the specific products they refer to.

In data management, enterprises are, in addition, confronted with problems that reach beyond data coordination problems. They refer to the differences in enterprises' ability to collect, communicate and use data, known as e-readiness (Bryceson, 2006). While one enterprise might be able to rely on a sophisticated digital data processing and communication ability, another one might still rely on paper-based data management. FMEA could help identify areas of weakness. The level of data communication along the chain is determined by the lowest level of e-readiness in any of the participating enterprises. In making improvements to communications, it may be necessary to agree the costs and resources required to bring all enterprises up to the level of efficiency, transparency and trust all members of the chain require.

4.2 Quality and environmental management

Quality and environmental management activities play an important role in agriculture and the food sector. Quality management assures the delivery of food to consumers, which is safe to eat and of the quality that trading partners and consumers expect. Environmental management serves the expectations of society in the protection of the environment and in an environmentally sustainable production.

The need for the coordination of activities along the food chain is quite similar in quality and environmental management. The coordination need is driven by regulatory requirements of public authorities and by pressures of markets. In consequence, the need to cooperate for coordination is much higher than in data coordination. Regulatory authorities have since many years and especially since the bovine spongiform encephalopathy (BSE) crises in

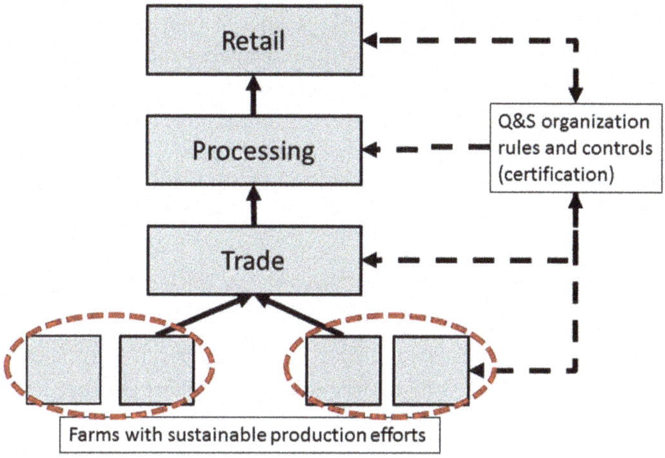

Figure 8 A basic quality coordination scheme. Source: author.

Europe enforced coordination activities not only regarding the assurance of food safety but also regarding the avoidance of environmentally damaging activities (Fig. 8).

As a result, there are a number of coordination models that are being tried out in real-life 'experimental settings'. They all build on a continuous coordination model, as requirements of policy and markets on food quality and environmental protection are continuously changing. All of them have advantages and disadvantages, which qualify them as decision alternatives, companies and food chains could select from. Differences concern the structural organization, the level of information exchange, the tackling of problems dealing with differences in the allocation of costs and benefits along the chain and the engagement of monitoring and auditing schemes for assuring companies' adherence to the coordination model agreed upon.

It is important to note that all alternatives do not build on a coordinated data exchange along stages of food chains throughout the sector, which allowed for the dynamics in food chain trade relationships but find opportunities to 'simulate' data exchange through a variety of means. A coordinated data exchange throughout the sector that could serve the needs of all stakeholders concerned are still far into the future.

A basic approach is realized through external quality management audit schemes as exemplified by the schemes Q&S in Germany (https://www.q-s.de/en/) and IKB in the Netherlands (http://www.prnp.nl/public/doc/PRLT_4230_IKBdiervoeder.pdf). The coordination is not between individual enterprises but through an external organization. It specifies production requirements for each individual stage of the food value chain, which assures certain agreed levels of food quality and environmental protection. With this approach, companies do not have to exchange data on food quality or environmental issues as they 'know' what requirements have been fulfilled by its suppliers. In addition, they are not bound to a certain supplier but could choose between all suppliers that follow the scheme. This simplifies the implementation of a coordinated quality and environmental quality management scheme not only along a chain but within the whole food sector. It is obvious that such a scheme depends on a reliable monitoring and audit activity that assures the adherence of all companies to the scheme.

However, the scheme also has drawbacks. First, it is not considering coordination in the allocation of costs and benefits. Furthermore, as its focus is on assuring certain levels in quality and environmental protection across a whole sector, improvements in quality and environmental protection have to assure that the majority of enterprises are prepared to follow. The realization of this decision alternative is, therefore, most suitable for situations where a sector-wide basic level of quality assurance and environmental protection needs to be implemented.

At the other end of coordination intensity is a scheme that tries to eliminate the drawbacks but requires, on the other hand, a coordinated data collection and exchange for reaching higher levels of quality assurance and environmental protection, which supports the visibility of products in the markets and a clearer identification and separation of additional costs and benefits.

In this scheme implemented by a major retailer, the final consumer products must fulfill higher levels of quality assurance and environmental protection based on appropriate requirements on all stages of the chain. These higher levels will be communicated to consumers through specific brand names that will allow higher prices. The additional gains are partly channeled back to agriculture, which carries the major burden in reaching the quality and environmental requirements (Fig. 9).

In an implementation of this scheme in the market, the scheme involves the consideration of high levels of environmental issues that are captured in an index that allows some flexibility in production. The final products must reach a certain index level to be accepted, carrying the exclusive brand name. It is obvious that the formulation of such an index requires intensive data collection and data use at each stage of the food chain. However, as in the preceding example, the assurance of quality and environmental protection builds on the establishment of an intensive monitoring and audit scheme and not on the information exchange between enterprises. This pays tribute to the data coordination difficulties discussed earlier and the still poorly developed schemes for assuring reliability in data collection, processing and communication. New developments in data collection and communication technology including sensors, blockchain technology, smart contracts linked

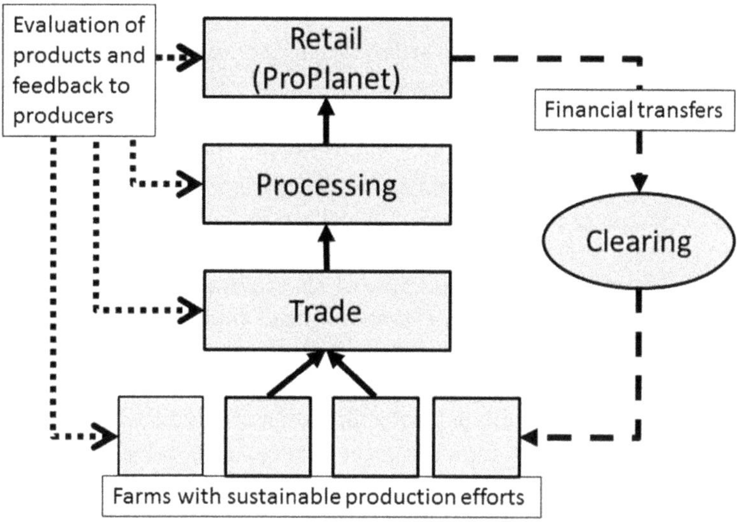

Figure 9 An advanced coordination scheme.

to blockchain technology and others will change the picture in the future and reduce the need for external monitoring and auditing activities.

5 Conclusion and future trends

5.1 Summary

Decision support in food chains is a complex issue as it needs to balance the interests of different actors with different objectives and interests. An approach used sometimes in literature involves the use of chain optimization models. However, these models were not well suited for arriving at solutions that capture the complexity and differences in the interests of the different participants in the chain. In the past, agriculture was very much dependent on decisions of processing and retail companies. With the need for an increase in food production without a similar increase in available resources, agriculture is improving its negotiation position in the chain. Decision making that considers the various partners in the chain as being of similar relevance needs to build on a number of coordination and negotiation activities. This is a difficult process that, however, can:

a) be supported by a number of easy-to-use decision tools, and
b) build on chain organization models that are being tried out in various implementations.

The chapter introduces two alternatives of decision tools that are well suited for considering the complexity in food chain coordination, decision tables and the SWOT analysis. Both tools allow within a rigid structural approach the consideration of individual opinions of food chain participants toward reaching a common decision outcome. However, decision activities require the availability of appropriate data and, in decisions with chain relevance, data from other stages of the food chain. The collection and communication of data along the chain is dealt with through varieties of organizational concepts. As of today, a seamless coordinated flow of all necessary data across the food chain is still limited to specific cases with fixed trading relationships, while the food sector is generally comprised of a network of enterprises where food chain relationships are developing dynamically through changing trade relationships. A coordinated data communication across the whole food sector is still a challenge for the future. The organizational relationships developed for quality and environmental management offer suitable alternatives.

5.2 Future trends and information support

The sector is continuously challenged to improve in serving changing needs of consumers while following sustainability and environmental expectations of society. The dynamics in changing consumer needs regarding products and production processes dealing with, for example, animal welfare, avoidance of pesticides, fair trade and related issues requires a continuous adaptation of decisions and data needs across the sector. These challenges cannot be solved by individual enterprises. The same is true for food chains that, as was discussed earlier, are not a stable construct but a dynamically evolving organizational scheme depending on actual trading activities. The challenge is for the sector as a whole that requires sector-wide coordination initiatives and the engagement of institutions who could act as facilitator or guidance by policy. Technological developments have reached a stage where further improvements in data management and decision coordination across food chains are no longer a technological but an institutional problem where the sector has to find suitable means for coordination and the balance of interests.

6 References

Bryceson, K. P. 2006. *E-Issues for Agribusiness: The What How and Why*. CABI Publishing, Oxfordshire.

Galvez, J. F., Mejuto, J. C. and Simal-Gandara, J. 2018. Future challenges on the use of blockchain for food traceability analysis. *Trends in Analytical Chemistry* 107, 222–32. doi:10.1016/j.trac.2018.08.011.

Grossi, M., Berardinelli, A., Sazonov, E., Beccaro, W. and Omaña, M. 2019. Sensors and embedded systems in agriculture and food analysis. *Journal of Sensors*. Special Issue 2019, 1-2. doi:10.1155/2019/6808674.

Liu, H., Liu, L. and Liu, N. 2013. Risk evaluation approaches in failure mode and effects analysis: a literature review. *Expert Systems with Applications* 40(2), 828-38. doi:10.1016/j.eswa.2012.08.010.

McFadden, D. 2001. Economic choices. *American Economic Review* 91(3), 351-78. doi:10.1257/aer.91.3.351.

Saaty, T. L. and Vargas, L. G. 2001. *Models, Methods, Concepts and Applications of the Analytical Hierarchy Process*. Kluwer Academic Publishers, Boston.

van Trijp, H. C. M., Steenkamp, J.-B. E. M. and Candel, M. K. J. M. 1997. Quality labeling as instrument to create product equity: the case of IKB in the Netherlands. In: Wierenga, B., van Tilburg, A., Grunert, K., Steenkamp, J.-B. E. M. and Wedel, M. (Eds), *Agricultural Marketing and Consumer Behavior in a Changing World*. Springer, Boston, pp. 201-15.

Chapter 11

Developing decision support systems for optimizing livestock diets in farms

Marina Segura, Concepción Maroto, Baldomero Segura and Concepción Ginestar, Universitat Politècnica de València, Spain

1 Introduction

The increasing demand for meat and other products provided by the livestock sector presents ever greater challenges worldwide, in both developing and developed countries, which should involve researchers from different disciplines and other stakeholders, such as managers, industry, farmers, policy-makers and consumers. Livestock production needs to provide sufficient amounts of safe and affordable food for society, but in an environmentally sustainable way that takes account of greenhouse gas (GHG) emissions and other pollutants. Modelling and decision support systems (DSSs) provide a way of optimizing production to make it more efficient and more sustainable. Gouttenoire et al. (2011), Ghahramani and Moore (2013), Rose et al. (2016), Eastwood et al. (2016) and Cabrera (2018) have reviewed a wide range of models and DSS applied to improve livestock farming systems in areas such as optimizing the nutritional

http://dx.doi.org/10.19103/AS.2020.0069.21

value and cost of feed, achieving more sustainable grazing management (e.g. by establishing optimum stocking densities), improving reproductive performance, assessing the impact of disease, as well as minimizing waste and other types of environmental impact. As feed represents around 70% of the total cost of livestock production, linear programming (LP) and other models have made a significant contribution since the 1960s to providing affordable meat by minimizing the cost of animal diets in areas such as pig and poultry production (Gous and Fisher, 2017; Brossard et al., 2017). LP and other models have also evolved to integrate advances in genetics, nutrition science, as well as animal welfare environmental concerns.

DSSs to calculate sustainable feed for livestock face a number of challenges. They rely on comprehensive, accurate and regularly updated databases recording availability and characteristics of raw materials and need to take into account that feeding requirements of animals depend on species, breed, stage of growth, type of production system and climate/weather variables among others. The availability of good quality data is a challenge particularly in developing countries. In developed countries, there are good sources, such as NRC in the USA, CVB in Holland, INRA in France, Atlas PREMIER in UK and FEDNA (2019) in Spain, although the quality of data can be improved by obtaining some of it in real farm conditions rather than in the laboratory. The availability and price of raw materials needs to regularly be updated if DSSs are to provide reliable recommendations. The food supply chain of animal products should assure the traceability of all ingredients and additives used in livestock diets to evaluate and improve food safety. Another challenge is ingredient availability, which can make it necessary to reformulate diets. Animal nutrition is changing from minimum cost diets to animal profitability optimization related to new economic models (Roembke, 2018).

The chapter is organized as follows. Section 2 provides a review of the use of mathematical programming models on diet formulation in livestock production. A complete LP model is explained in Section 3, after a brief explanation of the graphical solution method for a simple diet problem, which is relevant to know in order to avoid some misunderstanding that can be found in the literature due to the multidisciplinary nature of this problem. Currently, sustainable diet formulation needs to integrate knowledge from animal nutrition, operations research, software development and environmental impact assessment. Section 4 explains the use of goal programming (GP) for the calculation of sustainable diets. Section 5 summarizes the data and models to allow DSSs to make useful decisions. The chapter includes two case studies focussed on broiler and pig production with, finally, a summary and review of future trends in sustainable diet research.

2 Mathematical programming models for livestock production: a review

The contribution of mathematical programming models in decision making for farm planning covers both crop and livestock production. In both cases, the traditional objective has been economic, such as profit maximization or cost minimization. Livestock management depends on variables such as the type of livestock (e.g. pigs, poultry, cattle, sheep), the production system (intensive or extensive) and the product (meat, eggs or milk). The majority of the models developed in the 1970s focussed on feedstuffs, although some addressed other areas such as breeding, livestock replacement and waste disposal (Glen, 1987).

Stigler formulated the diet problem for human nutrition as an LP model in 1945. In 1947, Dantzig developed the simplex algorithm which made it possible to achieve the optimal solution for the minimum cost diet (Dantzig, 1990). This approach has been applied to animal nutrition for the last five decades, reducing the production cost in the livestock sector considerably and being one important factor in providing affordable meat and other livestock products worldwide. As this suggests, the importance of feed costs in the overall cost of livestock production has made diet and ration formulation a key objective in modelling. In Glen (1987), 'a diet is defined by the proportions of constituent foodstuffs, while a ration is defined by the quantities of constituent foodstuffs'. Glen's review includes diet and ration formulation problems for beef and dairy cattle, pigs and turkeys by applying LP models. In the 1970s, the majority of models were developed as research or teaching tools, as a prior step to improving decision making at the farm level. In the 1980s, with the availability of microcomputers, a large number of software programmes were developed for ration formulations which have continued to evolve with improvements in computer science. Some commercial software has been built up over 40 years, such as Bestmix (ADIFO software, 2019). The range of commercial software programmes has been reviewed by Saxena (2010). The benefits of LP models in optimizing nutrient use to improve farm profitability have been reviewed in various studies (Tedeschi et al., 2004; Guevara, 2004; Saxena and Chandra, 2011).

The first generation of LP-based ration formulation models have been criticized for being unable to account for the range of complexities in ration formulation, such as allowing for the variability in ingredients such as forage and cost of feed ingredients, as well as for producing nutrient imbalances in final recommendations to farmers (Rahman et al., 2010). Recent developments in LP models to address some of these shortcomings include Alqaisi et al. (2017) and Dooyum et al. (2018). Alqaisi et al. (2017), for example, have proposed a multi-stage LP model for broiler diet in order to take into account price variability of raw materials and its influence in diet composition. Their model makes it possible to assess the use of non-conventional by-product feed sources such as distiller's dried grains with solubles (DDGS) to substitute other raw materials (such as corn) in animal diets.

The limitations of LP models have led to the investigation of alternative approaches such as GP, multiobjective goal programming (MGP/MOP), multiobjective fractional programming (MOFP), nonlinear programming (NLP), as well as stochastic approaches such as chance constrained programming (CCP), quadratic programming (QP), risk formulation (RF) and genetic algorithms (GAs). The role of MOP has, for example, been reviewed by Pomar et al. (2007) and stochastic programming by Ramana et al. (2018). A review by Ghosh et al. (2014) highlights the important role of LP models in optimizing feed costs in mixing plants and livestock farms from the 1960s, as well as how other approaches such as GP can help overcome some of its limitations.

From the initial work of Rehman and Romero (1984, 1987), GP has emerged as the main alternative approach to LP in calculating feed rations, integrating additional criteria to cost minimization, such as nutrient imbalance and environmental objectives (GHG emissions and excretion of nitrogen and phosphorus). Developments in the use of GP in livestock production models are discussed by Lara and Romero (1992), Lara (1993), Bailleul et al. (2001), Tozer and Stokes (2001), Zhang and Roush (2002), Castrodeza et al. (2005), Pomar et al. (2007) and Moraes et al. (2015).

As these developments suggest, a major challenge for models is the need to account for a much wider range of variables such as the environmental impact of livestock feed. Focussing on poultry nutrition, Leeson (2008) highlights some important issues in the last decade and for the future, such as the need for traceability of animal products such as meat and eggs and environmental concerns, as well as the role of animal diets in reducing the quantity of nitrogen (N) and phosphorus (P) in livestock manure. At the end of the food supply chain, consumer preferences are affecting livestock feed and diet formulation. Examples include reductions in antibiotic use and the increasing demand for new feed additives, such as probiotics and prebiotics, to enhance animal immune systems and health (Roembke, 2018). A review of the views of key stakeholders by Makkar and Anker (2014) has identified three dimensions of sustainable animal diets (economic, environmental and social). These diets need to provide economically viable animal products which are affordable and safe for consumers, and which use natural resources efficiently and minimize pollution.

As an example of incorporating environmental issues into feed models, Tallentire et al. (2017) have quantified the environmental impact of broiler production in the UK and the USA using life cycle assessment (LCA) methodology. They assessed seven diets using LP models taking into account a set of minimum nutritional requirements. Other key objectives were to minimize cost, global warming potential, fresh water eutrophication potential, marine eutrophication potential, terrestrial acidification potential, non-renewable energy use and agricultural land use. The seven diets obtained for the UK and the USA used specific data for each region (EU legislation does not permit some ingredients allowed in the USA), nutrient content and prices of ingredients, as well as maximum and minimum limits due to other issues such as palatability. They found a higher diet cost when environmental factors were considered. This study is an interesting approach to evaluating the global impact of diets from livestock production based on average data at the country level. Nevertheless, to determine a real balance at the farm level, it is necessary to use specific farm-level data and to use a multicriteria method, such as GP, in order to balance cost and environmental objectives.

Hou et al. (2016) have proposed a uniform methodology, based on an aggregate LP model, to estimate feed use and nitrogen excreta from livestock in the European Union at the national level to improve comparisons between countries. This model has a high level of aggregation in both animal and feed categories. Segura et al. (2018) have developed a more detailed LP model to estimate real consumption of feed by animals and have validated it in the Spanish pig sector (the principal pig producer in the EU). This LP model is easy to implement for other species in other countries and also at the farm level. This optimization model provides the foundation for a DSS to analyse the effects of diet on pollutants and GHG emissions.

3 Linear programming (LP) models to minimize feed costs: solutions and sensitivity analysis

The formulation of diets and feed for livestock on farms is currently very complicated due to the need to integrate knowledge from different disciplines such as optimization methodologies, nutrition, environmental and economic science. Given that the feed composition has for decades been based on LP models, and the forecast that LP will continue to be a fundamental tool in the future, this section reviews the basic concepts of these models and illustrates them with a simple example.

Figure 1 illustrates the problem of determining the composition of the feed in a broiler farm, as well as the LP model. In order to represent the problem graphically, only two ingredients will be considered, corn and soybean meal. The problem is to calculate the optimal mix of corn and soybean meal that covers the nutritional needs of the animals while minimizing the cost. The feed must provide at least the amount of metabolizable energy (ME) and crude protein (CP) required.

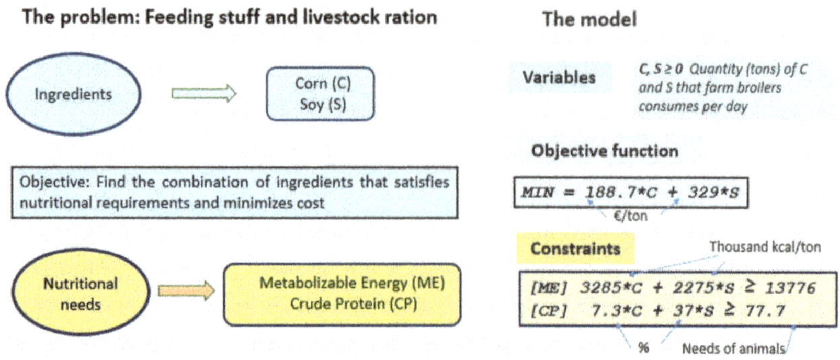

Figure 1 Problem and model to minimize feed cost per day of broilers in finisher stage.

Figure 1 also includes the model that solves this problem for a farm with 30000 broilers. The optimization models include three fundamental concepts: variables, objective function (OF) and constraints. The variables are the unknowns of the problem, in this case the quantity (tons) of corn (C) and soybean meal (S) that the broilers consume per day. The OF is the cost of the feed to be optimized; it is a linear function that depends on the price of corn and soybean meal. Finally, the constraints of the model guarantee that the feed provides the minimum needs for ME and CP that the broilers require.

In summary, the model in Fig. 1 allows the value of the variables C and S to be calculated, which minimize the cost of the feed and cover the ME and CP needs of the animals. The coefficients of the variables in the OF are the prices of corn and soybean meal expressed in euros/ton. The variables are measured in tons/day. The value of the OF represents the cost of the feed in euros/day.

The coefficients of the variables in the ME restriction represent 1000 kcal/ton of corn and soybean meal, while the right-hand side (RHS) shows the daily needs of ME for a farm with 30000 broilers. In the second CP restriction, the coefficients of the variables are the CP percentages of corn and soybean meal, while the RHS has been estimated taking into account that the feed has a minimum of 18.5% of CP, since feed for broilers in the finisher stage is being calculated.

This model can be solved graphically, as illustrated in Fig. 2. The optimal solution is point B, which is the combination of corn and soybean meal that satisfies the needs of broilers with the minimum cost. The relationship between the prices of corn and soybean meal determines the slope of the line of the OF and therefore the optimum composition of the feed. The LP model seeks the combination of ingredients among all solutions that meet the restrictions (feasible region) that represent the lowest cost. It is important to highlight that the feasible mixtures of corn and soybean meal that cover the needs of the animals depend solely on the constraints, not on the costs of the raw materials,

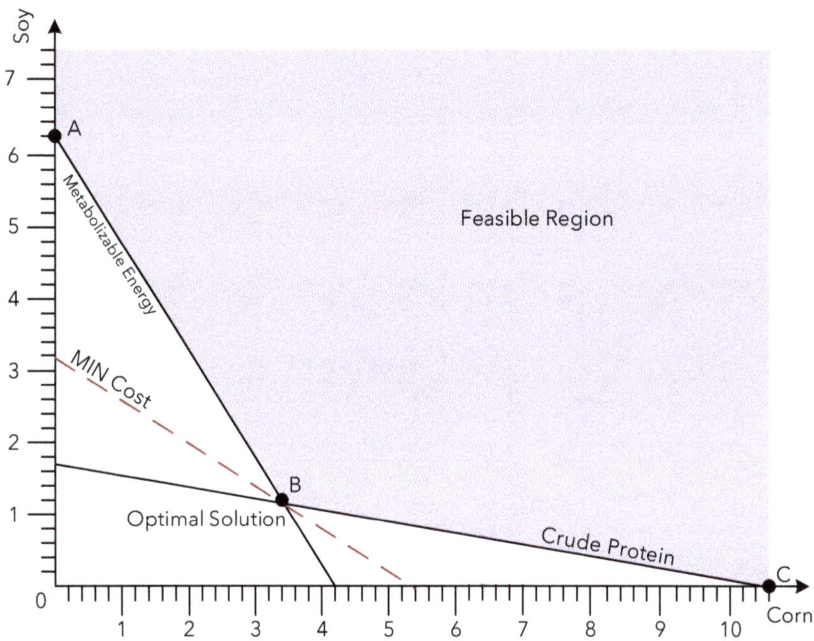

Figure 2 Graphical solution of the model in Fig. 1.

which only have an effect in determining lower cost solutions. More information on the formulation and resolution of LP models can be found in Maroto et al. (2014) and/or Hillier and Lieberman (2014).

In the previous model, limits on the availability of raw materials were not taken into account. Suppose that the farm has no more than 3 tons of corn and 4 tons of soybean meal a day. There would then be two bound constraints that represent this maximum availability of raw materials to make the feed. As shown in Fig. 3, by adding these two constraints on maximum levels of the variables, the feasible combinations of C and S that cover the needs of the animals have been reduced considerably and the optimal solution changes. By adding more constraints to the model, the cost of the feed increases, as shown in Fig. 4.

In practice, farms use more than three raw materials, so the problem cannot be solved graphically and it is necessary to use optimization software to calculate feed rations. Figure 4 shows models 1 and 2 and the solutions obtained with LINGO (LINDO Systems, 2019). The values of the variables (C and S) and the OF can be seen in Fig. 4.

The solutions of the LP models also provide other relevant information such as the amount of energy and protein contained in the ration. Slack or surplus concepts show the difference between the left-hand side (LHS) and the RHS of the restriction. Slack is the difference when the constraint is less or equal.

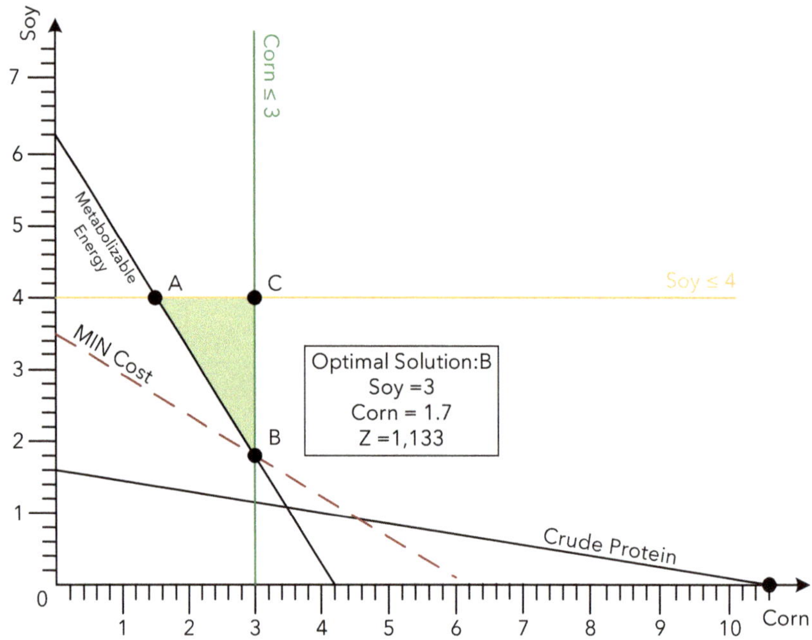

Figure 3 Graphical solution of model 2 in Fig. 4.

```
MODEL 1:
! Variables: Corn (C) and Soybean meal (S),
  ton/day for 30000 broilers in finisher stage

Objective: Minimize total cost. Prices: euros/ton;

[OF] MIN = 188.7*C + 329*S;

[ME] 3285*C + 2275*S >= 13776;
[CP]  7.3*C + 47*S >= 77.7;

OPTIMAL SOLUTION

Objective value:   1013.964

Variable     Value           Reduced Cost
C          3.416163            0.000000
S          1.122596            0.000000

Row    Slack or Surplus       Dual Price
ME         0.000000          -0.4693604E-01
CP         0.000000          -4.728096
```

```
MODEL 2:
! Variables: Corn (C) and Soybean meal (S),
  ton/day for 30000 broilers in finisher stage

Objective: Minimize total cost. Prices: euros/ton;

[OF] MIN = 188.7*C + 329*S;

[ME] 3285*C + 2275*S >= 13776;
[CP]  7.3*C + 47*S >= 77.7;
@BND(0,C,3);@BND(0,S,4);

OPTIMAL SOLUTION

Objective value: 1133.137

Variable     Value           Reduced Cost
C          3.000000           -286.3615
S          1.723516             0.000000

Row    Slack or Surplus       Dual Price
ME         0.000000          -0.1446154
CP        25.20527            0.000000
```

Figure 4 Solution of models 1 and 2 obtained by the optimization software LINGO.

Surplus is the difference when the constraint is greater or equal than in the case of ME and CP. Model 1 gives a daily ration for 30 000 broilers of 3.416 tons of corn and 1.123 tons of soybean meal, the cost of which is €1013.964. This ration provides exactly the minimum energy and protein required in the model. The optimal solution of model 2 provides different amounts of corn and soy, as well as higher cost due to restrictions on the availability of raw materials. In this case, what is really limiting is the amount of corn available, the surplus of the CP constraint is 25.205, indicating that the daily ration provides 0.252 tons more protein than the minimum required. Notice that the constraint is expressed as the percentage of CP multiplied by tons of each ingredient.

Figure 5 shows the model with other bounds related to availability of corn and soy, as well as the interpretation of basic concepts of LP such as reduced cost and dual price, which provide relevant information to design optimal diets. One of the advantages of using LP models for the calculation of rations and animal feed is having information about the modifications that would take place if the model coefficients such as prices and nutritional needs changed. The sensitivity analysis provides the ranges of variation of the prices and RHS of the constraints that does not change the mix of ingredients and production strategy, respectively, as can be seen in Fig. 5.

The general formulation of the LP model for feed ration is as follows. The model has the same number of variables as the number of ingredients to make the feed.

X_j is the quantity (kg, tons ...) of ingredient j in the ration.

```
MODEL 3:
! Variables: Corn (C) and Soybean meal (S),
 ton/day for 30000 broilers in finisher stage
Objective: Minimize total cost. Prices: euros/ton;

[OF] MIN = 188.7*C + 329*S;

[ME] 3285*C + 2275*S >= 13776;
[CP]  7.3*C + 47*S >= 77.7;
 @BND(0,C,2.8);@BND(0,S,3);

Objective value: 1190.409
Variable      Value            Reduced Cost
C             2.800000          -286.3615
S             2.012308           0.000000
Row       Slack or Surplus      Dual Price
ME            0.000000          -0.1446154
CP            37.31846           0.000000

                  Objective Coefficient Ranges:
           Current            Allowable          Allowable
Variable  Coefficient         Increase           Decrease
C           188.7000           286.3615           INFINITY
S           329.0000           INFINITY           198.3174

                     Righthand Side Ranges
           Current             Allowable          Allowable
Row          RHS               Increase           Decrease
ME         13776.00            2247.000           1806.372
CP         77.70000            37.31846           INFINITY
```

Improvement of cost (because the value of reduced cost is negative) when relaxing maximum availability of corn

Cost increase (because the value of dual price is negative) when ME Right-Hand Side increase in one unit

Sensitivity analysis of ingredient prices

Sensitivity analysis of nutritional needs

Figure 5 Solution and sensitivity analysis of model 3 obtained by the optimization software LINGO.

$$X_j \geq 0 \quad \text{for } j = 1,2,\ldots n$$

The OF is to minimize animal feeding cost, where Z is the cost of the livestock feeding and P_j is the price of ingredient j.

$$Z = \sum P_j X_j \quad \text{for } j = 1,2,\ldots n$$

The constraints are minimum and/or maximum nutrition requirements depending on factors, such as species, production cycle, age, genotype and management strategies. Examples of minimum nutrition requirements can be constraints related to energy, protein, cereal feed, protein-rich feed, starch, calcium and methionine among others.

$$\sum a_{ij} X_j \geq b_i \quad \text{for } j = 1,2,\ldots n \quad \text{and} \quad i = 1,2,\ldots m$$

Examples of maximum nutrition requirements can be constraints related to protein, cereal feed, dry matter, ether extract and calcium, amongst others.

$$\sum a_{ij} X_j \leq b_i \quad \text{for } j = 1,2,\ldots n \quad \text{and} \quad i = 1,2,\ldots m$$

Where a_{ij} is the quantity of the ith nutrient in ingredient j and b_i represents the needs of nutrient i in the ration.

This basic model can be extended with other constraints, such as bounds of the variables due to maximum availability of some ingredients or due to recommendations from nutrition experts.

$$X_j \leq \text{Bound } j$$

This model can be extended to determine several rations at a time for a complete feeding programme for broilers by defining variables X_{jk} as the quantity of ingredient j in the ration for bird stage k. Following the latest recommendations of FEDNA (2018), a feeding programme with four stages for broilers includes starter from 1 to 10 days, grower from 11 to 21 days, finisher from 22 to 35 days and withdrawal from 36 to 42 days. As another example, a multiformulation model for pigs allows the calculation of diets for all animal categories in the farm, such as piglets, fattening pigs, sows and boards.

4 Goal programming (GP) models: balancing costs and environmental impact

Sustainable diets involve a range of economic, environmental and social criteria. The model approach should, therefore, focus on animal nutrition that satisfies livestock requirements and balances cost and emissions (GHGs and other pollutants) from livestock production.

MOP models can be applied in order to find efficient solutions, which are those diets in which costs cannot be further reduced without increasing contamination levels. To obtain an efficient solution, a LP model would have to be solved, such as the one explained in Section 3, including the cost in the OF and the emissions as a parametric restriction, in which for each RHS value there would be an efficient solution. Many efficient solutions can then be calculated and the preferred solution chosen. Another approach is to apply GP models, which balance economic and environmental objectives by using an extended LP model as explained later with a small example (Maroto et al., 2014).

An environmental issue in intensive poultry production is the need to reduce N and P output in manure. As Leeson (2008) highlights, it is cheaper to overformulate these nutrients and adding maximum constraints to LP models cause higher costs. Adding phytase enzyme to feed reduces P output in broiler manure and N can be reduced by decreasing CP intake. A GP approach is, therefore, used to balance cost and CP of feed to calculate sustainable diets.

Figure 6 illustrates a GP model to find sustainable rations in ton/day for a farm with 30000 broilers in the finisher stage. The available ingredients are corn, wheat, soybean meal (SOYBM), calcium carbonate (CCA) and methionine (MET). The variables of the model are the unknowns, that is, the quantity in tons of these raw materials to be mixed to prepare the feed: CORN, WHEAT, SOYBM, CCA and MET.

The structure of the goal for the CP of the ration is a linear function, which expresses the CP provided by ingredients, plus the negative deviation variable (N_CP), minus the positive deviation variable (P_CP) equal a value that represents the aspiration level for the CP. This aspiration level is the recommended value of CP in the feed.

The negative deviation variable quantifies the lack of achievement of a goal with respect to its aspiration level. The positive deviation variable quantifies the surplus of achievement of a goal with respect to its aspiration level.

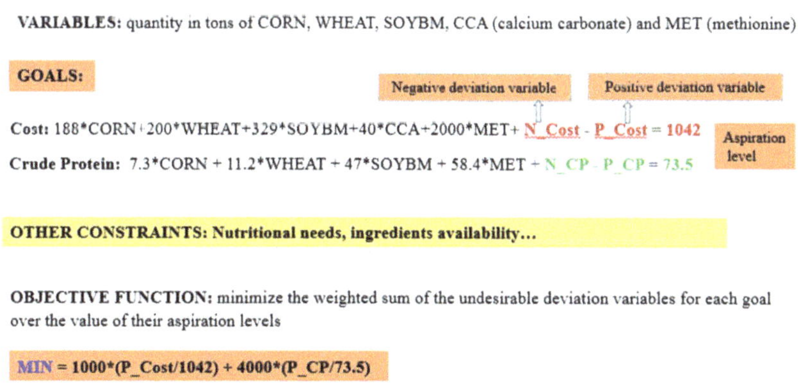

Figure 6 An example of goal programming model for sustainable livestock rations.

Goals for other important attributes of animal feed can also be defined, such as cost. In this case, a cost goal can be the expression to multiply prices (euro/ton) by variables (tons), plus the negative deviation variable (N_Cost), minus the positive deviation variable (P_Cost), equals the aspiration level, the minimum ration cost.

As Fig. 6 shows, the OF of a GP model is to minimize the weighted sum of the undesirable deviation variables for each goal over the value of their aspiration levels. In this case, the positive deviation variable for cost is to be minimized, as the lower the cost the better. For CP, the positive deviation variable is also minimized. As both goals, cost and CP, are measured in different units, it is necessary to divide each deviation variable by its aspiration level in order to have a homogeneous OF. The deviation variables are multiplied by weights according to the importance of the goal. Different optimal solutions can be generated by changing weights of goals.

In general, when the goal is of the lower, the better type, the positive deviation variable has to be minimized. If the goal is the higher, the better type, the negative deviation variable has to be minimized. Finally, when the best decision is to be as close as possible to a specific value, the GP model should minimize the sum of both negative and positive deviation variables.

Note that the structure of a goal with both negative and positive deviation variables is a way to include a flexible constraint in a model, whose LHS value can take a greater or a smaller value than the RHS. This strategy is useful, for instance, to add the P constraint, and the optimal solution shows the quantity of P that is provided by ingredients and the amount to be added to feed, for example with phytase enzyme. A complete GP model is presented in the case study (Section 6).

5 Decision support systems and data management for sustainable diets

The feed composition for livestock depends on many factors, such as the nutrient needs of the animals, which are related to species and production category (e.g. piglets, fattening pig, sows), the prices and availability of ingredients, genotype, management strategies, farm location, climate conditions etc. These need to be accounted for in LP models.

Figure 7 shows the main data needed for diet calculation, for instance, the nutrient amount of ingredients, as well as the nutrient needs by species. The main sources of these data are the benchmark publications of NRC in the USA (1994, 2012), CVB in Holland, INRA in France, Atlas PREMIER in UK and FEDNA in Spain, which are regularly updated. Applegate and Angel (2014) highlight the importance of these databases and the agreement between the scientific community and feed industry about the need to update them. The quality of

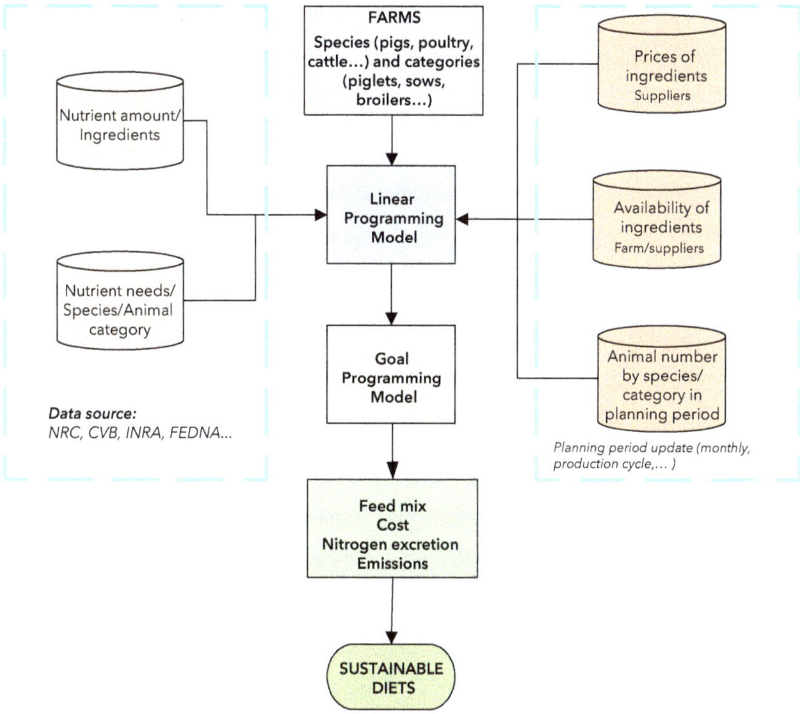

Figure 7 Decision support system to determine sustainable diets for livestock.

these data can be improved by obtaining some of data in real farm conditions, instead of in the laboratory. As data should take into account the farm context, it is necessary to develop a feed library for livestock production in other regions, such as South Asia, with the relevant information about the feed and fodder available in this region, which provides updated scientific data to livestock farmers and other stakeholders involved (Ramana et al., 2018).

In addition to information about ingredients, for the calculation of diets, it is necessary to know their real prices and availability, as well as the number of animals on the farm. Finally, livestock management has to apply the most current data from both research (animal genetics, productive and management systems, etc.) and the market.

An appropriate technique to minimize and balance cost and environmental problems (for instance, N excretion) is GP, as explained in Section 4. According to the Intergovernmental Panel on Climate Change (IPCC) Guidelines to determine emissions at the country level, the general approach is to multiply activity data by emission factor, which quantifies the emission per unit of activity. Depending on the available information, there are three methodologies. Tier 1 applies a linear relation between activity data from statistical information and default emission factors. In Tier 2, the emission factors are country specific. Tier 3 is based on more complex models and/or data from facility level. The emissions from livestock only use Tier 1 or 2 approaches, based on manure management, even though the influence of animal diets is well known. GP provides advanced models, which can be considered as a Tier 3 approach for improving estimations of emissions and pollutants at both the farm and country level (Segura et al., 2018).

6 Case study 1: sustainable rations for intensive broiler production

In this section, LP and GP models are used to show the strengths and weaknesses of these approaches to find sustainable rations for broiler production. A LP model to minimize cost of the rations for a farm with 30000 broilers in the finisher stage could be as follows. The model includes real data and the main ingredients are used in Spain, which are corn, wheat, soybean meal, calcium carbonate and methionine. The monthly prices are those for November 2017. The technical coefficients of the models and animal needs are from expert recommendations (FEDNA, 2018).

```
[OF] MIN = 187.8*CORN + 207.5*WHEAT + 311.5*SOYBM + 40*CCA +
2000*MET;

!Constraints;
[MEMIN] 3285*CORN + 3150*WHEAT + 2275*SOYBM + 4400*MET >= 13020;
[CPMIN] 7.3*CORN + 11.2*WHEAT + 47*SOYBM + 58.4*MET >= 73.5;
[CFMIN] 2.1*CORN + 2.4*WHEAT + 4.6*SOYBM >= 12.81;
[CFMAX] 2.1*CORN + 2.4*WHEAT + 4.6*SOYBM <= 18.06;
[METCIS] 0.31*CORN + 0.42*WHEAT + 1.38*SOYBM + 99*MET >= 3.192;
[CaMIN] 0.03*CORN + 0.05*WHEAT + 0.33*SOYBM + 38.6*CCA >= 3.15;
[CaMAX] 0.03*CORN + 0.05*WHEAT + 0.33*SOYBM + 38.6*CCA <= 3.57;
[PMIN] 0.25*CORN + 0.29*WHEAT +0.67*SOYBM + ND_P – PD_P = 2.352;

[PERWHEAT] WHEAT <= 0.60*FEED;
[PERCORN] CORN <= 0.40*FEED;
[TOTALFEED]FEED = CORN + WHEAT + SOYBM + CCA + MET;

[COSTVALUE]COST = 187.8*CORN + 207.5*WHEAT +311.5*SOYBM + 40*CCA
+ 2000*MET;
[MEVALUE] ME = 3285*CORN + 3150*WHEAT + 2275*SOYBM + 4400*MET;
[CPVALUE] CP = 7.3*CORN + 11.2*WHEAT + 47*SOYBM + 58.4*MET;
```

As explained in Section 3, the variables are the quantity in tons of these raw materials to be mixed: CORN, WHEAT, SOYBM, CCA and MET. The OF is to minimize the feed cost and the first group of constraints are the animal requirements related to ME, CP, crude fibre, the amino acids methionine and cystine and calcium. The phosphorus constraint is written as a goal to be able to determine the amount to be added as phytase. The next three constraints limit the percentages of corn and wheat in the feed. The last three constraints are there only to know the cost and the value of the ME and the CP of the ration, and by including these data in the results, make it easier to interpret the optimal solution.

After solving this model using LINGO software, the following optimal solution is obtained:

Variable	Value	Reduced cost
CORN	1.789585	0.000000
WHEAT	1.349922	0.000000
SOYBM	1.263491	0.000000
CCA	0.6766488E-01	0.000000
MET	0.3299431E-02	0.000000
ND_P	0.6665874	0.000000
FEED	4.473962	0.000000
COST	**1019.076**	0.000000
ME	**13020.00**	0.000000
CP	**87.75987**	0.000000

Row	Slack or surplus	Dual price
MEMIN	0.000000	−0.2526782E−01
CPMIN	14.25987	0.000000
CFMIN	0.000000	−48.88467
CFMAX	5.250000	0.000000
METCIS	0.000000	−19.05343
CAMIN	0.000000	−0.9706688
CAMAX	0.4200000	0.000000
PERWHEAT	1.334455	0.000000
PERCORN	0.000000	6.330466

The optimal solution of the LP model shows the amount of corn, wheat, soybean meal, calcium carbonate and methionine, as well as total feed and extra need for phosphorous (indicated by the negative deviation variable). The minimum cost and total values of energy and protein are also obtained. The slack/surplus of a constraint indicates how far the requirements are from satisfying their minimum or maximum value. When this is the case, the corresponding constraint has a dual price, which shows the effect on the cost when the RHS of the restriction is increased or decreased. For example, the ration provides the minimum quantity of energy and fibre, so if the needs are decreased the cost will also be reduced (dual price negative). Corn forms 40% of the feed, the maximum value allowed by the constraints. If this percentage could be increased, the ration cost would decrease (dual price positive).

To analyse how the feed composition changes according to price fluctuations during the year, monthly models taking into account the real prices in Spain from July 2017 to June 2018 have been solved. Figures 8–10 show the relationships between the evolution of prices, characteristics and composition of rations calculated monthly. In short, the rations provide the minimum energy and an excess of protein, except when the price of soybean meal increases a lot, which is the ingredient with the greatest variability in price. This result can be determined in advance by carrying out a sensitivity analysis of the prices in the previous model, whose results are the following:

Ranges in which the basis is unchanged:

Objective coefficient ranges:

Variable	Current Coefficient	Allowable Increase	Allowable Decrease
CORN	187.8000	6.341350	93.83059
WHEAT	207.5000	82.03207	5.543971
SOYBM	311.5000	45.21200	139.7801
CCA	40.00000	18461.79	37.47522
MET	2000.000	12942.78	1899.087

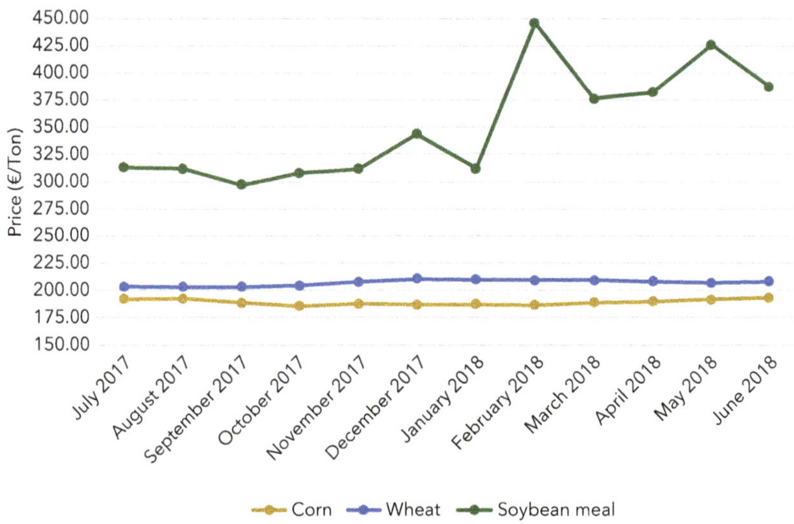

Figure 8 Price evolution of feed ingredients 2017–2018.

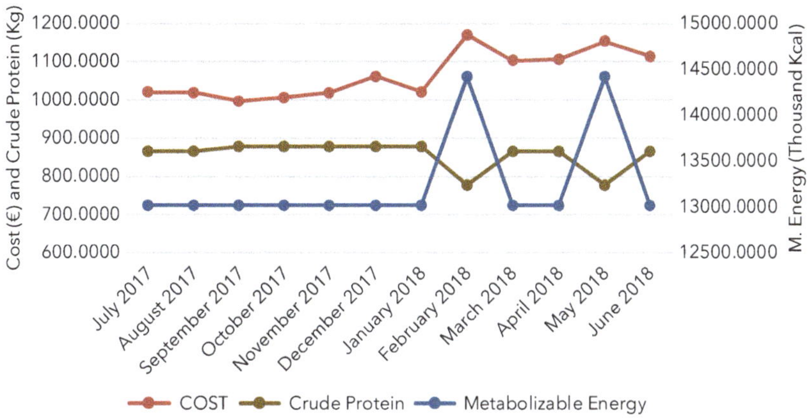

Figure 9 Cost, metabolizable energy and crude protein in daily ration.

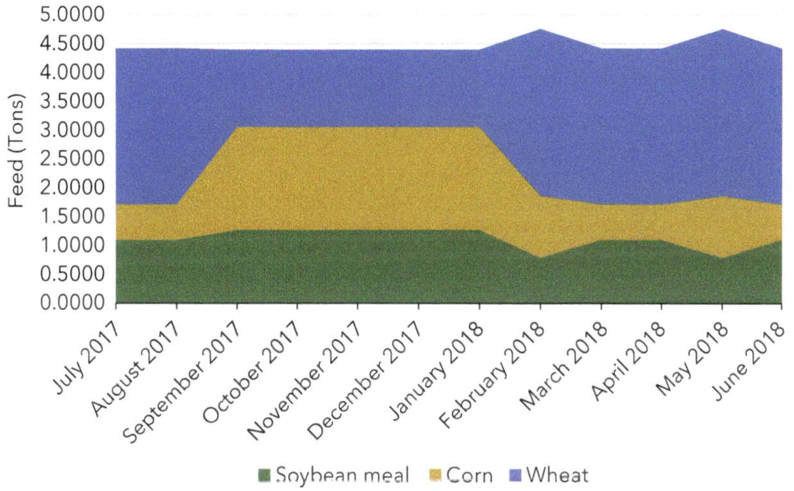

Figure 10 Evolution of amount of ingredients in the feed according to monthly prices.

The ranges provided by the sensitivity analysis indicate the allowable increase and decrease of the prices without causing changes in the composition for the minimum cost when only one price is modified at a time. The 100% rule can be used when several prices change at the same time as is usually the case. When the price of an ingredient decreases, the percentage that represents this change with respect to the allowable decrease is calculated. In the case that the price increases, the percentage is determined with respect to the allowable increase. Finally, if the sum of all of the percentages is lower than 100, the mix composition does not change, although the cost is modified according to the new prices. Applying this rule to the sensitivity analysis of the November model and considering the prices for December, it can be calculated that the mix would be the same, as the price of corn decreases by 1%, the price of wheat increases by 4% and the price of soybean meal goes up by 71%. The sum of these percentages is less than 100. When the soy price increases by more than €45.21, the mix for the minimum cost changes. This is the case for February 2018, as shown in Fig. 10.

With GP models, the cost, CP and energy of the ration can be balanced. The following model gives more weight to protein than to energy and cost in order to reduce the excess of protein in the feed.

```
[OF]MIN = 1000*(P_Cost/1019.076) + 4000*(P_CP/73.5)+ 2000*(P_ME/13020);

!Constraints:
[GOAL_COST] 187.8*CORN + 207.5*WHEAT + 311.5*SOYBM + 40*CCA +
2000*MET + N_Cost – P_Cost = 1019.076;
[GOAL_CP] 7.3*CORN + 11.2*WHEAT + 47*SOYBM + 58.4*MET + N_CP – P_
CP = 73.5;
[GOAL_ME] 3285*CORN + 3150*WHEAT + 2275*SOYBM + 4400*MET + N_
ME – P_ME = 13020;

[CFMIN] 2.1*CORN + 2.4*WHEAT + 4.6*SOYBM >= 12.81;
[CFMAX] 2.1*CORN + 2.4*WHEAT + 4.6*SOYBM <= 18.06;
[METCIS] 0.31*CORN + 0.42*WHEAT + 1.38*SOYBM + 99*MET >= 3.192;
[CaMIN] 0.03*CORN + 0.05*WHEAT + 0.33*SOYBM + 38.6*CCA >= 3.15;
[CaMAX] 0.03*CORN + 0.05*WHEAT + 0.33*SOYBM + 38.6*CCA <= 3.57;
[PMIN] 0.25*CORN + 0.29*WHEAT + 0.67*SOYBM + ND_P – PD_P = 2.352;

[PERWHEAT] WHEAT <= 0.60*FEED;
[PERCORN] CORN <= 0.40*FEED;

[TOTALFEED]FEED = CORN + WHEAT + SOYBM + CCA + MET;

[COSTVALUE]COST = 187.8*CORN + 207.5*WHEAT + 311.5*SOYBM +
40*CCA + 2000*MET;
[MEVALUE] ME = 3285*CORN + 3150*WHEAT + 2275*SOYBM + 4400*MET;
[CPVALUE] CP = 7.3*CORN + 11.2*WHEAT + 47*SOYBM + 58.4*MET;
```

Table 1 indicates the cost, the energy and the protein of the ration calculated by LP and weighted GP models (with prices from November 2017). When cost, energy and protein are goals with the same importance, the amount of protein decreases, the cost increases and the energy of the ration is the minimum required. By giving more importance to CP, in particular by minimizing the excess over the minimum needs, this goal is achieved, but only by increasing the cost, as well as the energy of the ration.

Figure 11 represents the feed composition calculated by the previous LP and GP models. This graph shows the reduction of CP in GP models compared

Table 1 Cost, energy (ME) and crude protein (CP) of the ration calculated by LP and GP models. Prices November 2017 (Spain)

Goals	LP model	GP model. Equal weights for three goals	GP model. Weight CP equal 4*weights cost and ME
Cost (euro)	1019.07	1026.49	1075.72
M. Energy (Thousand kcal)	13020.00	13020.000	15080.350
Crude protein (kg)	877.599	865.1153	735

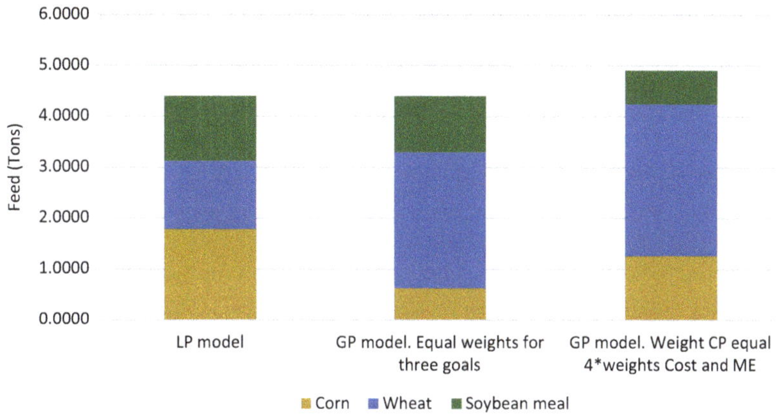

Figure 11 Composition of the broiler feed obtained by LP and weighted GP models. Prices November 2017 (Spain).

to the LP model. When more importance is given to reducing the excess of protein, the amount of feed is increased in order to satisfy all nutritional needs and other constraints. Finally, these models can be easily extended for a complete feeding programme as explained in Section 3, as well as to several cycles, according to the management system at the farm.

The previous results, together with concerns for animal welfare and increasing demand for safe food, show how complex diet and ration formulation for livestock is. Traceability of animal products requires accountability of the composition of livestock feed because of legislation and marketing strategies. There is a lot of room for future research that should integrate knowledge from several disciplines as nutrition science, MOP/GP and DSS development among others. All of this is necessary to make advances in satisfying society's demands and reducing emissions and pollutants by designing sustainable diets for livestock production in both developed and developing countries.

7 Case study 2: reducing emissions in pig production

The LP and GP models discussed in Sections 3 and 4 are also the core of the DSS for obtaining sustainable diets in farms for other livestock species, following the framework shown in Fig. 7. The models need to be adapted to take into account of factors such as animal numbers and needs in each stage of growth, as well as the appropriate ingredients for the species together with their availabilities and prices. Data should consider the type of production system (extensive/intensive farming) and environmental conditions. These DSSs can also provide the information needed to calculate emissions of GHGs and other pollutants, since feed intake is the most important variable in predicting these environmental impacts from livestock.

Recently, Hou et al. (2016) and Segura et al. (2018) have proposed LP models as a way to improve the assessment of emissions from livestock production, in part to meet annual reporting requirements for GHG emissions agreed internationally and regionally (e.g. through EU directives). Hou et al. (2016) cover a wide range of livestock species (dairy and beef cattle, sheep, pigs, goats, broilers and laying hens, goats, turkeys, duck and geese) and eight groups of ingredients (including cereals, protein-rich feed and brans). Segura et al. (2018) have focussed specifically on the pig sector in Spain which is the main European producer. This model includes all common ingredients in the country (21 in all), including cereals, protein-rich feeds and cereal by-products. To improve accuracy, the authors distinguish eight categories of pigs, as the needs of piglets, for example, are very different to those of fattening pigs and sows.

The variables of the aggregated model are the quantities in tons of each raw material that each animal category consumes in the reference year in a country. These variables have upper bounds depending on the type of feed and the animal category. The constraints are the minimum and/or maximum nutritional requirements of livestock, such as energy, protein or cereal feed. The available stock of raw materials in the country, calculated as production plus imports minus exports, in a reference year constitutes another group of constraints. Finally, the OF consists of minimizing the total animal feed cost by multiplying the unit prices by quantities of raw materials used for feeding all animals in the country. The main differences between LP models at the farm and country levels lie with the data requirements to be updated annually. Data to be updated include prices and stocks of feed ingredients, as well as animal numbers by category in a reference year. The quality and availability of these statistical data is the main challenge to making DSS a useful tool for policy-makers.

The model is able to provide a better estimate of the environmental impact (nitrogen excretion) of pig production, as it uses more precise technical data related to feed ingredients and the production cycle of animals. As an example, this model takes the specific nutrients of barley, wheat, corn and so forth into account instead of the average nutrients of cereals. Animal needs are related to specific categories (piglets, sows, etc.) instead of the average for the species. The model has been able to identify which categories of pigs produce the highest emissions as well as recommend target quantities of key ingredients such as cereals and protein-rich feeds to minimize emissions whilst meeting nutritional requirements and optimizing feed costs.

8 Summary and future trends

Livestock production should be able to provide sufficient safe, sustainable and affordable food to meet increasing demand worldwide. However, there are increasing concerns about health, animal welfare and environmental problems caused by livestock production. When the safety and sustainability of diets is improved, production costs usually increase as well. To resolve these conflicting objectives is a challenge, which must be overcome in the next few decades. An appropriate approach to solving this real problem is by involving all stakeholders, such as researchers from different disciplines, farmers, industry, consumers and policy-makers.

LP models have made a huge contribution to making meat affordable by reducing feed costs during the last four decades, with the pig and poultry sectors being very competitive examples. In fact, these optimization models are the core of DSSs, which calculate the ration composition for livestock. These DSSs should be integrated into systems, which also incorporate traceability and quality management information.

To understand the strengths and weaknesses of this powerful optimization technique in livestock diet/ration formulation, experts need to know some basic concepts and properties of LP. The diet problem consists of determining the proportions of feed ingredients, while the ration problem is defined by the quantities of feed ingredients which satisfy the requirements of animals with the objective of minimizing the total feed cost.

The simplex algorithm solves any LP model and provides the global optimal solution, if there is any solution that meets all model constraints. The constraints represent the minimum/maximum needs of animals, such as energy, protein, fibre, calcium, phosphorous and so forth. Among all of the feasible combinations of ingredients (which are usually infinite), the algorithm finds the best solution for the objective (in general, the ingredient combination with the minimum cost). In practice, farms have additional restrictions with respect to availability of cereals and other raw materials, which are also included as model constraints.

It is important to highlight that if a model does not have a solution, it is never due to the objective chosen by the decision maker. When there is no combination of ingredients that satisfies all constraints, the model has no solution. Farms and industry use additives (amino acids, minerals etc.) to avoid this problem, complying, as far as possible, with the recommendations of nutrition experts. Another property of the optimal solution from a LP model is that the number of ingredients to mix is as high as the number of constraints included in the model. One of the strengths of LP is the information which can be obtained from the sensitivity analysis. What happens to the feed formulation when the prices of the ingredients change and/or the animal needs or other restrictions increase or decrease can be calculated.

As society demands improvements to the sustainability of the agricultural production, diet formulation for livestock at farms and feed industries will extend LP models to GP in order to balance economic and environmental objectives. Goals are included as flexible constraints in the model and the objective of the decision maker is to get as close as possible to our aspiration levels in economic and environmental goals. This approach is easy to understand and to carry out in research studies, but it needs to be applied to sustainable production through appropriate DSSs.

DSS can help farmers to calculate diets for sustainable livestock production, based on LP/GP approaches. DSSs also require precise data on the characteristics, availability and prices of raw materials. Other necessary data are animal nutritional needs, which have to take into account species, production cycle, age, genotype and management strategies, as well as the farm context. Improving the quality of these data and maintaining them is essential to achieving sustainability in agriculture.

The composition of livestock diets is so related to the environment that it is the basis for estimating pollutants at the country level by using aggregated LP models, while other methodologies are based only on manure management. New multidisciplinary research studies have to support this trend and apply this approach to farms. In this context, DSS for diet formulation at the farm level would be key to documenting and achieving sustainable livestock production, which provides safe and affordable food for everyone.

9 Acknowledgements

The authors would like to thank the Regional Ministry of Education, Research, Culture and Sport for funding the research project 'Sustainability of the Food Value Chain from Production to Responsible Consumption' (grant number AICO/2017/066).

10 Where to look for further information

Book

Romero, C. and Rehman, T. 2003. *Multiple Criteria Analysis for Agricultural Decisions, Developments in Agricultural Economics* (vol. 11; 2nd edn.). Elsevier.

Software for feed formulation

BESTMIX Feed Formulation. https://www.adifo.com/en/brands/bestmix-feed-formulation.

Eco-Mix Feed Formulation Software. https://www.farm.com.np/company/eco-mix-feed-formulation-software/.

Optimix. http://www.optimix.com.br/ Software for the Animal Feed and Food Industry.

Conferences and websites

International Conference on Agro BigData and Decision Support Systems in Agriculture.

Feedipedia (https://feedipedia.org/) An open access information system on animal feed resources that provides information on nature, occurrence, chemical composition, nutritional value and safe use of nearly 1400 worldwide livestock feeds. A joint project of INRA (Institut National de la Recherche Agronomique, French National Institute for Agricultural Research), CIRAD (Centre de Coopération Internationale en Recherche Agronomique pour le Développement), French Agricultural Research Center for International Development, AFZ (Association Française de Zootechnie, French Association for Animal Production) and the FAO (Food and Agriculture Organization of the United Nations).

National Animal Nutrition Program (NANP), United States Department of Agriculture (USDA). National Institute of Food and Agriculture (NIFA). https://animalnutrition.org.

Research Institute, Wageningen Livestock Research. https://www.wur.nl/en/Research-Results/Research-Institutes/livestock-research.htm.

WATTAgNet. www.WATTAgNet.com News and analysis on the global poultry and animal feed industries.

11 References

ADIFO software. 2019. *BESTMIX Feed Formulation*. Available at: https://www.adifo.com/en/brands/bestmix-feed-formulation (accessed on 10 January 2019).

Alqaisi, O., Ndambi, O. A. and Williams, R. B. 2017. Time series livestock diet optimization: cost-effective broiler feed substitution using the commodity price spread approach. *Agric. Food Econ.* 5(1), 25. doi:10.1186/s40100-017-0094-9.

Applegate, T. J. and Angel, R. 2014. Nutrient requirements of poultry publication: history and need for an update. *J. Appl. Poult. Res.* 23(3), 567–75. doi:10.3382/japr.2014-00980.

Brossard, L., Dourmad, J.-Y., Garcia-Launey, F. and van Milgen, J. 2017. Modelling nutrient requirements for pigs to optimise feed efficiency. In: Wiseman, J. (Ed.), *Achieving Sustainable Production of Pig Meat Volume 2: Animal Breeding and Nutrition*. Burleigh Dodds Science Publishing, Cambridge, UK.

Cabrera, V. E. 2018. Helping dairy farmers to improve economic performance utilizing data-driven decision support tools. *Animal* 12(1), 134–44. doi:10.1017/S1751731117001665.

Castrodeza, C., Lara, P. and Pen, T. 2005. Multicriteria fractional model for feed formulation: economic, nutritional and environmental criteria. *Agric. Syst.* 86(1), 76–96. doi:10.1016/j.agsy.2004.08.004.

Dantzig, G. B. 1990. The practice of mathematical programming. *Interfaces* 20(4), 43-7. doi:10.1287/inte.20.4.43

Glen, J. J. 1987. Mathematical models in farm planning. *Oper. Res.* 35(5), 641-66. doi:10.1287/opre.35.5.641.

Dooyum, U. D., Mallipeddi, R., Trinadh Pamulapati, T., Parka, T., Kim, J., Woo, S. and Ha, Y. 2018. Interactive livestock feed ration optimization using evolutionary algorithms. *Comput. Electron. Agric.* 155, 1-11. doi:10.1016/j.compag.2018.08.031.

Eastwood, C., Dela Rue, B. and Gray, D. 2016. Using a 'network of practice' approach to match grazing decision-support design with farmer practice. *Anim. Prod. Sci.* 57(7), 1536–48.

FEDNA. 2018. *Necesidades Nutricionales para la avicultura: Normas FEDNA* (2nd edn.). FEDNA.

FEDNA. 2019. Available at: http://fundacionfedna.org/tablas-fedna-composicion-alimentos-valor-nutritivo (accessed on 01 February 2019).

Feedipedia. 2018. Available at: https://feedipedia.org/ (accessed on 02 December 2018).

Ghahramani, A. and Moore, A. D. 2013. Climate change and broadacre livestock production across southern Australia: adaptation options via grassland management. *Crop Pasture Sci.* 64(6), 615–30. doi:10.1071/CP13195.

Ghosh, S., Ghosh, J., Pal, D. T. and Gupta, R. 2014. Current concepts of feed formulation for livestock using mathematical modeling. *Anim. Nutr. Feed Technol.* 14, 205–23.

Gous, R. and Fisher, C. 2017. Using models to optimize poultry nutrition. In: Applegate, T. (Ed.), *Achieving Sustainable Production of Poultry Meat Volume 2: Breeding and Nutrition*. Burleigh Dodds Science Publishing, Cambridge, UK.

Gouttenoire, L., Counut, S. and Ingrand, S. 2011. Modelling as a tool to redesign livestock farming systems: a literature review. *Animal* 5(12), 1957–71. doi:10.1017/S175173111100111X.

Guevara, V. R. 2004. Use of nonlinear programming to optimize performance response to energy density in broiler feed formulation. *Poult. Sci.* 83(2), 147–51. doi:10.1093/ps/83.2.147.

Hillier, F. S. and Lieberman, G. J. 2014. *Introduction to Operations Research* (10th edn.). McGraw-Hill.

Hou, Y., Bai, Z., Lesschen, J. P., Staritsky, I. G., Sikirica, N., Ma, L., Velthof, G. L. and Oenema, O. 2016. Feed use and nitrogen excretion of livestock in EU-27. *Agric. Ecosyst. Environ.* 218, 232–44. doi:10.1016/j.agee.2015.11.025.

Jean dit Bailleul, P., Rivest, J., Dubeau, F. and Pomar, C. 2001. Reducing nitrogen excretion in pigs by modifying the traditional least-cost formulation algorithm. *Livest. Prod. Sci.* 72(3), 199–211. doi:10.1016/S0301-6226(01)00224-X.

Lara, P. 1993. Multiple objective fractional programming and livestock ration formulation: a case study for dairy cow diets in Spain. *Agric. Syst.* 41(3), 321–34. doi:10.1016/0308-521X(93)90007-O.

Lara, P. and Romero, C. 1992. An interactive multi-goal programming model for determining livestock rations: an application to dairy cows in Andalusia, Spain. *J. Oper. Res. Soc.* 43(10), 945–53. doi:10.1057/jors.1992.145.

Leeson, S. 2008. Predictions for commercial poultry nutrition. *J. Appl. Poult. Res.* 17(2), 315–22. doi:10.3382/japr.2007-00101.

LINDO Systems. 2019. Available at: https://lindo.com/ (accessed on 04 January 2019).

Makkar, H. P. S. and Anker, P. 2014. Towards sustainable animal diets: a survey-based study. *Anim. Feed Sci. Technol.* 198, 309–22. doi:10.1016/j.anifeedsci.2014.09.018.

Maroto, C., Alcaraz, J., Ginestar, C. and Segura, M. 2014. *Operations Research in Business Administration and Management*. Universitat Politècnica de València.

Moraes, L. E., Fadel, J. G., Castillo, A. R., Casper, D. P., Tricarico, J. M. and Kebreab, E. 2015. Modeling the trade-off between diet costs and methane emissions: a goal programming approach. *J. Dairy Sci.* 98(8), 5557–71. doi:10.3168/jds.2014-9138.

NRC. 1994. *Nutrient Requirements of Poultry* (9th edn.). Animal Nutrition Series, National Research Council.

NRC. 2012 *Nutrient Requirements of Swine* (11th rev. edn.). Animal Nutrition Series, National Research Council.

Pomar, C., Dubeau, F., Létourneau-Montminy, M.-P., Boucher, C. and Julien, P.-O. 2007. Reducing phosphorus concentration in pig diets by adding an environmental objective to the traditional feed formulation algorithm. *Livest. Sci.* 111(1–2), 16–27. doi:10.1016/j.livsci.2006.11.011.

Rahman, R., Ang, C. L. and Ramli, R. 2010. Investigating feed mix problem approaches: an overview and potential solution. *Int. J. Agric. Biosys Eng.* 4(10), 758.

Ramana, D. B. V., Selim, A. S. M. and Tedeschi, L. O. 2018. The necessity to develop a comprehensive feed library for livestock production in South Asia. *Curr. Sci. India* 115(7), 1270–5. doi:10.18520/cs/v115/i7/1270-1275.

Rehman, T. and Romero, C. 1984. Multiple-criteria decision-making techniques and their role in livestock ration formulation. *Agric. Syst.* 15(1), 23–49. doi:10.1016/0308-521X(84)90016-7.

Rehman, T. and Romero, C. 1987. Goal programming with penalty functions and livestock ration formulation. *Agric. Syst.* 23(1), 117–32.

Roembke, J. 2018. Livestock premixes evolve to meet consumer demands. *Feed Strategy*. Available at: https://www.feedstrategy.com/poultry-nutrition/livestock-premixes-evolve-to-meet-consumer-demands/ (accessed on 03 November 2018).

Rose, D. C., Sutherland, W. J., Parker, C., Lobley, M., Winter, M., Morris, C., Twining, S., Ffoulkes, C., Amano, T. and Dicks, L. V. 2016. Decision support tools for agriculture: towards effective design and delivery. *Agric. Syst.* 149, 165–74. doi:10.1016/j.agsy.2016.09.009.

Saxena, P. 2010. Feed formulation software: a comparative study. *All About Feed* 1(8), 1–4.

Saxena, P. and Chandra, M. 2011. Animal diet formulation models: a review 1950–2010. *CAB Rev.* 6(57), 1–6.

Segura, M., Maroto, C., Ginestar, C. and Segura, B. 2018. Optimization models to improve estimations and reduce nitrogen excretion from livestock production. *Sustainability* 10(7), 2362. doi:10.3390/su10072362.

Tallentire, C. W., Mackenzie, S. G. and Kyriazakis, I. 2017. Environmental impact trade-offs in diet formulation for broiler production systems in the UK and USA. *Agric. Syst.* 154, 145–56. doi:10.1016/j.agsy.2017.03.018.

Tedeschi, L. O., Fox, D. G., Chase, L. E. and Wang, S. J. 2000. Whole-herd optimization with the Cornell Net Carbohydrate and Protein System: predicting feed biological values for diet optimization with linear programming. *J. Dairy Sci.* 83(9), 2139–48. doi:10.3168/jds.S0022-0302(00)75097-1.

Tozer, P. R. and Stokes, J. R. 2001. A multi-objective programming approach to feed ration balancing and nutrient management. *Agric. Syst.* 67(3), 201–15. doi:10.1016/S0308-521X(00)00056-1.

Zhang, F. and Roush, W. B. 2002. Multiple-objective (goal) programming model for feed formulation: an example for reducing nutrient variation. *Poult. Sci.* 81(2), 182–92. doi:10.1093/ps/81.2.182.

Chapter 12

Developing decision-support systems for pasture and rangeland management

Callum Eastwood and Brian Dela Rue, DairyNZ, New Zealand

1 Introduction

Pasture and rangeland farming systems are inherently dynamic and complex due to multiple biological (e.g. plant, animal, soil) and climatic (e.g. rainfall, temperature, evapotranspiration) factors (Eastwood and Kenny 2009; Farrié et al. 2015; Flack 2016). When management (human) factors, market volatility, and society expectations are added in, the complexity only increases (Fulkerson and Lowe 2011; Roche et al. 2015). The ability of farmers to make accurate and timely decisions is also being progressively impacted by increasing farm sizes, meaning more stock and staff to manage (Eastwood et al. 2018) and also a higher stock-to-personnel ratio (Dela Rue et al. 2020). All of these pressures mean that collecting, managing, and interpreting information is more important than ever (French et al. 2015). Smart farming approaches are increasing the data available to farmers through a growing range of sensors (Wolfert et al. 2017), however this proliferation of data does not always match what farmers need for decision-making (Lokhorst et al. 2019). Whereas farmers have often relied on experiential and intuitive decision-making practices (Nuthall 2012), on many farms this approach is no longer sufficient for effective management. Decision-support tools represent an opportunity for farmers to better manage their dynamic and complex farms.

http://dx.doi.org/10.19103/AS.2020.0069.23

The design and use of decision-support tools in farming involves a range of applications from factsheet-style decision rules (Macdonald et al. 2010) through to complex farm system modelling software (Bryant et al. 2010). Often termed decision-support systems (DSSs), they have been previously defined as 'any tool or system designed to assist end users with decision-making' (Botha and Atkins 2006). The key issues for development and use of pasture and rangeland DSSs includes:

1 Balancing simplicity for users (e.g. farmers) with functionality;
2 Creating effective DSS tools within highly dynamic farm systems environments;
3 Ongoing development, maintenance, and relevance of DSSs;
4 Finding the right value proposition for farmers in a domain that is often managed with experiential knowledge;
5 Simplifying the data integration required to support DSS functionality; and
6 Linking the DSS with other tools on farm (e.g. sensors, software, or compliance programmes) to avoid double-entry.

There is a variety of audiences for these tools, including farmers, farm advisors (and other rural professionals), and farm system researchers. The needs of the different end users must be incorporated in tool design, for example, the complexity of farm system interactions desired by researchers may not be needed by farmers who are seeking operational or tactical decision-making guidance. A general lack of inclusion of the end user in the design process has seen few DSSs considered as successful (Parker and Sinclair 2001; McCown 2012; Eastwood et al. 2017b). This chapter outlines the different types of DSSs, examines the decision-making processes of farmers, and then discusses key features of DSS development for pasture and rangeland farm systems.

2 Decision-support systems (DSSs) in pasture and rangeland management

There are a range of tools that may fit into the definition of DSSs for pasture and rangeland management. Several authors have provided definitions of DSSs, and Botha and Atkins (2006, p. 1) and categorised them as:

- A simple definition of 'computer-based technologies that aid decision-making';
- interactive computer programs that utilise analytical methods for developing models to help decision-makers formulate alternatives, analyse the impacts, and interpret and select appropriate options for implementation; or

- easy-to-use software on a computer readily accessible to a manager to provide interactive assistance in the manager's decision process.

As seen from the definition of Botha and Atkins (2006), most authors focus on the use of computer software as a key feature of DSSs, and this has been broadened by some to include phone apps, email alerts, and paper-based guidance (Rose et al. 2016). In an industry report, Morrison (2006) classified DSSs into four types, with contemporary examples provided:

- Small tactical tools, for example Spring Rotation Planner and Autumn Planner (Neal et al. 2017b), Pasture.io, PastureMap, FarmIQ, AgriNet, FeedPlan (Dobos et al. 2004);
- Versatile simulators, for example Farmax (Bryant et al. 2010), Udder, PGSUS pasture growth simulation model (Romera et al. 2010), Integrated system dynamics model (Ibáñez and Martínez-Valderrama 2018);
- Research models, for example Cornell animal model (Fox et al. 1995), Whole Farm Model (Beukes et al. 2008), DairyMod (Johnson et al. 2008), FeedSmart (Dobos and Fulkerson 2004); and
- Structured information frameworks, for example PastureBase Ireland (Hanrahan et al. 2017), DairyBase (Ma et al. 2018).

More than 20 years ago, Cox (1996) identified confusions in the purpose of DSSs that are still common today. He suggested that organisations were committing to development of DSSs but 'confusing process models (for professional research) with model-based DSSs (intended as a guide to practical action)' (p355). The different uses of DSSs in pasture and rangeland management are outlined in Section 3.

The division in the purpose of DSSs, as identified by Cox (1996), involves use for either on-farm decision-making or to support research and learning (e.g. Bashari et al. 2008), and there is some overlap between the two categories, according to Botha and Atkins (2006). This overlap occurs where farmers use DSSs (e.g. 'versatile simulators') as scenario planning tools for helping them gain a better understanding of different actions – rather than directly for making decisions.

3 Decision-making processes of pasture and rangeland farmers

It is important to understand the adoption (Schaak and Musshoff 2018) and decision-making processes of pasture-based farmers to develop effective DSSs (Eastwood et al. 2017b). Literature on farmer decision-making shows this is influenced by problem definition, information collection, analysis of alternatives, decision on appropriate action, action taking and monitoring and evaluation (Rougoor et al. 1997). Strategic decision-making has been shown to be strongly influenced by family norms and attitudes, and the level of perceived behavioural control of decision-makers (Koropp et al. 2014). The process of strategic change has been described as a process of assessing current state, deciding on change, and managing change (Eastwood et al. 2017a). These phases align with other important studies of decision-making that describe a cycle of planning, implementation, and control (Rougoor et al. 1997; Gray 2001), which can be observed at an operational, tactical or strategic level (Gray 2001). This decision-making process, and the potential roles of DSSs in each decision level, is demonstrated in Table 1. Dairy farmers chose between sub-processes within the overall plan, called the decision points, which included planned activities, along with alternative tactics and contingencies to help cope with inter-year variations in climatic conditions.

Information is a central component in the decision-making process and searching for information occurs throughout the decision-making phases. Farmers often use a qualitative 'quick and simple' approach where information is collected through incremental implementation, and there can be two broad decision-making approaches - intuitive or analytical (Ohlmer et al. 1998). In a US study, Wilmer and Fernández-Giménez (2015) highlighted that ranchers 'construct decision-making processes through non-linear and complex

Table 1 Decision support systems in the context of Gray's (2001) planning, implementation, and control cycle for dairy farmers

Processes (Gray 2001)	Sub processes (Gray 2001)	Role of decision-support systems
Planning	Goals for the planning period Predictive schedule of events Targets Contingency plans Decision rules	Gathering historical information Presenting information on current status and trends Potentially predictive information/ modelling of future trends
Implementation		Providing farmers with decision options Directing automation and control devices to carry out plans
Control	Monitoring performance indicators and external environment Data storage and processing Decision point recognition Control response selection Evaluation and learning	Real-time monitoring of animal and plant resources Automated setting of alerts for when decision points are reached Automated decision rules to guide actions when decision points reached Use of data to evaluate success/ failure of processes

consideration of multiple ranch dynamics, through engagement with multiple ways of knowing and across different timescales' (p. 517). In these multiple ways of knowing, farmers can switch between use of 'informal and qualitative' and 'formal and quantitative' approaches (Nuthall and Old 2018), where the latter involves written planning while the former does not. McCown et al. (2012) describe farmer cognitive processes that involve a continuum between intuition (holistic to arbitrary) and analysis (casual to probabilistic). The concept of 'intuitive ability' has been identified as a critical part of farmer decision-making, and requires a mix of experience, technical knowledge, and skills in anticipation (Nuthall and Old 2018). This use of intuition, or heuristics (e.g. 'rules of thumb'), maintains currency in modern agricultural systems because it provides farmers the flexibility and speed to match the decision-making demands in dynamic and uncertain situations (Eastwood and Kenny 2009; Nuthall 2019).

It is within this management decision-making context that DSSs need to operate. Although the use of smart farming-type tools, underpinned by DSSs, are predicted to increase in agriculture (Shepherd et al. 2018), the current low usage suggests barriers to adoption by farmers (Rose et al. 2016; Gargiulo et al. 2018; Dela Rue et al. 2020). Informal systems often prevail because they are timely, suitably accurate and low cost in terms of time and capital (Gray 2001). Time and resource constraints are seen by others as reasons why farmers do not adopt formal decision-making tools (Rose et al. 2018).

The use of data and DSSs in decision-making can depend on the ability to use that information to address uncertainty in the decision-making environment (Eastwood and Kenny 2009). The lower the potential impact on uncertainty, the lower the value ascribed to the information by farmers. Lynch et al. (2000) noted that the use of planning aids depended on the degree to which farmers perceive they can exert control over the performance of their farming operations. Uncertainty of production in dryland areas due to variable and unpredictable rainfall was determined to be a major issue for management (McCown et al. 2009).

The main operational areas where decisions are made on a dairy farm have been described in literature (van Asseldonk et al. 1999; Holmes et al. 2002; Hansson and Öhlmér 2008; Macdonald et al. 2010). In assessing the potential for use of information technology in management, van Asseldonk et al. (1999) divided operational management into six main areas: reproduction; health care; milk production; nutrition; cattle replacement; and roughage (pasture and crop) production. Some dairy system DSSs seek to include all of these aspects, particularly in the research domain (e.g. DairyMod (Johnson et al. 2008), Whole Farm Model (Beukes et al. 2008)), but the most common DSS tools used by dairy farmers in New Zealand are focussed on the aspects of feeding and pasture production.

3.1 Seasonal decision-making cycles

Pasture and rangeland farmers are faced with managing their farm systems within the constraints of seasonal climatic variation. In interviews with dairy farmers, farm consultants and farming systems experts, the processes used to manage pasture and grazing throughout the year on seasonal-calving farms were examined (Eastwood et al. 2017b). A main outcome was the description of grazing management as a dynamic and seasonal process where achieving high performance involves the ability to adapt to seasonal changes. This may seem obvious, but interview participants noted that many commercial DSS tools did not sufficiently account for differences in seasonal management, or provide the tools to help farmers adapt to changing conditions. In New Zealand dairy systems, the pivotal point of the annual grazing management cycle was identified as 'balance date', the point when pasture supply meets herd pasture demand in early spring. Most farmers and rural professionals interviewed worked backwards from this date when planning for next season.

Dairy system planning typically started in autumn to manage pastures through the end of lactation, winter grazing, and the calving period. Rural professionals identified key performance indicators that they used with their clients such as pasture harvested per hectare, average farm pasture cover, grazing rotation speed, current post-grazing residuals, up-to-date monthly growth rates and tonnes of supplementary feed used. The seasonal phases, critical success factors, and main tools and information used, are incorporated into Fig. 1. Incorporating these phases, and critical success factors, into commercial DSSs is important to ensure the software matches farmer needs and therefore can provide value for their decision-making processes.

4 Development of effective decision-support tools

The factors that underpin successful development of DSSs have been the subject of several studies during the last couple of decades. Cox (1996, p. 355) outlined design issues as involving: 'the need for an analytical phase in DSS development to deconstruct professional models; resolution, validation and appropriateness in relation to intended purpose, in order to avoid over-engineered solutions; its contribution to communication practice; and the role of professional responsibility'. Also, in a study of computer software use in agriculture, Alvarez and Nuthall (2006) identified important factors such as DSS features that matched a farmer's working environment and decision context, farmer characteristics, pricing and investment of learning time post purchase, as well as ease of use.

In their review of pasture-based DSSs development, Botha and Atkins (2006) identified key considerations as: defining the purpose, focus on user's needs and funders needs, goal achievement, documenting properly for future

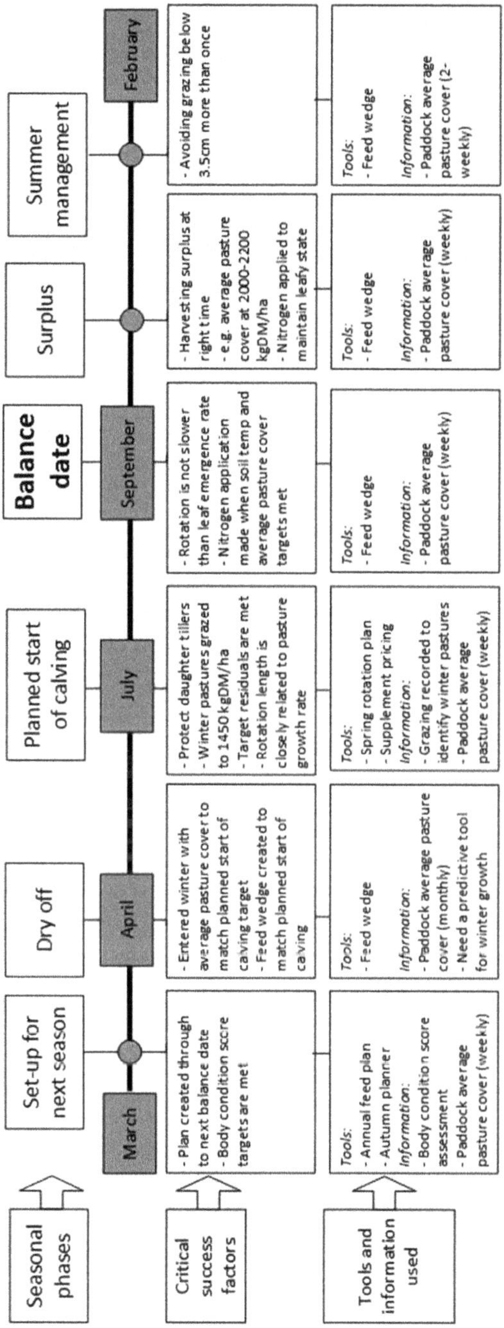

Figure 1 A dairy farmer perspective of the annual grazing management cycle, including information and tool needs and critical success factors (example from a non-irrigated spring calving dairy system in Waikato, New Zealand) (Eastwood et al. 2017b).

development and coding, overall cost, and strength of the leadership team. Also of significant importance was the involvement of the end user. The concept of involving users from the outset of DSS development, and throughout the process, seems logical but is often overlooked or only partially implemented (Botha and Atkins 2006; Nguyen et al. 2007).

Participatory design approaches are outlined by Cerf et al. (2012), and they suggest that participation should be based on a dialogical design process where the underlying assumptions of how users make decisions, and the respective agricultural paradigms of actors involved, are questioned. This process of user-centred design of DSS is seen by Rose et al. (2018) as involving six main steps:

- Who is the user? This step is focussed on identifying the end user and how they make decisions.
- Why should they want to use it? Determine the benefits for the end user, if any. Will the system allow for better decisions than the user can achieve currently?
- Can they use it? This involves investigating whether aspects such as rural infrastructure or other system dependencies exist to enable tool use. Is the Internet connection sufficient for the tool to operate properly?
- Is it easy for me? Involves testing of the DSS with end users (rather than other developers or researchers) to determine ease of use.
- Is there a delivery plan? This should consider peer-to-peer networking and integration with trusted advisory networks.
- What is the legacy? Determining how the system will be maintained long term, and whether it will rapidly become redundant.

In a smart farming framework, Eastwood et al. (2019a) broaden out these design categories to include the factors of (1) characteristics of the target population and market, (2) technology design and innovation, and (3) capability requirements and knowledge exchange. Although the framework is aimed at smart farming innovation in general, it is applicable to DSS development as a subset of smart farming. Through a series of 14 questions, the three factors above are examined to explore the roles of technology suppliers, farmers, and research and extension actors.

A benefit of participatory design is the co-learning that can result from the development process (Jakku and Thorburn 2010). In the case of more complex simulation tools, the co-learning (not just farmers but other actors such as advisors, researchers, and policy personnel) may be more valuable than the output of the tool itself. In his study, McCown (2012) found that this co-learning was contextual, working best when farmers could oscillate between intuition and analysis through simulation DSSs that used farmers' own soil and weather data. The role of DSSs as a focal point for communication among farmers was identified by Nguyen et al. (2007). The use of DSSs in this way has been termed a 'boundary object' (Eastwood et al. 2012), or as a 'device for integration' (Ayre and Nettle 2015), where actors use the DSS to connect and share knowledge around issues and challenges.

The adoption of agricultural technologies, such as DSSs, has also been widely studied for decades, including the early 'diffusion of innovations' work by Rogers (1962). Current thinking on adoption has broadened from diffusion to studies on innovation systems (Meijer et al. 2007), adoption of decision-support tools (Rose et al. 2016), and uptake of new agricultural practices (Kuehne et al. 2017). These studies are summarised in Table 2 (originally presented in Eastwood and Dela Rue (2017a)).

The studies summarised in Table 2 highlight key features to consider when developing DSS tools, explored further in the case study below. One of the main aspects is the relative advantage of the tool, along with how easy it is to trial without investing too much time, money, and data input. Also important is how

Table 2 Summary of studies identifying factors influencing adoption of agricultural technologies (Eastwood and Dela Rue 2017a)

Rogers (1962)	Meijer et al. (2007)	Kuehne et al. (2017)	Rose et al. (2016)
Relative advantage Observability Trialability Compatibility Reinvention Complexity	Technological uncertainty Resource uncertainty Competitive uncertainty Supplier uncertainty Consumer uncertainty Political uncertainty	**Relative Advantage for the Population**: profit, environmental and risk orientation; enterprise scale; management horizon; short-term constraints. **Learnability Characteristics of the Innovation**: Trialable; Innovation complexity; Observability. **Learnability of Population**: Advisory support; Group involvement; Relevant existing skills & knowledge; Innovation awareness. **Relative Advantage of the Innovation**: Relative upfront cost of innovation; Reversibility of innovation; Profit benefit in years that it is used; Future profit benefit; Time until any future profit benefits are likely to be realised; Environmental costs and benefits; Time to environmental benefit; Risk exposure; Ease and convenience.	Performance Ease of use Peer recommendation Trust Cost Habit Relevance to user Farmer-adviser compatibility Age Scale of business Farming type IT education Facilitating conditions Compliance Level of marketing

observable the results of using the tool are, which can be difficult for farmers to determine because positive results from using DSS tools may be masked by fluctuations elsewhere in the business (e.g. climate, input/output prices). Uncertainty surrounding the use of new tools can be a barrier to adoption, and Meijer et al. (2007) describe this in an innovation systems context, where uncertainty can be associated with more than just technology performance. It may also involve uncertainty in the political system (e.g. for DSS associated with national resource management outcomes (Wheeler et al. 2013)), or competitive uncertainty (e.g. how the tool interacts with other products on the market). This uncertainty can be addressed in part by the aspects of trialability and learnability (how easy it is to learn to use). Kuehne et al. (2017) consider the ability of the target population to learn about an innovation, and Rose et al. (2016) also considers this by citing factors such as age, information technology education, and support from advisors.

4.1 What stimulates or inhibits farmers to use decision-support system (DSS) tools?

A study was conducted with commercial and research providers of grazing management software in New Zealand to understand the factors for farmers using, or not using, grazing management software (Eastwood and Dela Rue 2017a). Workshop participants identified a wide range of factors motivating or demotivating users, and non-users, of grazing management software. The main *stimulating* and *inhibiting* factors have been categorised in Fig. 2. Important stimulating factors included:

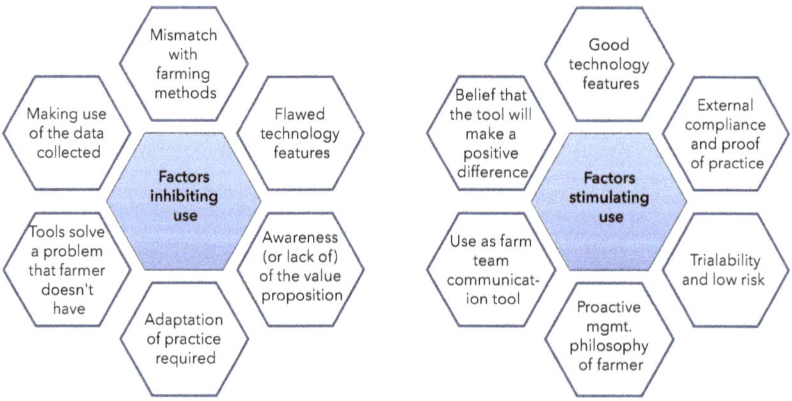

Figure 2 Summary of the factors inhibiting and stimulating use of grazing management software and tools.

- *Good technology features* that made it easy for farmers to use pasture data, and integrate it with other data sources. Features included ease of use and learnability.
- *External compliance and proof of practice* influences, such as collecting data for compliance and for milk processors, agri-finance organisations, and corporate governance. Reporting on key performance indicators to farm investors is becoming increasingly important, as is data collection for compliance software such as Overseer™ (Wheeler et al. 2013).
- Ability to *trial the software* on their own farm, without wasting a farmer's time. Farmers should be able to easily cease use and export data to a standard format if required.
- *A proactive management philosophy* of the farmer or farm owner - for example, needing data to make proactive decisions, striving for excellence, competition with other farmers (benchmarking), identifying feed surpluses early, simplifying tasks to get more time away from the farm, and making the most of the pasture resource. Also, farmers who did not trust intuition and were comfortable with structured and organised information on pasture performance might be more comfortable with the software.
- The role of software as a *communication and learning tool for the farm team*, for example providing feedback to staff on the consequences of grazing decisions. Good software can be used to communicate plans to staff, manage multiple staff, enable jobs to be tracked and to facilitate learning for new staff. Features such as smartphone apps have been shown to be important for farmers in this role (Eastwood et al. 2017b).
- Importantly, there needed to be a *belief that use of the software would make a positive difference* to the farming operation. Underlying this was the need to see (and understand) the value proposition, and to see how using the data and software would lead to capture of these benefits. This belief was also seen to be underpinned by an attitude toward technology, and experience of a tangible (or emotional) reward from using software.

Non-adoption, or dis-adoption, by farmers was seen to be influenced by inhibiting factors including:

1 *Mismatch of technology design with farmer practice*. For example, workshop participants said that farmers could be deterred by software that was too complex to be usable, or too simple to be meaningful. Farmers often also feel time spent at a computer is not part of 'real farming' and that they know their farm well and 'don't need to be told what to do by a computer'. Participants identified the range of dairy farming 'segments' in New Zealand and noted that farmers often felt

that current tools were not flexible enough to match their specific needs (for example operating with many staff, or across multiple farms, or with a range of feed inputs).

2 *Flaws or misunderstandings of technology performance*. According to workshop participants these included: functionality across operating systems (Windows PC, Apple OS), not linking with other management tools, and the time requirement to set up the farm characteristics (paddocks, herd information). Other issues identified were companies stopped developing a product further, and/or frustration around installing upgrades. One bad experience could be an inhibitor to a farmer trying software in the future.
3 *Lack of a value proposition (or lack of awareness of the value)*. Although scientists and technology developers believed there was a value proposition in investing time and money in more objective decision-making, farmers often did not perceive a sufficient return on their investment. This was underpinned by factors such as confusion about which software option was best for their farm, or a general lack of awareness about the grazing software available or an awareness of the available features.
4 *Adaptation of practice*. Participants raised several examples, including a reluctance to enhance skills needed to use information technology, not trusting the data collected, and farmers feeling that pasture measurement took too long and/or was not a priority.
5 *Tools solving a problem that farmers did not recognise* (e.g. a problem they do not think they have, for example taking a 'science-based' objective mindset rather than a more outcome focussed 'farm-practitioner' mindset). Participants felt that many farmers still prefer to trust their own judgement and use a visual non-measured estimate of pasture mass. Participants felt farmers often did not understand that managing pasture quality was more important than short-term quantity, and that 'just in time' management processes were needed to manage pasture effectively. There were also comments that too many farmers believe they are doing better than average '80% of farmers think they are in the top 20%' so this infers a belief that there is no room for improvement.
6 *A lack of knowledge on how to make best use of the data* was linked to the perception of a value proposition. If farmers were not comfortable with understanding the key performance indicators to look for in their operation, then it could be difficult to know how to best apply grazing software. Participants felt some farmers lacked confidence in using the tools, and therefore preferred to stay with their current, more intuitive methods. The potential role of farm consultants was mentioned, with

some participants feeling that there was a disincentive for consultants to help farmers use software tools as it could replace their core business.

A better understanding of approaches toward grazing management will help to guide design and marketing these DSSs. The findings from this study indicate that there are a range of possible inhibiting and stimulating factors influencing the adoption of decision-support software. Software developers cannot hope to design products that are 'all things to all people', however, it was obvious from this study that software needs to be flexible enough to be relevant to a range of farmer perspectives and farm business structures. This aligns with findings from a study of grazing management among New Zealand dairy farmers (Eastwood et al. 2017b). A different value proposition may be required for different farmer segments, for example the functionality needed by large multi-farm or corporate operations may be very different from that needed by individual owner-operators.

Understanding the general psychology around the use of precision farming tools was also seen as important by workshop attendees. Participants noted that often the reasons for adoption or non-adoption were 'hidden' and that this was manifested in many farmers always finding reasons not to use precision tools. Therefore, there was a need to better understand the different motivators for farmers and their key influences. Including farmers, and other stakeholders such as farm consultants, in the design process is an important method of incorporating these 'hidden' psychological factors and determining what role rural professionals have in the adoption process (Douthwaite et al. 2001).

Among the inhibiting factors there was little discussion of the cost of grazing software. This reflects the view from participants (commercial and research providers of grazing management software) that the price of software (ranging from NZD$100–$1000 p.a.) represented a minimal cost for farmers. In other studies, on the use of agricultural decision-support software, cost has been shown to be important for farmers when the value proposition, or return on investment, is not clear (Rose et al. 2016). Where there are so many potential inhibiting factors, including the need for farmer practice change from experienced-based decision-making to data-driven decision-making, even a small cost may prove a barrier to use. There will also be other costs for farmers (perceived and real) including the time taken to set up software, learn to use, and engage staff. The importance of peer influence (farmer to farmer knowledge exchange), highlighted as important in other studies (Kilpatrick and Johns 2003; Rose et al. 2016), was not identified as a major inhibiting or stimulating factor. This may be because the participants were commercial technology and research providers and therefore not cognisant of all factors important to farmers.

Participants in the study of Eastwood and Dela Rue (2017a) also discussed a range of roles in the DSS innovation system, including: technology developers needing greater knowledge about designing and marketing tools that fit the adoption factors (for example Fig. 2); universities and primary industry training organisations in lifting understanding and skills in DSS use, and conducting social research into adoption psychology; industry organisation roles in elucidating the value proposition for DSSs and building agreement and awareness of key performance indicators for pasture management; roles of rural professionals (e.g. farm consultants, banks, irrigation and fertiliser companies) in incorporating data and cloud-based software systems in their business models; and roles of farmers/farming companies to train their staff and build modern data-driven business structures. Also, the potential for industry organisations to lead precision grazing uptake was highlighted where the combination of benchmarking within the farmer community, training, social media, and discussion groups has been shown to be effective stimulating factors for uptake of grazing software in Ireland (Hanrahan et al. 2017)

5 Case studies of decision-support system (DSS) development in pasture and rangeland management

In this section, three case studies are discussed to provide examples of the development and use of DSSs in pasture and rangeland management. The first case study examines research focussed on the development of grazing DSSs that meet the needs of users, pasture-based dairy farmers. The second case study describes the use of DSS for sustainability outcomes. The final case study considers DSS development in the context of socio-ethical perspectives that are increasingly being applied to smart farming approaches.

5.1 Case study 1: Decision-support development for New Zealand dairy systems

For Case study 1, the results of research examining the development of DSSs for pasture-based dairy farming in New Zealand are outlined. The aim is to provide a case study focussed on how to understand the needs of end users and then design for these needs in grazing management DSS. More details of the research used for this case can be found in several recent publications (Eastwood and Dela Rue 2017a,b; Eastwood et al. 2017b; Neal et al. 2017a; Dela Rue et al. 2020). The research involved assessments of current grazing DSS, interviews with farmers about their grazing management practices, interviews with farm consultants about their grazing management principles and interactions with clients, detailed work with a panel of farming systems experts, and a workshop with designers of grazing DSS.

5.1.1 Key features of a grazing decision-support system (DSS) in pasture-based dairy

The following section discusses high level 'attributes' and more specific 'features' of DSS identified in the study (Fig. 3, Table 3). Among the farmers and farm consultants interviewed, there was a variety of commercial grazing DSS tools in use. Although some DSS were identified as better than others, there were differences in what farmers needed from a DSS depending on their farm system. For example, family farmers who only owned one property with few staff might only need a simple tool to record pasture and feed availability; this could even be smartphone-based. However, farmers who owned multiple farms, or farms that were owned in corporate arrangements, had needs for DSS that could link data across farms, allow benchmarking between farms, and facilitate tracking and communication of key performance indicators (KPIs). A need for DSS to be 'scalable' for different farming businesses was identified (Fig. 3).

Aspects, such as the DSS tool having 'seasonal' features, being 'simple', and being 'trialable' (Fig. 2), were also reinforced in the interviews. The attribute of 'flexibility' was identified by participants, which refers to the ability of grazing DSS to be relevant for different farm system types (e.g. irrigated v non-irrigated, different growing seasons, different season of calving). As with many other smart technologies (Bahlo et al. 2019), the attribute of 'integration' was important so that data could be exchanged with other DSS tools, or easily imported from sensors.

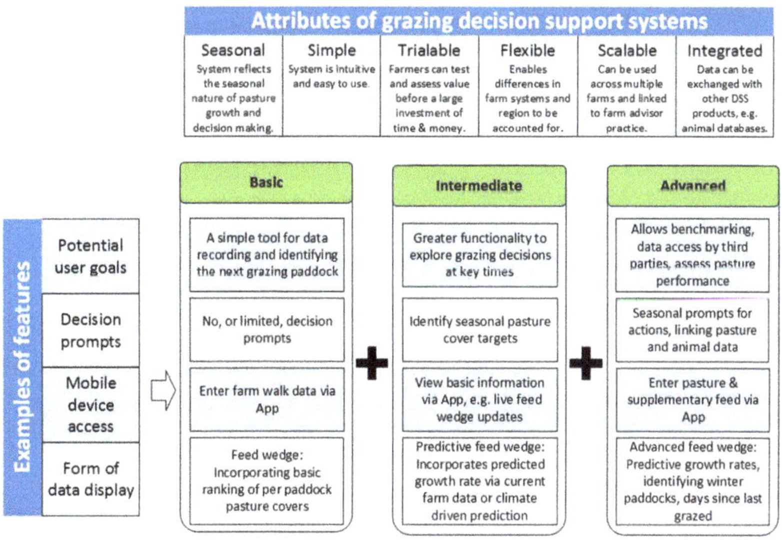

Figure 3 Desirable attributes of dairy farm grazing decision-support tools (Eastwood et al. 2017b).

Table 3 Specific features suggested for grazing software products

Attribute	Specific feature
General	
	PC based or Cloud-based
	Mobile App (Android or iOS)
	Uses digital farm maps
Input	
	Data entry and presentation
Standard features	Displays & prints paddock list in customised order (e.g. farm walk order)
	Data download for analysis in spreadsheets or other (.xls or .csv files)
	Grazing dates can be entered (with/without pre- and post-grazing covers)
Advanced features	Software uploads external pasture data (e.g. from pasture measuring device)
	Direct data entry from the paddock (via Smartphone App etc.)
	Remote access to data and planning tools (e.g. Feed Wedge)
	Displays data on digital farm map
	Data upload from non-pasture sources (e.g. milk production)
	Fertilizer applications by paddock
Strategic pasture management	
	Annual feed budget
Standard features	Calculation of monthly feed supply and demand
	Set seasonal targets (e.g. target pasture covers at key times)
	Accounts for different stock types (milkers/dry/heifers etc.)
	Graph: Seasonal average pasture cover target - planned vs. actual
	Annual key performance indicators
Advanced features	Calculates annual farm growth rates (t DM/ha)
	Calculates annual individual paddock growth rates (t DM/ha)
	Calculates annual number of grazings per paddock
	Calculates milk solids from pasture
	Calculates annual fertilizer use by farm and by paddock

Attribute	Specific feature
Tactical/Operational management	
Calving to balance date	**Spring Rotation Planner**
Standard features	Allocate grazing area to multiple mobs to match total planned area
	Calculates extra supplement required per mob and total herd
	Graph: Rotation length - planned vs. actual
	Graph: Avg. pasture cover - planned vs. actual
	Graph: Accumulated area grazed - planned vs. actual
Advanced features	Spring feed plan accounting for mobs (milkers, springers, dry cows etc.)
	Account for cows coming back on to farm from grazing throughout Spring
	Adjustable predicted pasture growth
	Calculates supplement required until balance date
	Graph: Avg pasture cover - planned vs. predicted vs. actual
Balance date to Autumn	**Feed wedge for paddock grazing order**
Standard features	Accepts manual entry and uploads from an external file
	Accepts partial farm walk data (e.g. farm walk split into different days)
	Re-ordering of paddocks
	Conservation: exclude paddocks from rotation, for example, for silage/crops
	Displays days since last grazed
	Printable & exportable as table and graph
Advanced features	Separate feed wedge for mobs
	Pasture growth rate since last grazing by paddock
	Pasture growth rate prediction/forecast by farm and by paddock
	Estimates growth from farm walk date until grazing date (when recorded)
	Records fertilizer applications by paddock
	Estimate of feed quality
	Estimate of energy supply/demand for grazing paddock
	Displays weekly milk production (kg MS/cow/day)

(*Continued*)

Table 3 (Continued)

Attribute	Specific feature
Balance date to Autumn	**Pasture and supplement allocation**
Standard features	Grazing area by herd demand
	Grazing area by feed supply (round length)
	Calculates extra supplement required per mob and total herd
Advanced features	Calculates number of cows required to graze the paddock
	Accounts for different stock types (milkers/dry/heifers etc.)
	Allows a grazing session to be split between two paddocks
Summer to Planned Start of Calving (PSC)	**Autumn management Planning**
Standard features (Autumn rotation planner)	Allocate grazing area to multiple mobs to match total planned area
	Calculate extra supplement required per mob and total herd
	Graph: Round length - planned vs. actual
	Graph: Avg. pasture cover - planned vs. actual
	Graph: Accumulated area grazed - planned vs. actual
Advanced features (Autumn management tool)	Calculation of weekly feed supply and demand to PSC
	Calculation of supplement requirements to PSC including feed on hand
	Calculation of milk revenue less feed and operational costs
	Considers the impact of once-a-day milking by mob from a set date
	Considers the impact of drying off by mob from a set date
	Considers the impact of reducing stock on the milking platform by mob from a set date
	Graphs feed demand by mob from current date to PSC
	Graphs feed supply vs. demand from current date to PSC
Farm business requirements	
	Business and communication functions
	Integration with other products for analysis
	Multiple user accounts, for example, for owner or farm consultant access

Attribute	Specific feature
	Incorporates multiple farm data
	Benchmarking against own or regional farms
	Data security
General attributes	
	Usability and Functionality
	Ease of use
	Farm team usability
	Flexibility for different farm systems
	Match seasonal decision points
	Scalable for different farm sizes
	Support and training
	On-call support
	Self-help support and training
	Formal training

Some of the farm consultants were using commercial DSS tools, while others had built their own spreadsheets to enable specific functionality and reporting which was not present in commercial DSS offerings. All the consultants interviewed used optimisation DSSs to determine strategic options for farmer clients. These tools were seen as valuable at a strategic level, although the consultants noted that few farmers used them directly. This may be because farmers can use simpler tools year-round and pay a consultant to do an annual strategic plan using their collected data rather than having to up-skill. Overall, the need for easier information sharing between different software and databases was acknowledged, with continuous data transfer required. Some farm consultants would access and review pasture data prior to client meetings, to identify key points for discussion and for more efficient use of their time - and this required good cloud-based systems with the ability to download data to spreadsheets.

Participants noted that the use of grazing DSSs required skilled and motivated farm staff for regular capture of good quality data. Farmers need to provide training for staff in how to use data effectively for better grazing management decisions. Dairy trainees were being made aware of basic tools during training courses, however, they were not being trained in more sophisticated decision-making processes. Mobility, for example being able to use the system on a device in the field, was a highly desirable feature and managers in large multi-farm businesses used smartphone and online platform functionality to track performance across the business, and as a means of engaging with staff.

5.2 Case study 2: Decision-support systems (DSSs) for pasture and rangeland sustainability

While much of the development of commercial DSSs for pasture and rangeland management has been focussed on feed management, there has been considerable development of DSSs targeted at improving decision-making for farm sustainability. Controlling soil erosion has been a historical focus, but more recently optimised management of nutrients and managing drought impacts, and now greenhouse gas (GHG) mitigation, are important uses for DSSs. In this case study, some examples of DSS development for sustainability outcomes are reviewed.

A dilemma for grazed rangeland ecosystems was identified by Roche et al. (2015) where society demands goods from livestock and forage, but also ecosystem services such as high-quality water. The need for these multi-functional rangelands is adding increasing complexity, and sometimes ambiguity, to land manager decision-making (Duru et al. 2015). DSSs can aid farmers with these sometimes-conflicting issues by modelling potential outcomes from different management decisions and helping farmers identify the most positive trajectories. Such DSS need to be developed with consideration of the complex decision-making experiences and different timescales of farmers (Bosch et al. 1996; Bashari et al. 2008; Wilmer and Fernández-Giménez 2015). In their review of aspects related to the practical application of agroecological principles for more sustainable agriculture, Duru et al. (2015) identified two types of decision-support tools: those involving knowledge bases built on scientific data and experiential knowledge, and model-based games. They also highlighted that many DSSs are limited mechanistic models that cannot deal with uncertainty or adequately interact with farm management processes.

Creative and flexible mitigation strategies are required through use of partnerships to address potential conflicts between stakeholders related to rangeland sustainability (Roche et al. 2015). Social learning and group decision-making have been shown to be important in these contexts to develop collective understanding (Wilmer et al. 2018). In this context, sustainability-focussed DSSs can provide an effective centre-point, or boundary object (Pigford et al. 2018), around which multi-stakeholder social learning can occur. Examples of large-scale DSSs that could be used for group learning include G-Range, a decision-support model that provides information on large-scale impacts of rangeland management on ecosystem services such as carbon sequestration based on climate and CO_2 changes (Sircely et al. 2019). This model can work at regional and national scales to guide rangeland systems that can better withstand climate changes. Another DSS, the Great Plains Framework for Agricultural Resource Management (GPFARM), has functionality that includes environmental impact analysis (Andales et al. 2005). DSSs can also be highly interactive, for example the use of a computer model-based board game called 'Rangeland Rummy' (Farrié et al. 2015). This tool was specifically designed to be used with farmer groups allows them to collectively design and assess rangeland-based systems in the context of weather variability and volatile input prices. Farrié et al. (2015) found that use of such DSS can build the adaptive capacity of farmers and stimulate discussions on different management strategies. Model-based games have also been shown to enable iterative design among farmers, scientists, and other stakeholders for implementation of biodiversity-based agriculture (Duru et al. 2015).

The ability to access timely and accurate information is a key to driving complex environmental and sustainability-focussed DSSs. The use of the Internet for provision of up-to-date data for predictive DSSs is growing. In a US study, Ghajar et al. (2019) found that rangeland managers use the Internet to access weather forecasts and other information. Changing climate patterns may increase the need for predictive tools to help farmers interpret more variable weather information. Recent examples of DSSs for short-medium term weather trends are described in studies on a Drought Calculator (Dunn et al. 2013) and Soil and Water Outlook tool (Eastwood et al. 2019a). Results from these studies highlighted the need for simple tools that were based on readily available information, and that were adequately supported by advisors who could help farmers understand the DSS predictions.

The development of DSSs has evolved over recent decades from spreadsheet-based models, to online cloud-based systems, and more recently the use of smartphone-based apps. An example of a rangeland sustainability app, the Land-Potential Knowledge System (LandPKS) app, is aimed at helping users understand their soils and plants (Herrick et al. 2017). Smartphones represent a major opportunity to make DSSs more useful and usable for pasture and rangeland managers, due to the ability to combine GPS, mobility, image capture, and real-time interaction with cloud-based data storage, and ever increasing in-device processing capability (Herrick et al. 2017). The capture of real-time interactions between users and an app-based DSS will also allow designers to adapt functionality to actual use profiles. In the future, the combination of augmented reality and smartphone-based DSS will enable sustainability decision-support tools that more closely match the intuitive and visual preferences of farmers.

5.3 Case study 3: Responsible innovation, decision-support systems (DSSs) and smart farming

While DSSs, as part of the smart farming ecosystem, have the potential to guide agricultural decision-making to increase productivity and improve sustainability outcomes (Shepherd et al. 2018), concerns have also been raised about a lack of attention on social implications of using such data-driven and techno-centric approaches (Eastwood et al. 2019b; Rose and Chilvers 2018). These implications have been identified as the interaction of DSS and technology use with animal care and ethics (Holloway et al. 2014), farmers (Jago et al. 2013) employees (Rotz et al. 2019) and farm advisory services (Eastwood et al. 2019a). Use of DSSs require a particular skill set and farming paradigm (e.g. a data-driven mindset), which may potentially advantage some farmers/staff over others. For example, the complexity and cost associated with implementing sensors and DSSs for pasture and rangeland management may be more cost-effective for larger farming operations, or prohibitive for small operations. Also, the skills needed to use computerised DSSs may exclude people with lower skills or confidence with information technology, such as unskilled migrants. In this case study, the concept of responsible research and innovation is applied to the development of DSSs.

Responsible research and innovation (RRI) is a concept that was outlined by Stilgoe et al. (2013) to apply the dimensions of anticipation, inclusion, reflexivity, and responsiveness to innovation processes. The overall purpose of RRI was to develop a process to anticipate potential implications of research and innovation, with a focus on enhancing responsiveness to societal concerns. Many studies concerning RRI have been published since 2013, although most are at the conceptual level and few have tested RRI principles in practice (Rose and Chilvers 2018). There has been little coordinated implementation of the RRI dimensions in smart farming developments, with a need for more focus on the requirements of end users and their advisors, and the potential broader consequences of such technology in a data-driven and artificial intelligence enabled future.

In Table 4, previously identified dimensions and indicators of RRI in smart farming have been adapted to help guide the responsible development of DSSs. Some of the activities suggested include: using surveying approaches to better understand the need for DSSs before commencing development; understanding the potential implications of DSSs by using scenario building and foresighting exercises; using human-centred design to include the relevant stakeholders; and using flexible development processes such as stage-gates to enable changes in direction rather than staying fixed to development trajectories which are found to be irrelevant. Not all DSS development projects will be of sufficient scale or scope to require consideration of RRI principles, but increasingly DSS tools in pasture and rangeland management will have integrated multi-platform and artificial intelligence functionality. DSSs that are focussed on making and even implementing decisions for farmers within a 'black box' configuration, where farmers are unable to fully understand the algorithms driving the DSS-based decision-making, are highly possible within the next few decades. This functionality will bring with it questions about

Table 4 Using the framework of responsible research and innovation to guide development of decision-support systems for pasture and rangeland management

RRI Dimensions	Potential RRI indicators for DSS development	Activities for responsible DSS development
Anticipation	Foresighting exercises Scenario building of possible futures	Use of Delphi surveys to identify advantages and disadvantages of a particular DSS (e.g. Eastwood and Dela Rue, 2017b). Establishing baselines of DSS use by farmers to have 'real' data about levels of adoption, also including farmer perceptions of technology (e.g. Edwards et al., 2015). Using scenario building to understand the future role of technologies such as DSSs and artificial intelligence-based decision-making (e.g. Romera et al., 2020).
Inclusion	Involvement of relevant actors Private sector engagement Use of new methods for societal engagement	Using human-centred design to develop DSSs in conjunction with end users and actors in their knowledge networks (e.g. Rose et al., 2018). Increasing societal engagement in DSS development, where appropriate, using tools such as systematic mapping, comparative case studies and meta-analysis (e.g. Macnaghten and Chilvers, 2014).
Reflexivity	Reflexive guidance Structures to guide second-order reflexivity	During larger DSS development projects, using reflexive monitors (people dedicated to facilitating reflection on project progress) (e.g. Fielke et al., 2017). Facilitating second-order reflexivity through codes of conduct or guidelines for best practice in DSS development, for example data privacy and sharing standards (Bahlo et al., 2019).
Responsiveness	Potential to adapt development timelines Open research processes and access to research data	In larger DSS development projects, having the ability to change direction in the development process based on stakeholder feedback, or even halt development if needed. This can be achieved through building in project stage-gating and including stakeholder interactions throughout projects (e.g. McCown et al., 2009). Improved responsiveness can also be achieved through taking a 'lean start-up' approach where development is progressed though iterative experiments to allow ongoing feedback and adjustment (Long et al., 2019). Responsible DSS innovation may also require a focus on open access to research results and data, and ongoing open data exchange.

Source: adapted from Eastwood et al. (2019b), Bronson (2018), Rose and Chilvers (2018).

the ethics of automated processes for animals and people, what it means to be a farmer (Burton et al. 2012), and what societal expectations are of food-producing enterprises (Eastwood et al. 2019b). Future development of DSSs, particularly those linked to national policies and regulations or integrating with smart farming technologies, will need to integrate the RRI principles into the development process.

6 Conclusion and future trends

There have been several decades of development of DSSs for pasture and rangeland systems; however there is still limited uptake of such tools by farmers. This chapter has shown that this is partly due to confusion related to the intended audience, that is either farmers, or for learning and research purposes. This confusion can impact the end use of the DSS tool, with tools historically designed by researchers and over-engineered when used in a farming context. It is important to focus on both the end user, and the actual need for the tool, early in the development process (prior to working on the technical development). In this chapter, the decision-making process of pasture and rangeland farmers was reviewed, highlighting the continued role of intuition and experience. This is an important consideration for DSS development, as the importance of intuitive decision-making should not be undervalued, as prior failures of DSS have involved an assumption that objective-based tools can replace intuition for farmers.

Future trends for pasture and rangeland DSSs include:

1 Increased use of cloud-based online systems: Many commercial systems are now accessible online, allowing more security with data backups, easier access by farm teams and advisors, and regular system updates by DSS providers. It also facilitates easier off-farm data sharing and traceability. One caveat to this development is the limitations of rural Internet access and speeds.
2 Greater use of real-time data streams and middleware: Existing DSS development has focussed on relatively static models, required manual or semi-automated inputs, often requiring more time investment from farmers. Opportunities for improvement will be focussed on the automation of real-time data streams into farm system DSS, and linkages to smart middleware.
3 Expanded DSS application on smartphone platforms to facilitate the connection with GPS location, image capture and analysis, cloud-based storage, and user mobility.
4 Better features to align with farmer decision-making processes and worldviews, rather than researcher-driven worldviews as in the past.

5 The greater use of data science: With increased data there is also the potential for advanced analytics and artificial intelligence (Smith 2020) where DSS tools not only learn from feedback loops, but also from decision-making input from farmers – thereby linking intuitive and explicit paradigms (Nuthall and Old 2018).

Further research is needed to:

1 Develop DSS that learn from individual farmer priorities and is more flexible to adapt to different farming enterprises (e.g. differences in user goals, decision-making, mobility, and farm business structure);
2 Better understand effective co-design of DSS with farmers and other stakeholders to merge scientific and practical knowledge;
3 Understand and build policies related to data ownership and use of DSS data to improve on-farm compliance practices and product provenance; and
4 Assess the positive and negative impacts of widespread DSS use in pasture and rangeland farming, and to incorporate aspects of responsible innovation into future design (Bronson 2018).

7 Where to look for further information

7.1 Further reading and online resources

1 *Agricultural Systems* vol. 74, no. 1 (October 2002). This issue covers the development and use of decision support tools and is still very relevant.
2 Holmes, C. W., Brookes, I. M., Garrick, D. J., Mackenzie, D. D. S., Parkinson, T. J. and Wilson, G. F. 2002. *Milk Production from Pasture*. Massey University, Palmerston North, New Zealand. Available at: https://www.amazon.com/Production-Pasture-Butterworths-agricultural-books/dp/0409702315.
3 Flack, S. 2016. *The Art and Science of Grazing: How Grass Farmers Can Create Sustainable Systems for Healthy Animals and Farm Ecosystems*. Chelsea Green Publishing Co, White River Junction, USA.
4 Decision-support tools developed by CSIRO in Australia 'intended to help farmers to make decisions about their farm management principally in grazing enterprises' https://grazplan.csiro.au.
5 A summary of decision-support tools for grazing land management in Australia, particularly for rangeland beef systems https://futurebeef.com.au/knowledge-centre/decision-support-tools-grazing-land-management/.
6 For guidance on how to think about design of new tools from an end user adoption mindset, see the ADOPT model https://adopt.csiro.au/.

7.2 Links to current commercial decision-support systems (DSSs) used for pasture and rangeland management

1 Pasture.io (Australia) https://pasture.io/;
2 Pasture Map (USA) https://pasturemap.com/;
3 PastureBase (Ireland), https://pasturebase.teagasc.ie;
4 FarmIQ (NZ), https://farmiq.co.nz/;
5 Agrinet (Ireland) https://www.agrinet.ie/;
6 Pasture Coach (NZ) https://www.pasturecoachnz.co.nz/; and
7 Farmax (NZ) http://www.farmax.co.nz/.

7.3 Key journals/conferences

- *Agricultural Systems;*
- *Computers and Electronics in Agriculture;*
- *Environmental Modelling and Software;*
- *Rangeland Ecology and Management;*
- International Grasslands Conference;
- Asian-Australasian Conference on Precision Agriculture (ACPA); and
- European Conference on Precision Livestock Farming.

8 References

Alvarez, J. and Nuthall, P. 2006. Adoption of computer based information systems: the case of dairy farmers in Canterbury, NZ, and Florida, Uruguay. *Computers and Electronics in Agriculture* 50(1), 48–60. doi:10.1016/j.compag.2005.08.013.

Andales, A. A., Derner, J. D., Bartling, P. N. S., Ahuja, L. R., Dunn, G. H., Hart, R. H. and Hanson, J. D. 2005. Evaluation of GPFARM for simulation of forage production and cow-calf weights. *Rangeland Ecology and Management* 58(3), 247–55. doi:10.2111/1551-5028(2005)58[247:EOGFSO]2.0.CO;2.

Ayre, M. and Nettle, R. 2015. Doing integration in catchment management research: insights into a dynamic learning process. *Environmental Science and Policy* 47, 18–31. doi:10.1016/j.envsci.2014.10.011.

Bahlo, C., Dahlhaus, P., Thompson, H. and Trotter, M. 2019. The role of interoperable data standards in precision livestock farming in extensive livestock systems: a review. *Computers and Electronics in Agriculture* 156, 459–66. doi:10.1016/j.compag.2018.12.007.

Bashari, H., Smith, C. and Bosch, O. J. H. 2008. Developing decision support tools for rangeland management by combining state and transition models and Bayesian belief networks. *Agricultural Systems* 99(1), 23–34. doi:10.1016/j.agsy.2008.09.003.

Beukes, P. C., Palliser, C. C., Macdonald, K. A., Lancaster, J. A. S., Levy, G., Thorrold, B. S. and Wastney, M. E. 2008. Evaluation of a whole-farm model for pasture-based dairy systems. *Journal of Dairy Science* 91(6), 2353–60. doi:10.3168/jds.2007-0728.

Bosch, O., Allen, W., Williams, J. and Ensor, A. 1996. An integrated approach for maximising local and scientific knowledge for land management decision-making in the New Zealand high country. *The Rangeland Journal* 18(1), 23–32. doi:10.1071/RJ9960023.

Botha, N. and Atkins, K. 2006. The design, utility and adoption of decision support systems in the New Zealand pastoral industry, *Proceedings of APEN International Conference 2006*. Beechworth, Victoria, Australia, 6–8 March 2006. Available at: http://www.regional.org.au/au/apen/2006/refereed/5/2869_bothan.htm.

Bronson, K. 2018. Smart farming: including rights holders for responsible agricultural innovation. *Technology Innovation Management Review* 8(2), 7–14. doi:10.22215/timreview/1135.

Bryant, J. R., Ogle, G., Marshall, P. R., Glassey, C. B., Lancaster, J. A. S., García, S. C. and Holmes, C. W. 2010. Description and evaluation of the Farmax Dairy Pro decision support model. *New Zealand Journal of Agricultural Research* 53(1), 13–28. doi:10.1080/00288231003606054.

Burton, R. J. F., Peoples, S. and Cooper, M. H. 2012. Building 'cowshed cultures': a cultural perspective on the promotion of stockmanship and animal welfare on dairy farms. *Journal of Rural Studies* 28(2), 174–87. doi:10.1016/j.jrurstud.2011.12.003.

Cerf, M., Jeuffroy, M.-H., Prost, L. and Meynard, J.-M. 2012. Participatory design of agricultural decision support tools: taking account of the use situations. *Agronomy for Sustainable Development* 32(4), 899–910. doi:10.1007/s13593-012-0091-z.

Cox, P. G. 1996. Some issues in the design of agricultural decision support systems. *Agricultural Systems* 52(2–3), 355–81. doi:10.1016/0308-521X(96)00063-7.

Dela Rue, B. T., Eastwood, C. R., Edwards, J. P. and Cuthbert, S. 2020. New Zealand dairy farmers preference investments in automation technology over decision-support technology. *Animal Production Science* 60(1), 133–7. doi:10.1071/AN18566.

Dobos, R. C. and Fulkerson, W. J. 2004. A database program to assist in the allocation of pasture and supplements to grazing dairy cows. *Environmental Modelling and Software* 19(6), 581–9. doi:10.1016/j.envsoft.2003.08.013.

Dobos, R. C., Ashwood, A. M., Moore, K. and Youman, M. 2004. A decision tool to help in feed planning on dairy farms. *Environmental Modelling and Software* 19(10), 967–74. doi:10.1016/j.envsoft.2003.11.013.

Douthwaite, B., Keatinge, J. D. H. and Park, J. R. 2001. Why promising technologies fail: the neglected role of user innovation during adoption. *Research Policy* 30(5), 819–36. doi:10.1016/S0048-7333(00)00124-4.

Dunn, G. H., Gutwein, M., Green, T. R., Menger, A. and Printz, J. 2013. The drought calculator: decision support tool for predicting forage growth During drought. *Rangeland Ecology and Management* 66(5), 570–8. doi:10.2111/REM-D-12-00087.1.

Duru, M., Therond, O., Martin, G., Martin-Clouaire, R., Magne, M.-A., Justes, E., Journet, E.-P., Aubertot, J.-N., Savary, S., Bergez, J.-E. and Sarthou, J. P. 2015. How to implement biodiversity-based agriculture to enhance ecosystem services: a review. *Agronomy for Sustainable Development* 35(4), 1259–81. doi:10.1007/s13593-015-0306-1.

Eastwood, C. and Dela Rue, B. 2017a. Precision grazing management-understanding farmer uptake of grazing software. *Proceedings of the 1st Asian-Australasian Conference on Precision Pastures and Livestock Farming*, Hamilton, New Zealand.

Eastwood, C. R. and Dela Rue, B. 2017b. Identification of performance attributes for pasture measuring devices. *Journal of New Zealand Grasslands* 79, 17–22.

Eastwood, C. R. and Kenny, S. 2009. Art or science? Heuristic versus data driven grazing management on dairy farms. *Extension Farming Systems Journal* 5, 95–102.

Eastwood, C. R., Chapman, D. F. and Paine, M. S. 2012. Networks of practice for co-construction of agricultural decision support systems: case studies of precision dairy farms in Australia. *Agricultural Systems* 108, 10–8. doi:10.1016/j.agsy.2011.12.005.

Eastwood, C., White, T., Sheridan, J., Manning, M. and Mashlan, K. 2017a. Skills required by dairy farmers when strategically adapting their farm system. *Rural Extension and Innovation Systems Journal* 13, 22–31.

Eastwood, C. R., Dela Rue, B. T. and Gray, D. I. 2017b. Using a 'network of practice' approach to match grazing decision-support system design with farmer practice. *Animal Production Science* 57(7), 1536–42. doi:10.1071/AN16465.

Eastwood, C. R., Greer, J., Schmidt, D., Muir, J. and Sargeant, K. 2018. Identifying current challenges and research priorities to guide the design of more attractive dairy-farm workplaces in New Zealand. *Animal Production Science* 60(1). doi:10.1071/AN18568.

Eastwood, C., Ayre, M., Nettle, R. and Dela Rue, B. 2019a. Making sense in the cloud: farm advisory services in a smart farming future. *NJAS – Wageningen Journal of Life Sciences* 90–91. doi:10.1016/j.njas.2019.04.004.

Eastwood, C. R., Klerkx, L., Ayre, M. and Dela Rue, B. 2019b. Managing socio-ethical challenges in the development of smart farming: from a fragmented to a comprehensive approach for responsible research and innovation. *Journal of Agricultural and Environmental Ethics* 32(5–6), 741–68. doi:10.1007/s10806-017-9704-5.

Edwards, J. P., Dela Rue, B. T. and Jago, J. G. 2015. Evaluating rates of technology adoption and milking practices on New Zealand dairy farms. *Animal Production Science* 55(6), 702–9. doi:10.1071/AN14065.

Farrié, B., Jouven, M., Launay, F., Moreau, J. C., Moulin, C. H., Piquet, M., Taverne, M., Tchakérian, E., Thénard, V. and Martin, G. 2015. Rangeland Rummy – a board game to support adaptive management of rangeland-based livestock systems. *Journal of Environmental Management* 147, 236–45. doi:10.1016/j.jenvman.2014.08.018.

Fielke, S., Nelson, T., Blackett, P., Bewsell, D., Bayne, K., Park, N., Rijswijk, K. and Small, B. 2017. Hitting the bullseye: learning to become a reflexive monitor in New Zealand. *Outlook on Agriculture* 46(2), 117–24. doi:10.1177/0030727017708490.

Flack, S. 2016. *The Art and Science of Grazing : How Grass Farmers Can Create Sustainable Systems for Healthy Animals and Farm Ecosystems*. Chelsea Green Publishing Co, White River Junction.

Fox, D. G., Barry, M. C., Pitt, R. E., Roseler, D. K. and Stone, W. C. 1995. Application of the Cornell Net Carbohydrate and Protein model for cattle consuming forages. *Journal of Animal Science* 73(1), 267–77. doi:10.2527/1995.731267x.

French, P., O'Brien, B. and Shalloo, L. 2015. Development and adoption of new technologies to increase the efficiency and sustainability of pasture-based systems. *Animal Production Science* 55(7), 931–5. doi:10.1071/AN14896.

Fulkerson, W. J. and Lowe, K. F. 2011. Forages and pastures | grazing management. In: Fuquay, J. W. (Ed.), *Encyclopedia of Dairy Sciences* (2nd edn.). Academic Press, San Diego, pp. 594–601.

Gargiulo, J. I., Eastwood, C. R., Garcia, S. C. and Lyons, N. A. 2018. Dairy farmers with larger herd sizes adopt more precision dairy technologies. *Journal of Dairy Science* 101(6), 5466–73. doi:10.3168/jds.2017-13324.

Ghajar, S., Fernández-Giménez, M. E. and Wilmer, H. 2019. Home on the digital range: ranchers' web access and use. *Rangeland Ecology and Management* 72(4), 711-20. doi:10.1016/j.rama.2018.12.009.

Gray, D. I. 2001. The tactical management processes used by pastoral-based dairy farmers: a multiple-case study of experts. A thesis presented in partial fulfilment of the requirements for the degree of Doctor of Philosophy in Farm Management at Massey University.

Hanrahan, L., Geoghegan, A., O'Donovan, M., Griffith, V., Ruelle, E., Wallace, M. and Shalloo, L. 2017. PastureBase Ireland: a grassland decision support system and national database. *Computers and Electronics in Agriculture* 136, 193-201. doi:10.1016/j.compag.2017.01.029.

Hansson, H. and Öhlmér, B. 2008. The effect of operational managerial practices on economic, technical and allocative efficiency at Swedish dairy farms. *Livestock Science* 118(1-2), 34-43. doi:10.1016/j.livsci.2008.01.013.

Herrick, J. E., Karl, J. W., McCord, S. E., Buenemann, M., Riginos, C., Courtright, E., Van Zee, J., Ganguli, A. C., Angerer, J., Brown, J. R., Kimiti, D. W., Saltzman, R., Beh, A. and Bestelmeyer, B. 2017. Two new mobile apps for rangeland inventory and monitoring by landowners and land managers. *Rangelands* 39(2), 46-55. doi:10.1016/j.rala.2016.12.003.

Holloway, L., Bear, C. and Wilkinson, K. 2014. Re-capturing bovine life: robot-cow relationships, freedom and control in dairy farming. *Journal of Rural Studies* 33, 131-40. doi:10.1016/j.jrurstud.2013.01.006.

Holmes, C. W., Brookes, I. M., Garrick, D. J., Mackenzie, D. D. S., Parkinson, T. J. and Wilson, G. F. 2002. *Milk Production from Pasture*. Massey University, Palmerston North, New Zealand.

Ibáñez, J. and Martínez-Valderrama, J. 2018. Global effectiveness of group decision-making strategies in coping with forage and price variabilities in commercial rangelands: a modelling assessment. *Journal of Environmental Management* 217, 531-41. doi:10.1016/j.jenvman.2018.03.127.

Jago, J., Eastwood, C. R., Kerrisk, K. and Yule, I. 2013. Precision dairy farming in Australasia: adoption, risks and opportunities. *Animal Production Science* 53(9), 907-16. doi:10.1071/AN12330.

Jakku, E. and Thorburn, P. J. 2010. A conceptual framework for guiding the participatory development of agricultural decision support systems. *Agricultural Systems* 103(9), 675-82. doi:10.1016/j.agsy.2010.08.007.

Johnson, I. R., Chapman, D. F., Snow, V. O., Eckard, R. J., Parsons, A. J., Lambert, M. G. and Cullen, B. R. 2008. DairyMod and EcoMod: biophysical pasture-simulation models for Australia and New Zealand. *Australian Journal of Experimental Agriculture* 48(5), 621-31. doi:10.1071/EA07133.

Kilpatrick, S. and Johns, S. 2003. How farmers learn: different approaches to change. *Journal of Agricultural Education and Extension* 9(4), 151-64. doi:10.1080/13892240385300231.

Koropp, C., Kellermanns, F. W., Grichnik, D. and Stanley, L. 2014. Financial decision making in family firms:an adaptation of the theory of planned behavior. *Family Business Review* 27(4), 307-27. doi:10.1177/0894486514522483.

Kuehne, G., Llewellyn, R., Pannell, D. J., Wilkinson, R., Dolling, P., Ouzman, J. and Ewing, M. 2017. Predicting farmer uptake of new agricultural practices: a tool for research, extension and policy. *Agricultural Systems* 156, 115–25. doi:10.1016/j.agsy.2017.06.007.

Lokhorst, C., de Mol, R. M. and Kamphuis, C. 2019. Invited review: Big Data in precision dairy farming. *Animal* 13(7), 1519–28. doi:10.1017/S1751731118003439.

Long, T. B., Blok, V., Dorrestijn, S. and Macnaghten, P. 2019. The design and testing of a tool for developing responsible innovation in start-up enterprises. *Journal of Responsible Innovation*, 1–31. doi:10.1080/23299460.2019.1608785.

Lynch, T., Gregor, S. and Midmore, D. 2000. Intelligent support systems in agriculture: how can we do better? *Australian Journal of Experimental Agriculture* 40(4), 609–20. doi:10.1071/EA99082.

Ma, W., Renwick, A. and Bicknell, K. 2018. Higher intensity, higher profit? Empirical evidence from dairy farming in New Zealand. *Journal of Agricultural Economics* 69(3), 739–55. doi:10.1111/1477-9552.12261.

MacDonald, K. A., Glassey, C. B. and Rawnsley, R. P. 2010. The emergence, development and effectiveness of decision rules for pasture based dairy systems. *Proceedings of the 4th Australasian Dairy Science Symposium*, Lincoln, New Zealand, 31 August – 2 September 2010. Caxton Press, Christchurch.

Macnaghten, P. and Chilvers, J. 2014. The future of science governance: publics, policies, practices. *Environment and Planning C: Government and Policy* 32(3), 530–48. doi:10.1068/c1245j.

McCown, R. L. 2012. A cognitive systems framework to inform delivery of analytic support for farmers' intuitive management under seasonal climatic variability. *Agricultural Systems* 105(1), 7–20. doi:10.1016/j.agsy.2011.08.005.

McCown, R. L., Carberry, P. S., Hochman, Z., Dalgliesh, N. P. and Foale, M. A. 2009. Re-inventing model-based decision support with Australian dryland farmers. 1. Changing intervention concepts during 17 years of action research. *Crop and Pasture Science* 60(11), 1017–30. doi:10.1071/CP08455.

McCown, R. L., Carberry, P. S., Dalgliesh, N. P., Foale, M. A. and Hochman, Z. 2012. Farmers use intuition to reinvent analytic decision support for managing seasonal climatic variability. *Agricultural Systems* 106(1), 33–45. doi:10.1016/j.agsy.2011.10.005.

Meijer, I. S. M., Hekkert, M. P. and Koppenjan, J. F. M. 2007. The influence of perceived uncertainty on entrepreneurial action in emerging renewable energy technology; biomass gasification projects in the Netherlands. *Energy Policy* 35(11), 5836–54. doi:10.1016/j.enpol.2007.07.009.

Morrison, J. 2006. *Decision Support System Review*. Hamilton, New Zealand.

Neal, M., Dela Rue, B. and Eastwood, C. 2017a. Defining the value proposition for using technology to improve pasture management and harvest more pasture. In: *1st Asian-Australasian Conference on Precision Pasture and Livestock Farming*. Hamilton, New Zealand, 6pp.

Neal, M., Kay, J., Peel, S. and McCarthy, S. 2017b. A decision-support tool for autumn management in a spring-calving pasture-based dairy system. *Animal Production Science* 57(7), 1343–9. doi:10.1071/AN16468.

Nguyen, N. C., Wegener, M. K. and Russell, I. W. 2007. Decision support systems in Australian agriculture: state of the art and future development. *AFBM Journal, Australasian Farm Business Management Network* 4, 1–7.

Nuthall, P. L. 2012. The intuitive world of farmers – the case of grazing management systems and experts. *Agricultural Systems* 107, 65–73. doi:10.1016/j.agsy.2011.11.006.

Nuthall, P. 2019. A review of the intuition literature relative to a recent quantitative study of the determinants of farmers' intuition. *International Journal of Agricultural Management* 8, 31–40.
Nuthall, P. L. and Old, K. M. 2018. Intuition, the farmers' primary decision process. A review and analysis. *Journal of Rural Studies* 58, 28–38. doi:10.1016/j.jrurstud.2017.12.012.
Ohlmer, B., Olson, K. and Brehmer, B. 1998. Understanding farmers' decision making processes and improving managerial assistance. *Agricultural Economics* 18(3), 273–90. doi:10.1016/S0169-5150(97)00052-2.
Parker, C. and Sinclair, M. 2001. User-centred design does make a difference. The case of decision support systems in crop production. *Behaviour and Information Technology* 20(6), 449–60. doi:10.1080/01449290110089570.
Pigford, A.-A. E., Hickey, G. M. and Klerkx, L. 2018. Beyond agricultural innovation systems? Exploring an agricultural innovation ecosystems approach for niche design and development in sustainability transitions. *Agricultural Systems* 164, 116–21. doi:10.1016/j.agsy.2018.04.007.
Roche, L. M., Schohr, T. K., Derner, J. D., Lubell, M. N., Cutts, B. B., Kachergis, E., Eviner, V. T. and Tate, K. W. 2015. Sustaining working rangelands: insights from rancher decision making. *Rangeland Ecology and Management* 68(5), 383–9. doi:10.1016/j.rama.2015.07.006.
Rogers, E. M. 1962. *Diffusion of Innovations*. Free Press of Glencoe, New York.
Romera, Á. J., Beukes, P., Clark, C., Clark, D., Levy, H. and Tait, A. 2010. Use of a pasture growth model to estimate herbage mass at a paddock scale and assist management on dairy farms. *Computers and Electronics in Agriculture* 74(1), 66–72. doi:10.1016/j.compag.2010.06.006.
Romera, Á. J., Neal, M. and Bos, B. 2020. Use of reflexive interactive design to address challenges to New Zealand dairy farming. *Animal Production Science* 60(1), 121–6. doi:10.1071/AN18597.
Rose, D. C. and Chilvers, J. 2018. Agriculture 4.0: broadening responsible innovation in an era of smart farming. *Frontiers in Sustainable Food Systems* 2. doi:10.3389/fsufs.2018.00087.
Rose, D. C., Sutherland, W. J., Parker, C., Lobley, M., Winter, M., Morris, C., Twining, S., Ffoulkes, C., Amano, T. and Dicks, L. V. 2016. Decision support tools for agriculture: Towards effective design and delivery. *Agricultural Systems* 149, 165–74. doi:10.1016/j.agsy.2016.09.009.
Rose, D. C., Parker, C., Fodey, J. O. E., Park, C., Sutherland, W. J. and Dicks, L. V. 2018. Involving stakeholders in agricultural decision support systems: improving user-centred design. *International Journal of Agricultural Management* 6, 80–9.
Rotz, S., Gravely, E., Mosby, I., Duncan, E., Finnis, E., Horgan, M., LeBlanc, J., Martin, R., Neufeld, H. T., Nixon, A., Pant, L., Shalla, V. and Fraser, E. 2019. Automated pastures and the digital divide: how agricultural technologies are shaping labour and rural communities. *Journal of Rural Studies* 68, 112–22. doi:10.1016/j.jrurstud.2019.01.023.
Rougoor, C. W., Dijkhuizen, A. A., Huirne, R. B. M., Mandersloot, F. and Schukken, Y. H. 1997. Relationships between technical, economic and environmental results on dairy farms: an explanatory study. *Livestock Production Science* 47(3), 235–44. doi:10.1016/S0301-6226(96)01408-X.

Schaak, H. and Musshoff, O. 2018. Understanding the adoption of grazing practices in German dairy farming. *Agricultural Systems* 165, 230–9. doi:10.1016/j.agsy.2018.06.015.

Shepherd, M., Turner, J. A., Small, B. and Wheeler, D. 2018. Priorities for science to overcome hurdles thwarting the full promise of the "digital agriculture" revolution. *Journal of the Science of Food and Agriculture*. doi:10.1002/jsfa.9346.

Sircely, J., Conant, R. T. and Boone, R. B. 2019. Simulating rangeland ecosystems with G-Range: model description and evaluation at global and site scales. *Rangeland Ecology and Management* 72(5), 846–57. doi:10.1016/j.rama.2019.03.002.

Smith, M. J. 2020. Getting value from artificial intelligence in agriculture. *Animal Production Science* 60(1), 46–54. doi:10.1071/AN18522.

Stilgoe, J., Owen, R. and Macnaghten, P. 2013. Developing a framework for responsible innovation. *Research Policy* 42(9), 1568–80. doi:10.1016/j.respol.2013.05.008.

van Asseldonk, M. A. P. M., Huirne, R. B. M., Dijkhuizen, A. A., Beulens, A. J. M. and Udink ten Cate, A. J. 1999. Information needs and information technology on dairy farms. *Computers and Electronics in Agriculture* 22(2–3), 97–107. doi:10.1016/S0168-1699(99)00010-1.

Wheeler, D. M., Ledgard, S. F. and Boyes, M. 2013. Farm-specific carbon footprinting to the farm gate for agricultural co-products using the OVERSEER® model. *Animal: an International Journal of Animal Bioscience* 7(Suppl. 2), 437–43. doi:10.1017/S1751731113000797.

Wilmer, H. and Fernández-Giménez, M. E. 2015. Rethinking rancher decision-making: a grounded theory of ranching approaches to drought and succession management. *The Rangeland Journal* 37(5), 517–28. doi:10.1071/RJ15017.

Wilmer, H., Derner, J. D., Fernández-Giménez, M. E., Briske, D. D., Augustine, D. J. and Porensky, L. M. 2018. Collaborative adaptive rangeland management fosters management-science partnerships. *Rangeland Ecology and Management* 71(5), 646–57. doi:10.1016/j.rama.2017.07.008.

Wolfert, S., Ge, L., Verdouw, C. and Bogaardt, M.-J. 2017. Big Data in smart farming – a review. *Agricultural Systems* 153, 69–80. doi:10.1016/j.agsy.2017.01.023.

Index

Lightning Source UK Ltd.
Milton Keynes UK
UKHW021847121021
392072UK00001B/30